# METHODS IN COMPUTATIONAL PHYSICS

*Advances in Research and Applications*

Volume 17

General Circulation Models of the Atmosphere

**Methods in Computational Physics**

*Advances in Research and Applications*

1 Statistical Physics

2 Quantum Mechanics

3 Fundamental Methods in Hydrodynamics

4 Applications in Hydrodynamics

5 Nuclear Particle Kinematics

6 Nuclear Physics

7 Astrophysics

8 Energy Bands of Solids

9 Plasma Physics

10 Atomic and Molecular Scattering

11* Seismology: Surface Waves and Earth Oscillations

12* Seismology: Body Waves and Sources

13* Geophysics

14 Radio Astronomy

15† Vibrational Properties of Solids

16‡ Controlled Fusion

17§ General Circulation Models of the Atmosphere

* Volume Editor: Bruce A. Bolt.
† Volume Editor: Gideon Gilat.
‡ Volume Editor: John Killeen.
§ Volume Editor: Julius Chang.

# METHODS IN COMPUTATIONAL PHYSICS

*Advances in Research and Applications*

*Series Editors*

BERNI ALDER

*Lawrence Livermore Laboratory*
*Livermore, California*

SIDNEY FERNBACH

*Lawrence Livermore Laboratory*
*Livermore, California*

MANUEL ROTENBERG

*University of California*
*La Jolla, California*

## Volume 17

## General Circulation Models of the Atmosphere

*Volume Editor*

**JULIUS CHANG**

*Lawrence Livermore Laboratory*
*University of California*
*Livermore, California*

1977

ACADEMIC PRESS NEW YORK SAN FRANCISCO LONDON

A Subsidiary of Harcourt Brace Jovanovich, Publishers

ACADEMIC PRESS, INC.
111 Fifth Avenue, New York, New York 10003

*United Kingdom Edition published by*
ACADEMIC PRESS, INC. (LONDON) LTD.
24/28 Oval Road, London NW1

LIBRARY OF CONGRESS CATALOG CARD NUMBER: 63-18406

ISBN 0-12-460817-5

PRINTED IN THE UNITED STATES OF AMERICA

# Contents


CONTRIBUTORS . . . . . . . . . . vii
PREFACE . . . . . . . . . . ix


## COMPUTATIONAL ASPECTS OF NUMERICAL MODELS FOR WEATHER PREDICTION AND CLIMATE SIMULATION

*Akira Kasahara*


I. Introduction . . . . . . . . . . 2
II. Basic Equations of the Atmosphere . . . . . . . . . . 5
III. Principal Procedure in Numerical Prediction . . . . . . . . . . 10
IV. Physical Processes in Prediction Models . . . . . . . . . . 17
V. Numerical Integration Methods . . . . . . . . . . 27
VI. Initial Conditions . . . . . . . . . . 54
VII. Future Outlook . . . . . . . . . . 56
References . . . . . . . . . . 60




## UNITED KINGDOM METEOROLOGICAL OFFICE FIVE-LEVEL GENERAL CIRCULATION MODEL

*G. A. Corby, A. Gilchrist, and P. R. Rowntree*


I. Introduction . . . . . . . . . . 67
II. Coordinate System and Grid . . . . . . . . . . 69
III. Basic Equations and Finite Difference Approximations . . . . . . . . . . 70
IV. Dissipation Terms (Lateral Eddy Viscosity) . . . . . . . . . . 74
V. Large-Scale Precipitation and Latent Heating . . . . . . . . . . 75
VI. Simple Boundary Layer Parameterization . . . . . . . . . . 76
VII. Representation of Surface Exchange Processes . . . . . . . . . . 82
VIII. Treatment of Land and Ice Surfaces . . . . . . . . . . 86
IX. Convective Interchange . . . . . . . . . . 87
X. Radiation Scheme . . . . . . . . . . 92
XI. Model's January and July Simulations . . . . . . . . . . 96
List of Symbols . . . . . . . . . . 107
References . . . . . . . . . . 108

## A DESCRIPTION OF THE NCAR GLOBAL CIRCULATION MODELS

*Warren M. Washington and David L. Williamson*


I. Origin and Development of the NCAR Global Circulation Models .......... 111
II. Continuous Equations .......... 115
III. Numerical Approximation .......... 131
IV. Application of NCAR Models .......... 164
References .......... 169


## COMPUTATIONAL DESIGN OF THE BASIC DYNAMICAL PROCESSES OF THE UCLA GENERAL CIRCULATION MODEL

*Akio Arakawa and Vivian R. Lamb*


I. Outline of the General Circulation Model .......... 174
II. Principles of Mathematical Modeling .......... 176
III. Finite Difference Schemes for Homogeneous Incompressible Flow .......... 179
IV. Basic Governing Equations .......... 207
V. The Vertical Difference Scheme of the Model .......... 213
VI. The Horizontal Difference Scheme of the Model .......... 236
VII. Vertical and Horizontal Differencing of the Water Vapor and Ozone Continuity Equations .......... 251
VIII. Time Differencing .......... 260
IX. Summary and Conclusions .......... 262
References .......... 264


## GLOBAL MODELING OF ATMOSPHERIC FLOW BY SPECTRAL METHODS

*William Bourke, Bryant McAvaney, Kamal Puri, and Robert Thurling*


I. Introduction .......... 268
II. Spectral Algebra .......... 271
III. Multilevel Spectral Model .......... 285
IV. Numerical Weather Prediction via a Spectral Model .......... 302
V. General Circulation via a Spectral Model .......... 311
VI. Conclusion .......... 319
Appendix .......... 320
References .......... 323

AUTHOR INDEX .......... 325

SUBJECT INDEX .......... 330

CONTENTS OF PREVIOUS VOLUMES .......... 333

# Contributors

Numbers in parentheses indicate the pages on which the authors' contributions begin.

AKIO ARAKAWA, *Department of Atmospheric Sciences, University of California, Los Angeles, California* (173)

WILLIAM BOURKE, *Australian Numerical Meteorology Research Centre, Melbourne, Australia* (267)

G. A. CORBY, *Meteorological Office, Bracknell, England* (67)

A. GILCHRIST, *Meteorological Office, Bracknell, England* (67)

AKIRA KASAHARA, *National Center for Atmospheric Research, Boulder, Colorado* (1)

VIVIAN R. LAMB, *Department of Atmospheric Sciences, University of California, Los Angeles, California* (173)

BRYANT MCAVANEY, *Australian Numerical Meteorology Research Centre, Melbourne, Australia* (267)

KAMAL PURI, *Australian Numerical Meteorology Research Centre, Melbourne, Australia* (267)

P. R. ROWNTREE, *Meteorological Office, Bracknell, England* (67)

ROBERT THURLING, *Australian Numerical Meteorology Research Centre, Melbourne, Australia* (267)

WARREN M. WASHINGTON, *National Center for Atmospheric Research, Boulder, Colorado* (111)

DAVID L. WILLIAMSON, *National Center for Atmospheric Research, Boulder, Colorado* (111)

# Preface

THE NECESSITY FOR NUMERICAL computations to describe the behavior of the earth's atmosphere had been recognized well before Richardson's (1922) pioneering calculation. The construction in the mid-1940s of an electronic computer was initiated by von Neumann primarily to facilitate the computational task of numerical weather forecasting. Ever since, general circulation models of the atmosphere have absorbed the existing computational capabilities and spearheaded the drive for faster and larger computers in order to achieve still higher resolution and to allow for still more complex physical processes. Thus, the general circulation models illustrate well both the potential and the limitations of present-day computing. As such, their presentation should be of general interest to those who attempt calculations with other three-dimensional models of complex physical systems.

The existing general circulation models are increasingly employed in solving practical problems related to the environment and for that reason alone it seems desirable to introduce the subject to a larger audience. Accordingly, the first article by A. Kasahara is a survey that describes the various options in modeling physical processes and in computational procedures. The next two reviews illustrate how, in any actual application, practical considerations influence the compromise between a detailed physical description and reasonable computing time. The article by G. A. Corby, A. Gilchrist, and P. R. Rowntree describes a basic model that is economical, while the article by W. M. Washington and D. L. Williamson includes more physical details resulting in greater flexibility. The computational details of two different numerical schemes for general circulation models are examined in the next two reviews. The first of these by A. Arakawa and V. R. Lamb provides an in-depth analysis of finite difference methods by proceeding from general considerations of homogeneous incompressible flow to the fine details of the particular numerical scheme. In contrast, in the final article by W. Bourke, B. McAvaney, K. Puri, and R. Thurling the fundamentals of the alternative spectral method are developed and implemented for a multilevel spectral model that illustrates the capability of that approach.

J. CHANG<br>
B. ALDER<br>
S. FERNBACH

# Computational Aspects of Numerical Models for Weather Prediction and Climate Simulation

AKIRA KASAHARA

NATIONAL CENTER FOR ATMOSPHERIC RESEARCH*
BOULDER, COLORADO

I. Introduction . . . . . . . . 2
II. Basic Equations of the Atmosphere . . . . . . . . 5
A. Equations of Motion . . . . . . . . 5
B. Simplification of Equations of Motion . . . . . . . . 7
C. Equation of Mass Continuity . . . . . . . . 8
D. Thermodynamic Energy Equation . . . . . . . . 9
III. Principal Procedure in Numerical Prediction . . . . . . . . 10
A. Time Extrapolation . . . . . . . . 10
B. Primitive Equations . . . . . . . . 11
C. Vertical Coordinates . . . . . . . . 12
IV. Physical Processes in Prediction Models . . . . . . . . 17
A. General Consideration . . . . . . . . 17
B. Solar and Terrestrial Radiation . . . . . . . . 18
C. Prediction of Clouds and Precipitation . . . . . . . . 19
D. Atmospheric Boundary Layer . . . . . . . . 21
E. Influence of the Oceans . . . . . . . . 23
F. Parameterization of Subgrid-Scale Motions . . . . . . . . 24
G. Effects of Upper Boundary Conditions . . . . . . . . 26
V. Numerical Integration Methods . . . . . . . . 27
A. Types of Differential Equations . . . . . . . . 27
B. Time-Differencing Schemes . . . . . . . . 29
C. Finite-Difference Method . . . . . . . . 32
D. Quadratic Conserving Schemes . . . . . . . . 35
E. Problems in Mapping Flows over a Sphere . . . . . . . . 40
F. Spectral Methods . . . . . . . . 43
G. Implicit Methods . . . . . . . . 47
H. Nonlinear Instability . . . . . . . . 48
I. Discretization in the Vertical . . . . . . . . 51
VI. Initial Conditions . . . . . . . . 54
A. Initialization . . . . . . . . 54
B. Four-Dimensional Data Assimilation . . . . . . . . 55
VII. Future Outlook . . . . . . . . 56
A. Atmospheric Predictability . . . . . . . . 56

* The National Center for Atmospheric Research is sponsored by the National Science Foundation.

B. Climate Sensitivity . . . . . . . . . . . . . . . . . . . . . . 58
C. Conclusions . . . . . . . . . . . . . . . . . . . . . . . . . . 59
References . . . . . . . . . . . . . . . . . . . . . . . . . . . . 60

## I. Introduction

THERE IS NO NEED to stress the economic and social importance of understanding the behavior of the atmosphere and predicting weather changes and climatic trends. Steady progress has been made toward achieving "man's dream," as L. F. Richardson (1922) put it, and it is useful to review from time to time our prospects and problems.

Predicting the behavior of the earth's atmosphere is a real challenge to scientists. Motions of the atmosphere are governed by physical laws that may be expressed as equations of hydrodynamics and thermodynamics. From this point of view alone, prediction of atmospheric behavior would be more feasible than prediction of, for instance, an economic system for which governing principles are difficult to establish quantitatively. Although we can rule out man's speculation as a reason for weather change, a quantitative description of the atmosphere becomes exceedingly complex with motions of all scales and associated weather phenomena. The scientific discipline dealing with motions of the atmosphere and oceans is called *geophysical fluid dynamics*, distinguished from usual fluid dynamics because of their scales and mechanisms of driving motions.

Laboratory experiments have been useful in understanding conventional fluid dynamics, but a faithful simulation in the laboratory of geophysical fluid motions is difficult because of their unique characteristics. With the advent of fast electronic computers, it is possible to solve numerically the prediction equations as an initial value problem and to perform experiments by changing initial conditions and various parameters in the governing equations. The system of governing equations is called a *mathematical model* and when numerical approximations are applied to the system, it is termed a *numerical model.*

A concrete idea for predicting atmospheric behavior by solving a numerical model of the atmosphere was set forth by Richardson (1922) who, in fact, performed numerical calculations using the meager meteorological data then available. It is no surprise that he was unsuccessful; several major technological and theoretical breakthroughs were necessary before today's successful weather forecasts could be produced. The real value of Richardson's work was that it crystallized the essential problems to be faced in this field and laid a thorough groundwork for their solution (Charney, 1951; Platzman, 1968).

In the mid-1940's, John von Neumann, of the Institute of Advanced Studies at Princeton, began building an electronic computer primarily for weather forecasting. At the Institute, a meteorological research group was established to attack the problem of numerical weather prediction through a step-by-step investigation of models designed to approximate more closely the real properties of the atmosphere (Thompson, 1976). Within a few years, results of the first successful numerical prediction were published by Charney *et al.* (1950).

From 1950 to 1960, active research in numerical weather prediction sprang up in many countries, and various prediction models were formulated. One noteworthy achievement in this period was the first successful simulation of the general circulation of the atmosphere by numerical integration of simplified atmospheric equations (Phillips, 1956).

The large-scale motions of the atmosphere are predominately horizontal and approximately balanced geostrophically. The large-scale horizontal wind blows approximately parallel to the isobars, clockwise around a high-pressure area and counterclockwise around a low-pressure area in the Northern Hemisphere (N.H.), and in opposite directions in the Southern Hemisphere (S.H.). The pressure force, directed at right angles to the isobars from high to low pressure, is approximately balanced by the *Coriolis force* (an apparent force resulting from the earth's rotation and acting at a right angle to the flow, left to right facing downstream in the N.H. and upstream in the S.H.). This state of large-scale motion is referred to as *quasi-geostrophic balance.*

Numerical prediction models from 1950 to 1960 incorporated the quasi-geostrophic approximation in their basic equations. After considerable experimentation and practice, it became evident that quasi-geostrophic models exhibited certain deficiencies in describing atmospheric behavior. These deficiencies were caused by limitations in the quasi-geostrophic assumption in dealing with global-scale motions and in handling frictional effects and heat sources. In spite of research efforts to improve prediction models incorporating nongeostrophic aspects, the attempts had only limited success. Meanwhile, the use of primitive equations without the quasi-geostrophic assumption became more promising in predicting global-scale motions (Smagorinsky, 1958).

The *primitive equations* are the Eulerian equations of motion, under the influence of the earth's rotation, which are simplified by the assumption of hydrostatic balance (Section III, B). Primitive equation models contain less dynamical constraints than quasi-geostrophic models, but are more time-consuming in calculating a one-day forecast. Owing to the greatly improved speed and storage capacity of electronic computers and the progress in numerical techniques needed to solve hydrodynamical problems, many advanced forecasting and simulation models of the atmosphere are now based on primitive equations.

Let us look at the progress in numerical weather prediction over the past 20 years. Figure 1 is the yearly averaged record of forecasting skill based on the $S_1$ score of 36-hr forecasts over North America at 500 mb (approximately 6 km) compiled by the National Meteorological Center, National Oceanic and Atmospheric Administration (Shuman, 1972). The $S_1$ score is a normalized average absolute error of height gradient. We see a general upward trend indicating steady improvement in forecasting skill. Without a forecast model, we must rely on the continuity of weather, or *persistency*. This is equivalent to saying that today's weather continues forever. Even this simple rule has a finite skill of 27% and, therefore, the forecasting score using any means should be judged against this lowest skill score. The average score during 1954 to 1958 was better than that of persistency owing to subjective procedure by experienced forecasters. The improvements from 1958 to 1966 were achieved by the successive introduction of various quasi-geostrophic models, first a model neglecting variation with height, then including vertical variation. In 1966, a six-layer primitive equation model (Shuman and Hovermale, 1968) became operational and has been under continuous development. The better skill score of the primitive equation model may substantiate the fact that removal of the quasi-geostrophic assumption is a step toward improving atmospheric modeling. Yet considerable improvement is needed to achieve the best forecast, and this is the question to which we often address ourselves—How can we further improve short-range forecast skill?

There are three aspects to the formulation of numerical models for weather prediction and climate simulation—mathematical, physical, and

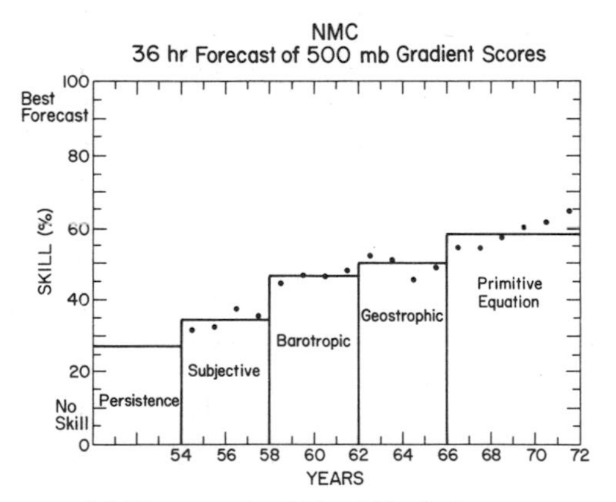


FIG. 1. Annual average "skill" scores for 36 hr, 500 mb forecasts over North America by National Meteorological Center. Data are based on Shuman (1972).

computational. In this article, we review the basics in numerical modeling with emphasis on the mathematical and computational aspects as an introduction to various numerical models detailed in other chapters in this volume.

## II. Basic Equations of the Atmosphere

### A. Equations of Motion

We first present the basic atmospheric equations for large-scale flows. Lagrange's equations of motion in generalized spatial coordinates fixed in space may be given by (Lamb, 1932; Ertel, 1938)

$$\frac{d}{dt}\left(\frac{\partial K}{\partial \dot{q}_k}\right) - \frac{\partial K}{\partial q_k} = -\frac{\partial \Phi^*}{\partial q_k} - \frac{1}{\rho}\frac{\partial p}{\partial q_k} + F_k; \qquad k = 1, 2, 3 \tag{2.1}$$

where $q_k$, $k = 1, 2, 3$ denote the three components of generalized spatial coordinates and $\dot{q}_k = dq_k/dt$; $k = 1, 2, 3$ represent the three components of generalized velocity. Here $t$ denotes time, $p$ pressure, $\rho$ density and $F_k$ the $k$th component of the frictional term. Quantity $K$ denotes kinetic energy referring to an absolute system fixed in space. Quantity $\Phi^*$ is the earth's gravitational potential. The substantial derivative $d/dt$ is expressed by

$$\frac{d}{dt} = \frac{\partial}{\partial t} + \sum_j \dot{q}_j \frac{\partial}{\partial q_j}, \tag{2.2}$$

where $\sum_j$ stands for the sum over $j$ from 1 to 3.

We now derive from (2.1) the equations of motion in spherical coordinates relative to the rotating earth. Spherical coordinates $(\lambda, \phi, r)$ are defined by longitude $\lambda$, latitude $\phi$, and radial distance $r$ from the center of the globe. We assume that the coordinate system is rotating about an axis through the poles with constant angular velocity $\Omega$. We express the generalized spatial coordinates $q_k$ and corresponding velocity components $\dot{q}_k$ by

$$\begin{aligned} q_1 &= \lambda + \Omega t, \qquad & q_2 &= \phi, \qquad & q_3 &= r, \\ \dot{q}_1 &= \dot{\lambda} + \Omega, \qquad & \dot{q}_2 &= \dot{\phi}, \qquad & \dot{q}_3 &= \dot{r}. \end{aligned} \tag{2.3}$$

In spherical coordinates, the velocity components in the direction of increasing $\lambda$, $\phi$, and $r$ are given by

$$u = r\cos\phi\dot{\lambda}, \qquad v = r\dot{\phi}, \qquad w = \dot{r}. \tag{2.4}$$

Here relative to the rotating frame, $u$, $v$, and $w$ are longitudinal, meridional, and vertical components of velocity, respectively. With reference to the absolute system, the longitudinal velocity component $u_a$ is

$$u_a = r \cos \phi(\dot{\lambda} + \Omega). \tag{2.5}$$

Thus, referring to the absolute system and using (2.4) and (2.5), we write kinetic energy $K$ as

$$\begin{aligned} K &= \tfrac{1}{2}(u_a{}^2 + v^2 + w^2) \\ &= \tfrac{1}{2}[r^2 \cos^2 \phi(\dot{\lambda}^2 + 2\dot{\lambda}\Omega) + r^2\dot{\phi}^2 + \dot{r}^2] + \Phi^* - \Phi, \end{aligned} \tag{2.6}$$

where

$$\Phi - \Phi^* = -\tfrac{1}{2}(\Omega r \cos \phi)^2 \tag{2.7}$$

and the right-hand side is *centrifugal potential.* The sum of the earth's gravitational potential $\Phi^*$ and centrifugal potential $-\frac{1}{2}(\Omega r \cos \phi)^2$ is called *geopotential* denoted by $\Phi$.

Considering that $\Phi^*$ is a function of $q_k$ only (actually a function of $r$ only), carrying out partial differentiations with respect to $\lambda$, $\phi$, $r$, $\dot{\lambda}$, $\dot{\phi}$, and $\dot{r}$, substituting the resultant derivatives into (2.1), and using the expression of velocity components (2.4), we obtain

$$\frac{du}{dt} + \frac{uw}{r} - \frac{uv}{r}\tan \phi - fv + \hat{f}w = -\frac{1}{\rho r \cos \phi}\frac{\partial p}{\partial \lambda} - \frac{1}{r \cos \phi}\frac{\partial \Phi}{\partial \lambda} + F_\lambda, \tag{2.8}$$

$$\frac{dv}{dt} + \frac{vw}{r} + \frac{u^2}{r}\tan \phi + fu = -\frac{1}{\rho r}\frac{\partial p}{\partial \phi} - \frac{1}{r}\frac{\partial \Phi}{\partial \phi} + F_\phi, \tag{2.9}$$

$$\frac{dw}{dt} - \left(\frac{u^2 + v^2}{r}\right) - \hat{f}u = -\frac{1}{\rho}\frac{\partial p}{\partial r} - \frac{\partial \Phi}{\partial r} + F_r, \tag{2.10}$$

where

$$f = 2\Omega \sin \phi, \qquad \hat{f} = 2\Omega \cos \phi$$

and the total derivative $d/dt$ relative to the rotating earth is

$$\frac{d}{dt} = \frac{\partial}{\partial t} + \frac{u}{r \cos \phi}\frac{\partial}{\partial \lambda} + \frac{v}{r}\frac{\partial}{\partial \phi} + w\frac{\partial}{\partial r}. \tag{2.11}$$

The second and third terms in (2.8) and (2.9) and the second term in (2.10) are apparent acceleration terms due to the curvature of the coordinate system. Terms $fv$ and $\hat{f}w$ in (2.8), $fu$ in (2.9) and $\hat{f}u$ in (2.10) are also apparent acceleration terms, but are the result of the rotation of the coordinate system. The apparent acceleration due to the rotation of the coordinate system is called *Coriolis acceleration.*

## B. Simplification of Equations of Motion

By scale analysis for large-scale meteorological motions, some terms in the equations of motion (2.8)–(2.10) are small compared to the remaining terms. To simplify consistently the equations of motion, we introduce an approximation due to shallowness of the atmosphere (Phillips, 1966; Hinkelmann, 1969). The vertical extent of the atmosphere of our interest is approximately 100 km above the earth's surface and, thus, the thickness of the atmosphere is very small compared to the earth's radius.

Let radial distance $r$ be represented by $a + z$, where $a$ is a mean radius of the earth and $z$ denotes altitude relative to radius $a$. The *shallowness approximation* is that any undifferentiated radial distance $r$ appearing in the expression of kinetic energy $K$ is replaced by $a$ and the differentiated form of $r$ by $dz$, namely,

$$r = a \qquad \text{and} \qquad dr = dz. \tag{2.12}$$

With the above approximation, the new form of kinetic energy $K$ and geopotential $\Phi$ becomes

$$K = \tfrac{1}{2}[a^2 \cos^2 \phi(\dot{\lambda}^2 + 2\dot{\lambda}\Omega) + a^2\dot{\phi}^2 + \dot{z}^2] + \Phi^* - \Phi \tag{2.13}$$

and

$$\Phi = \Phi^* - \tfrac{1}{2}(a\Omega \cos \phi)^2. \tag{2.14}$$

Substituting expression (2.13) into $K$ of (2.1) and specifying

$$\begin{aligned} q_1 &= \lambda + \Omega t, & q_2 &= \phi, & q_3 &= z, \\ \dot{q}_1 &= \dot{\lambda} + \Omega, & \dot{q}_2 &= \dot{\phi}, & \dot{q}_3 &= \dot{z}, \end{aligned} \tag{2.15}$$

we obtain the following simplified equations:

$$\frac{du}{dt} - \left(f + \frac{u \tan \phi}{a}\right) v = -\frac{1}{a \cos \phi}\left(\frac{1}{\rho}\frac{\partial p}{\partial \lambda} - \frac{\partial \Phi}{\partial \lambda}\right) + F_\lambda, \tag{2.16}$$

$$\frac{dv}{dt} + \left(f + \frac{u \tan \phi}{a}\right) u = -\frac{1}{\rho a}\frac{\partial p}{\partial \phi} - \frac{1}{a}\frac{\partial \Phi}{\partial \phi} + F_\phi, \tag{2.17}$$

$$\frac{dw}{dt} = -\frac{1}{\rho}\frac{\partial p}{\partial z} - \frac{\partial \Phi}{\partial z} + F_z, \tag{2.18}$$

where

$$\frac{d}{dt} = \frac{\partial}{\partial t} + \frac{u}{a \cos \phi} \frac{\partial}{\partial \lambda} + \frac{v}{a} \frac{\partial}{\partial \phi} + w \frac{\partial}{\partial z} \tag{2.19}$$

and

$$u = a \cos \phi \dot{\lambda}, \qquad v = a\dot{\phi}, \qquad w = \dot{z}. \tag{2.20}$$

We also assume that the surface of constant apparent gravity potential $\Phi$ is approximated by a sphere so that $\Phi$ depends only on the vertical coordinate $z$. We further assume that the thickness of the atmosphere of our interest is sufficiently small to justify a linear variation of $\Phi$ in $z$. Hence,

$$\partial\Phi/\partial\lambda = \partial\Phi/\partial\phi = 0, \qquad \partial\Phi/\partial z = g, \tag{2.21}$$

where $g$ denotes apparent acceleration of the earth's gravity and we use $g = 9.8 \text{ m sec}^{-2}$. The error introduced by treating the magnitude of $g$ as constant is about 3% at $z = 100$ km.

Comparing the system of simplified equations (2.16)–(2.18) with the original system (2.8)–(2.10), we find that terms $uw/r$ and $\hat{f}w$ in (2.8), $vw/r$ in (2.9), and $(u^2 + v^2)/r$ and $\hat{f}u$ in (2.10) are absent in (2.16)–(2.18). The neglected terms are small by scale analysis (Charney, 1948) compared to the remaining terms.

## C. Equation of Mass Continuity

The equation of motion deals with the relation between the acceleration of air and the fields of velocity and pressure. The rate of change of air density per unit time is determined from another physical law—the continuity equation of air.

In spherical coordinates, the mass continuity equation takes the form

$$\frac{\partial \rho}{\partial t} + \frac{1}{r \cos \phi}\left[\frac{\partial(\rho u)}{\partial \lambda} + \frac{\partial}{\partial \phi}(\rho v \cos \phi)\right] + \frac{\partial(\rho r^2 w)}{r^2 \, \partial r} = 0. \tag{2.22}$$

Introducing the shallowness approximation (2.12), we obtain

$$\frac{\partial \rho}{\partial t} + \frac{1}{a \cos \phi}\left[\frac{\partial(\rho u)}{\partial \lambda} + \frac{\partial}{\partial \phi}(\rho v \cos \phi)\right] + \frac{\partial(\rho w)}{\partial z} = 0. \tag{2.23}$$

This form of mass continuity equation is consistent with the system of equations (2.16)–(2.18).

### D. Thermodynamic Energy Equation

The equations of motion relate the dynamics of flow to the pressure and density fields. The equation of mass continuity determines the time rate of change in the density field in terms of kinematics of flow. We now need a relation between the time rate of change of pressure and density; this relationship is specified by the first law of thermodynamics. The atmosphere is considered an ideal gas so that internal energy is a function of temperature only and not density. We also assume for an ideal gas that the specific heat at constant volume $C_v$ is constant.

For an ideal gas or a mixture of ideal gases, the equation of state defines temperature $T$ in relation to pressure $p$ and density $\rho$:

$$p = R\rho T, \qquad (2.24)$$

where $R$ denotes the specific gas constant for the particular ideal gas under consideration and relates to $C_v$ through

$$C_p = R + C_v, \qquad (2.25)$$

which defines specific heat at constant pressure $C_p$.

The first law of thermodynamics appropriate to the atmosphere may be given by

$$C_v\, d(\ln p)/dt - C_p\, d(\ln \rho)/dt = Q/T, \qquad (2.26)$$

where $Q$ denotes the time rate of heating (or cooling) per unit mass and $d/dt$ represents the substantial derivative. Combining (2.24) with (2.26), we obtain

$$C_p\, dT/dt - (1/\rho)(dp/dt) = Q. \qquad (2.27)$$

This form of thermodynamic equation will appear often in atmospheric modeling.

Another convenient form is given concisely by

$$d(\ln \theta)/dt = Q/(C_p T) \qquad (2.28)$$

by defining *potential temperature* $\theta$ as

$$\theta = T(p_0/p)^{\kappa} \qquad (2.29)$$

with $\kappa = R/C_p$ $(\equiv 2/7)$ and $p_0 = 1013.25$ mb as a reference pressure.

## III. Principal Procedure in Numerical Prediction

### A. Time Extrapolation

In the system of equations (2.16)–(2.18), (2.23), (2.24), and (2.26), $\lambda$, $\phi$, $z$, and $t$ are called *coordinate variables* or *independent variables*, and $u$, $v$, $w$, $\rho$, $p$, and $T$ are *dependent variables*. $u$, $v$, $w$, $\rho$, and $p$ are *prognostic variables* since there are corresponding equations for the time rate of change of variables. On the other hand, $T$ is a diagnostic variable because once $p$ and $\rho$ are known, temperature can be computed from an equation that does not contain the time derivative.

If frictional terms $F_\lambda$, $F_\phi$, $F_z$ and heating term $Q$ are expressed as functions of dependent and independent variables, these equations, together with proper boundary conditions, form a complete set. Namely, the number of unknowns is the same as the number of equations, and the evolution of the dependent variables can be determined for $t > t_0$ once the fields of the dependent variables are prescribed at an initial moment $t = t_0$.

A method of numerically solving a system of prognostic partial differential equations is to replace the partial derivatives with finite differences in both space and time. Dependent variables are defined only at the points of a four-dimensional lattice in space and time. Let $\mathbf{W}(t) = (u_1 \cdots u_N v_1 \cdots v_N$, etc.) denote a column vector consisting of $N$ elements of prognostic variables $u$, $v$, $w$, $\rho$, and $p$ at time $t$ and let $\mathbf{W}(t_0)$ represent the values of $\mathbf{W}$ at $t = t_0$. Then the values of $\mathbf{W}$ at $t_0 + \Delta t$, where $\Delta t$ is a small time increment, may be calculated from

$$\mathbf{W}(t_0 + \Delta t) = \mathbf{W}(t_0) + (\partial \mathbf{W}/\partial t)_{t=\tau}\, \Delta t, \tag{3.1}$$

where $t_0 \leqslant \tau \leqslant t_0 + \Delta t$.

If the components of the time derivative $\partial \mathbf{W}/\partial t$ at $t = \tau$ were computed from (2.16)–(2.18), (2.23), and (2.26), the value of $\mathbf{W}$ at $t_0 + \Delta t$ could be computed from (3.1). However, because we do not know the value of $\tau$, we must approximate the values of $\partial \mathbf{W}/\partial t$ at $t = \tau$ by the value of $\partial \mathbf{W}/\partial t$ at, for example, $t_0$ and/or $t_0 + \Delta t$. When $\partial \mathbf{W}/\partial t$ is evaluated using the value of $\mathbf{W}$ at $t_0$, the time extrapolation is *explicit*. On the other hand, when $\partial \mathbf{W}/\partial t$ is evaluated at $t_0 + \Delta t$, the time extrapolation is *implicit*, because the evaluation of $\partial \mathbf{W}/\partial t$ at $t_0 + \Delta t$ requires the value of $\mathbf{W}$ at $t_0 + \Delta t$ which is still unknown at $t_0$. Hence an implicit scheme requires some kind of iteration or an inversion of matrix for the computation of $\partial \mathbf{W}/\partial t$ and generally is time consuming to solve. In any case, repetition of (3.1) will yield a prediction of $\mathbf{W}$ for any later time.

When the system of equations is solved by an explicit difference scheme, there is a constraint on the value of $\Delta t$. This constraint (Section V, C) states that $\Delta t$ must satisfy the condition $C_m \Delta t/\Delta x < 1$, where $\Delta x$ represents one of the space increments in the three-dimensional grid and $C_m$ denotes the maximum characteristic speed of the system.

The maximum characteristic speed in this system is that of sound, approximately 300 m $\sec^{-1}$. If a representative horizontal grid distance is 300 km, we find that $\Delta t$ is less than 1000 sec. Sound waves in this system, however, propagate not only in the horizontal but also in the vertical. If a representative vertical grid increment is 3 km, we find that $\Delta t$ must be less than 10 sec! To use such a small time step for the numerical integration of a system designed to predict weather patterns for several days is impractical, even with advanced high-speed computers.

Consequently, we face two alternatives: to use implicit finite-difference schemes in the vertical without computational constraints (Section V, G) or to modify the atmospheric equations to eliminate the vertical propagation of sound waves.

### B. Primitive Equations

For large-scale motions of the atmosphere, the horizontal extent of motion is much larger than the vertical, and we can show by scale analysis (Charney, 1948) that the vertical acceleration $dw/dt$ in (2.18) may be neglected. Since frictional term $F_z$ is, at most, of the order of $dw/dt$, by incorporating (2.21) for $g$ we may approximate (2.18) as

$$\partial p/\partial z = -\rho g \tag{3.2}$$

known as the hydrostatic equation.

The system of equations of horizontal motion (2.16), (2.17), hydrostatic equilibrium (3.2), mass continuity (2.23), thermodynamics (2.26), and the ideal gas law (2.24) is the hydrostatic prediction system or *primitive equations*. It turns out that the hydrostatic assumption modifies the normal modes (eigensolutions of small free oscillations) of the original prediction system by filtering out the propagation of sound waves in the vertical (Monin and Obukhov, 1959). All global prediction models in this volume are based on primitive equations.

Another significant aspect of primitive equations is that $w$ no longer becomes a prognostic variable because $dw/dt$ is not available. The hydrostatic prediction system consists of three prognostic equations for $u$, $v$, and $p$ or $\rho$ and two diagnostic equations of state (2.24) and hydrostatic equilibrium (3.2), supposedly for temperature $T$ and vertical velocity $w$. However, it is not obvious how $w$ can be obtained. In the hydrostatic system, the calculation of $\partial\rho/\partial t$ from the thermodynamic equation must always satisfy the hydrostatic condition. Since both equations for $\partial\rho/\partial t$ and $\partial p/\partial t$ contain $w$, vertical velocity $w$ should be determined to satisfy the hydrostatic constraint. A diagnostic equation to calculate $w$ was first derived by Richardson (1922).

## C. Vertical Coordinates

So far, we have adopted geometrically fixed coordinates to express prediction equations. Fixed coordinates are *Eulerian* in contrast to moving coordinates, which may be called *Lagrangian*. In Eulerian coordinates, the properties of a fluid are assigned to fixed points in space at a given time without identifying individual fluid parcels from one time to the next. On the other hand, in Lagrangian coordinates properties of a fluid are described in terms of the coordinates in space of the same fluid parcels following the movement of every fluid parcel. It is often convenient to adopt a hybrid system of coordinates in which two sets of coordinate surfaces are geometrically fixed and a third is a material surface. Starr (1945) proposed the so-called quasi-Lagrangian coordinate system where a Lagrangian coordinate is employed in the vertical and Eulerian coordinates in the horizontal.

The use of a Lagrangian coordinate in the vertical is not entirely suitable for extended-range numerical weather prediction because of the large distortion of vertical coordinate surfaces. Nevertheless, moving coordinates in the vertical are useful, and pressure has been adopted extensively as a vertical coordinate in numerical prediction models. We show here how to transform the primitive equations in a height coordinate to those in a vertical coordinate utilizing any meteorological variable that gives a single-valued monotonic relationship with height $z$ (Kasahara, 1974).

Let the $z$ system be the coordinate system with independent variables $x_1$, $x_2$, $z$, and $t$, and the $s$ system the generalized coordinate system with $x_1$, $x_2$, $s$, and $t$, where $s$ represents the generalized coordinate as a function of $x_1$, $x_2$, $z$, and $t$. We assume a single-valued monotonic relationship between $s$ and $z$ when $x_1$, $x_2$, and $t$ are held fixed. Here $x_1$ and $x_2$ are suitably defined orthogonal curvilinear horizontal coordinates, so that for any scalar function $A$ and horizontal velocity $\mathbf{V}$, with $x_1$ and $x_2$ components as $u$ and $v$, $\nabla A$ defines grad $A$ and $\nabla \cdot \mathbf{V}$ horizontal divergence.

The partial derivative of scalar function $A$ with respect to $c$, where $c$ can be $x_1$, $x_2$, or $t$, is generally different in the $z$ and $s$ systems and the following relationship exists:

$$\left(\frac{\partial A}{\partial c}\right)_s = \left(\frac{\partial A}{\partial c}\right)_z + \frac{\partial s}{\partial z}\left(\frac{\partial z}{\partial c}\right)_s \frac{\partial A}{\partial s}, \tag{3.3}$$

where subscripts $s$ and $z$ denote a particular vertical coordinate to be held constant for partial differentiation. If we choose $t$ for $c$, it follows that

$$\left(\frac{\partial A}{\partial t}\right)_s = \left(\frac{\partial A}{\partial t}\right)_z + \frac{\partial s}{\partial z}\left(\frac{\partial z}{\partial t}\right)_s \frac{\partial A}{\partial s}. \tag{3.4}$$

Similarly, choosing $x_1$ and $x_2$ for $c$, we obtain

$$\nabla_s A = \nabla_z A + (\partial s/\partial z)(\nabla_s z)(\partial A/\partial s). \tag{3.5}$$

The total derivative can be expressed in the $z$ system as

$$d/dt = (\partial/\partial t)_z + \mathbf{V} \cdot \nabla_z + w(\partial/\partial z) \tag{3.6}$$

and in the $s$ system as

$$d/dt = (\partial/\partial t)_s + \mathbf{V} \cdot \nabla_s + \dot{s}(\partial/\partial s), \tag{3.7}$$

where $\dot{s}$ is the generalized vertical velocity $\dot{s} = ds/dt$, which corresponds to $w$ in the $z$ system $w = dz/dt$. Using (3.4) and (3.5), we find that the relation between $w$ and $\dot{s}$ is

$$w = (\partial z/\partial t)_s + \mathbf{V} \cdot \nabla_s z + \dot{s}(\partial z/\partial s). \tag{3.8}$$

The horizontal equations of motion (2.16) and (2.17), incorporating (2.21), in the $s$ system may be expressed by

$$\frac{d\mathbf{V}}{dt} + f^*\mathbf{K} \times \mathbf{V} = -\frac{1}{\rho}\nabla_s p + \frac{1}{\rho}\left(\frac{\partial s}{\partial z}\right)(\nabla_s z)\frac{\partial p}{\partial s} + \mathbf{F}, \tag{3.9}$$

where

$$f^* = f + (u/a)\tan\phi$$

and $\mathbf{K}$ denotes a unit vector in the vertical and $\mathbf{F}$ represents the frictional term.

The hydrostatic equation (3.2) is written as

$$\rho\, \partial z/\partial s = -(1/g)(\partial p/\partial s). \tag{3.10}$$

Hence, in the hydrostatic system the horizontal equations of motion (3.9) are reduced to

$$d\mathbf{V}/dt + f^*\mathbf{K} \times \mathbf{V} = -(1/\rho)\nabla_s p - g\nabla_s z + \mathbf{F}. \tag{3.11}$$

Note that the pressure gradient is expressed by the sum of two gradient terms on the right-hand side; the first is the pressure gradient on the $s$ surface, and the second is the hydrostatic correction. If height $z$ is used as a coordinate, the hydrostatic correction vanishes. If pressure is used as a vertical coordinate, then the first term vanishes and the pressure gradient is expressed by the geopotential gradient.

To transform the mass continuity equation (2.23) into the $s$ system, note that from (3.8) we have

$$\begin{aligned}\frac{\partial w}{\partial z} &= \frac{\partial w}{\partial s}\frac{\partial s}{\partial z} \\ &= \frac{\partial s}{\partial z}\left[\frac{d}{dt}\left(\frac{\partial z}{\partial s}\right) + \frac{\partial \mathbf{V}}{\partial s}\cdot \nabla_s z\right] + \frac{\partial \dot{s}}{\partial s}. \end{aligned} \tag{3.12}$$

Also, with the aid of (3.3), we can show that

$$\nabla_z \cdot \mathbf{V} = \nabla_s \cdot \mathbf{V} - \left(\frac{\partial s}{\partial z}\right)(\nabla_s z)\cdot\frac{\partial \mathbf{V}}{\partial s}. \tag{3.13}$$

Substitution of (3.12) and (3.13) into (2.23) yields the mass continuity equation in the $s$ system

$$d \ln [\rho(\partial z/\partial s)]/dt + \nabla_s \cdot \mathbf{V} + \partial \dot{s}/\partial s = 0. \tag{3.14}$$

With the hydrostatic equation (3.10), we may rewrite (3.14) as

$$\frac{\partial}{\partial s}\left(\frac{\partial p}{\partial t}\right)_s + \nabla_s \cdot \left(\mathbf{V}\frac{\partial p}{\partial s}\right) + \frac{\partial}{\partial s}\left(\dot{s}\,\frac{\partial p}{\partial s}\right) = 0. \tag{3.15}$$

If pressure is used as the vertical coordinate, the first term vanishes and the continuity equation reduces to a diagnostic equation for $dp/dt$, which is a measure of the vertical motion. This makes analyses of large-scale motions simpler, especially with quasi-geostrophic models as demonstrated by Godart (Sutcliffe, 1947) and Eliassen (1949).

Any numerical model requires an upper boundary to limit its vertical extent, but the nature of upper boundary conditions is by no means obvious. Here we assume that there is no mass transport through coordinate surface $s = s_T =$ constant corresponding to the model top. This is stated as

$$\dot{s} = 0 \qquad \text{at} \quad s = s_T = \text{const.} \tag{3.16}$$

The lower boundary of the model is expressed, in general, by

$$s = s_H = s(x_1, x_2, H, t), \tag{3.17}$$

where $H$ denotes the altitude of the earth's surface above mean sea level $z = 0$. The value of $s$ at $z = H$ may vary with time and space. Because air at the earth's surface may move only along the earth's surface, we have the lower boundary condition

$$\dot{s} = (\partial s_H/\partial t) + \mathbf{V}_H \cdot \nabla s_H \qquad \text{at} \quad s = s_H, \tag{3.18}$$

where $\mathbf{V}_H$ denotes horizontal velocity at $z = H$.

If pressure is used as the vertical coordinate, the above lower boundary condition reads such that

$$\dot{p} = \partial p_H/\partial t + \mathbf{V}_H \cdot \nabla p_H \qquad \text{at} \quad p = p_H. \tag{3.19}$$

Because surface pressure can vary with time and space, it is awkward to handle the lower boundary condition in the prediction model. To circumvent this difficulty, Phillips (1957) introduced the so-called sigma-coordinate system

$$\sigma = (p - p_T)/P_* \quad \text{and} \quad P_* = p_H - p_T, \tag{3.20}$$

where $p_H$ denotes surface pressure at $z = H$ and $p_T$ a constant pressure corresponding to the model top. Thus, the model top is expressed by $\sigma = 0$ and the earth's surface by $\sigma = 1$ where no special procedure is required to deal with the lower boundary condition. Most numerical models discussed in this volume adopt the sigma-coordinate system.

We can formulate a transformed (or normalized) coordinate in which a transformed variable $\tilde{s}$ is defined by

$$\tilde{s} = (s - s_H)/(s_T - s_H), \tag{3.21}$$

so that $\tilde{s} = 0$ at the earth's surface $s = s_H$ and $\tilde{s} = 1$ at the model top $s = s_T$. Because the earth's surface coincides with a constant $\tilde{s}$ surface, the lower boundary condition (3.18) reduces to

$$\dot{\tilde{s}} = 0 \qquad \text{at} \quad s = s_H. \tag{3.22}$$

This simple boundary condition is an advantage in using a transformed coordinate.

By integrating (3.15) with respect to $s$ from some level $s$ to the model top $s_T$ and applying upper boundary condition (3.16), we obtain the pressure tendency equation

$$\left(\frac{\partial p}{dt}\right)_s = \frac{\partial p_T}{\partial t} - \dot{s}\frac{\partial p}{\partial s} + \int_s^{s_T} \nabla_s \cdot \left(\mathbf{V}\frac{\partial p}{\partial s}\right) ds, \tag{3.23}$$

where $\partial p_T/\partial t$ denotes the pressure tendency at the model top.

Kasahara (1974) examined the vertically integrated total energy equation in the $s$ system and found that the $z$ and $p$ systems are exceptional in that the upper and lower boundary conditions (3.16) and (3.18)—when properly expressed as $p$ or $z$—are sufficient to maintain energy conservation. [In transformed coordinates for $z$ and $p$, the lower boundary condition (3.22) should be used.] However, if a variable other than $z$ or $p$, such as potential temperature or density, is used as the vertical coordinate, we need an additional upper boundary condition to maintain total energy conservation. We have two choices:

(a) $\partial p_T/\partial t = 0$ at $s = s_T = \text{const.}$

(b) $z_T = \text{const.}$ for $s = s_T = \text{const.}$,

where $s_T$ is a constant value of $s$ corresponding to the model top. In (b), we must calculate $\partial p_T/\partial t$ on $s = s_T$. It is inconsistent to assume conditions (a) and (b) at the same time.

The thermodynamic equation (2.26) or (2.27) is written in terms of the total derivatives of dependent variables. Hence, the thermodynamic equation in the $s$ system takes the same form except that the total derivative in the $s$ system is given by (3.7).

To derive a particular prediction model, we must specify a value for $s$ as the vertical coordinate. So far, pressure $p$, transformed pressure such as defined by (3.20), and transformed height $z$ have been used as the vertical coordinate for the global circulation models in this volume. Another vertical coordinate variable especially convenient to describe adiabatic motions is potential temperature $\theta$. For adiabatic motions $Q = 0$, $\theta$ of an air parcel is conserved, as seen from (2.28). Hence, the pattern of isentropic (equal potential temperature) surfaces reflects the vertical structure of adiabatic motions. Prediction models with isentropic coordinate have been formulated by, for example, Eliassen and Rekustad (1971), Bleck (1974), and Shapiro (1975).

The form of prediction equations is considerably different depending on the variable selected as the vertical coordinate. It is by no means obvious which coordinate system is best for global circulation modeling. From the point of view of computer programming, it is advantageous to apply a transformed system so that the earth's surface becomes a coordinate surface. However, there is an inherent difficulty connected with the calculation of pressure gradient force in a transformed coordinate (Section V, I). Relative merits of various vertical coordinate systems are difficult to determine without considerable experimentation because they involve such practical matters as the vertical resolution of finite-difference schemes and the vertical structure of meteorological motions to be described with those systems.

## IV. Physical Processes in Prediction Models

### A. General Consideration

We have discussed the dynamical aspects of prediction models, but a prediction model is not complete without specific forms of heating $Q$ in the thermodynamic equation and frictional term $\mathbf{F}$ in the equations of horizontal motion. In the early days of numerical weather prediction, it was thought that for short-range forecasts the physical aspects of prediction models were of secondary importance to the dynamical aspects. This was because synoptic-scale motions such as cyclones and anticyclones are mostly transient, and the energy source for the development of systems is mainly the conversion from potential energy to kinetic energy. On the other hand, global-scale motions are quasi-stationary and generated mostly by external heating or cooling related to the continentality of the earth's surface and the dynamical forcing due to the earth's orography. Without proper specification of heating and frictional terms, prediction models deal only with transient aspects, and even for one-day forecasts errors can be serious for global-scale motions. Because climate describes the quasi-stationary aspects of atmospheric motions, the correct formulation of heating and frictional terms becomes even more important in the simulation of global climates.

The time rate of heating $Q$ in the thermodynamic equation may be expressed by

$$Q = Q_a + Q_b + Q_c, \tag{4.1}$$

where the three parts of the heating rate are due to solar and atmospheric radiation $Q_a$, vertical and horizontal diffusion of sensible heat $Q_b$, and the release of latent heat of condensation of water vapor $Q_c$.

Similarly, frictional term $\mathbf{F}$ in the equation of horizontal motion may be expressed by

$$\mathbf{F} = \mathbf{F}_\mathrm{L} + \mathbf{F}_\mathrm{s} \tag{4.2}$$

as the sum of the time rate of change of momentum per unit mass by large-scale friction $\mathbf{F}_L$ and by transport due to the scales of motion smaller than the computational grid increment $\mathbf{F}_s$. For example, the dissipation of momentum by frictional processes near the earth's surface falls in the category of $\mathbf{F}_L$, and the vertical transport of momentum by an ensemble of cumulus clouds belongs to $\mathbf{F}_s$. The integration of frictional force $\mathbf{V} \cdot \mathbf{F}$ over the entire atmosphere is negative, acting as an energy sink.

Specification of heating and frictional terms by dependent variables of a prediction model requires detailed knowledge of the physical processes involved. These details of the atmosphere are specialized branches of atmospheric science—atmospheric radiation, micrometeorology, hydrology, cloud physics, and so forth.

This section briefly describes the various physical processes in the atmosphere incorporated into prediction models. The formulation of physical processes in terms of large-scale flow variables as model parameters is called *parameterization* (Joint Organizing Committee, 1972).

## B. Solar and Terrestrial Radiation

All energy of atmospheric motions is ultimately derived from incoming solar radiation, in short called *insolation*. The insolation at the top of the atmosphere is approximately 1.95 cal $\text{min}^{-1}$ through a 1 $\text{cm}^2$ surface perpendicular to the solar beam when the earth is at mean distance from the sun. This value is known as *solar constant*. A mean flux of solar energy perpendicular to the earth's surface is about 0.5 cal $\text{cm}^{-2}$ $\text{min}^{-1}$, since the surface area is four times the cross-sectional area of the sphere. Of this, approximately 33% is reflected from the atmosphere, including clouds and portions of the earth's surface. This percentage of energy reflected back to space is referred to as the *planetary albedo* of the earth as a whole. Approximately 20% is absorbed into the atmosphere and 47% at the earth's surface.

Because the mean temperature of the earth and atmosphere does not change appreciably from one year to the next, the energy received by the earth must be sent back to space. This ultimate return is in the form of low-temperature infrared radiation from the earth–atmosphere system—*terrestrial radiation.*

The heating/cooling term due to radiational sources $Q_a$ in (4.1) may be divided into two parts:

$$Q_a = Q_{as} + Q_{al}, \tag{4.3}$$

where $Q_{as}$ denotes the time rate of heating due to absorption of insolation in the atmosphere and $Q_{al}$ is the time rate of heating/cooling due to infrared radiation emitted from the earth's surface and the atmosphere.

In radiative transfer calculations, we need to specify the amount of a given absorbing or emitting gas in a vertical column of unit cross section extending between two specific levels. This quantity is called *optical thickness*. We ordinarily consider three major absorbing gases in the atmosphere—water vapor, carbon dioxide, and ozone. The optical thickness of absorbing gas can be calculated by knowing its volume content as a function of height.

The time rate of heating due to absorption of insolation $Q_{as}$ in (4.3) may be expressed by

$$Q_{as} = \frac{\partial F_{as}(z)}{\rho\, \partial z}, \tag{4.4}$$

where $F_{as}$ denotes the downward flux of solar energy reaching some level $z$, modified by absorption by major absorbing gases in the atmosphere, and reflection due to scattering by atmospheric molecules, dust, and clouds.

The energy flux of insolation through the atmosphere can be determined by knowing its absorptivity and optical thickness. The flux of solar energy depends on the solar constant and the Sun's zenith angle. The reflection of solar energy by clouds is geometrically calculated by taking into account cloudiness as a function of height.

The time rate of heating/cooling due to the divergence of net infrared flux $Q_{al}$ in (4.3) may be expressed by

$$Q_{al} = -\frac{\partial}{\rho\, \partial z}\left[F^{\uparrow}(z) - F^{\downarrow}(z)\right], \tag{4.5}$$

where $F^{\uparrow}(z)$ and $F^{\downarrow}(z)$ denote the total upward and downward infrared fluxes, respectively, at height $z$. These fluxes under clear sky can be calculated knowing the absorptivity of an atmosphere containing water vapor, carbon dioxide, and ozone and the vertical distribution of temperature.

Since both upward and downward fluxes are involved in the infrared radiation calculation, the handling of cloud effect on infrared radiation becomes more complicated than the effect on insolation. The cloud effect is usually taken into account by first calculating separately infrared fluxes under clear sky and overcast sky. Then for a partial cloud cover the mean fluxes are calculated from the combination of cloudfree and overcast fluxes taken with proper weight of the fractional cloud amount.

### C. Prediction of Clouds and Precipitation

Clouds affect atmospheric motions by altering the distribution of heating/ cooling through absorption, reflection, and scattering in solar and terrestrial radiative transfer. Clouds also play an active role of interaction with atmospheric flow by transporting momentum, sensible heat, and water vapor and by providing the source/sink of latent heat due to the phase transition of water substance.

Clouds are formed by condensation, sublimation, and freezing. The precipitation from clouds occurs when the cloud droplets and crystals grow sufficiently large that they are no longer suspended but fall from the clouds and eventually reach the ground as rain, snow, or hail.

Prediction of clouds and precipitation is perhaps the most important aspect of weather forecasting. Unfortunately, the problem is also difficult because the time and space scales involved are much smaller than those of large-scale motions. Ideally, the problem should be treated separately by formulating prediction models with finer space and time resolutions and incorporating cloud dynamics and physics. Despite the inherent difficulty in predicting clouds, their collective effect as a source of condensation heating must be parameterized in prediction models.

The mixing ratio of water vapor $q$ is defined by

$$q = \rho_w/\rho, \tag{4.6}$$

where $\rho_w$ is the density of water vapor and $\rho$ is the density of dry air. When the mixing ratio exceeds the saturation mixing ratio $q_r$, the excess water vapor over $q_r$ ordinarily gives rise to condensation in the form of liquid or to sublimation in the form of ice. Latent heat of condensation or sublimation is liberated in the phase transition and becomes available to heat the air. The rate of heating from this process is designated by $Q_c$ in (4.1). Thus, the first step in the computation of $Q_c$ is to predict the large-scale moisture field.

The continuity equation of water vapor may be expressed by

$$\partial(\rho q)/\partial t + \nabla \cdot (\rho q \mathbf{V}) + \partial(\rho q w)/\partial z = M + \rho E, \tag{4.7}$$

where $M$ is the rate of condensation of water vapor per unit volume and $E$ is the time rate of change of water vapor content per unit mass due to the vertical and horizontal diffusion of water vapor.

If we assume that condensed water remains in the same air parcel as clouds, then we must set up two more equations dealing with the time rate of change of liquid water and ice content. Knowledge of cloud physics is needed to express the rate of phase transition between different forms of water substance. A part of the water substance will eventually fall from the air and the mechanism of precipitation must be considered. On the other hand, if we assume that all condensed and sublimated water substance falls immediately from the air, then the problem becomes much simpler. The calculation of condensation heating discussed in this volume usually assumes this simpler precipitation process. However, this assumption overestimates the amount of precipitation because evaporation of water droplets during the fall, thereby reducing the amount of precipitation, is not taken into account.

### D. Atmospheric Boundary Layer

The region from the earth's surface to an altitude approximately 1000 meters above is called the *atmospheric boundary layer*. Not only does kinetic energy dissipate by friction here, but the layer acts as an energy source transporting sensible heat and water vapor (latent heat) from the earth's surface to the interior of the atmosphere.

The region above the atmospheric boundary layer is the *free atmosphere*. Here frictional effects are generally neglected except intermittent turbulence on a subsynoptic scale caused by towering cumulus clouds, clear-air turbulence, and upward propagation of gravity waves.

The atmospheric boundary layer may be divided into two horizontal layers. The lowest layer, extending not more than 100 meters above the surface, is the *surface boundary layer*. Here the vertical fluxes of momentum, heat, and water vapor may be assumed independent of height and the atmospheric structure is primarily determined by characteristics of the earth's surface and thermal stratification. This layer is also referred to as the *constant flux* or *Prandtl layer*.

The domain between the surface boundary layer and the free atmosphere is the *planetary boundary layer*. Its atmospheric structure is influenced by pressure gradient force, Coriolis force, thermal stratification, and surface stress. It is usually called the *Ekman layer*, but lately is referred to as the *mixed layer*. The top of the mixed layer is ordinarily well-delineated by a solid or broken cloud cover having a stable free atmosphere above and below which turbulent motions are confined.

Frictional force $\mathbf{F}_L$ in (4.2) may be expressed by

$$\mathbf{F}_L = \rho^{-1}(\partial\tau/\partial z) + \mathbf{F}_H, \tag{4.8}$$

where $\mathbf{F}_H$ is the horizontal component of $\mathbf{F}_L$ and $\tau$ is the Reynolds stress given by

$$\tau = -\rho\overline{\mathbf{V}'w'}, \tag{4.9}$$

where the prime denotes the deviation of a quantity from the running time average indicated by the overbar.

The heating rate due to the vertical and horizontal diffusion of sensible heat $Q_b$ in (4.1) may be expressed by

$$Q_b = \rho^{-1}(\partial h/\partial z) + Q_H, \tag{4.10}$$

where $Q_H$ represents the horizontal component of eddy diffusion of sensible heat and $h$ denotes the vertical flux of sensible heat given by

$$h = C_p \rho \overline{\theta' w'}, \tag{4.11}$$

in which $\theta$ is potential temperature.

The time rate of change in water vapor content per unit mass due to eddy diffusion $E$ in (4.7) may be given by

$$E = -\rho^{-1}(\partial r/\partial z) + E_H, \tag{4.12}$$

where $E_H$ represents the horizontal component of diffusion of water vapor and $r$ denotes the vertical flux of water vapor, expressed by

$$r = \rho \overline{q' w'}. \tag{4.13}$$

To complete the formulation of the atmospheric boundary layer, we must specify the vertical fluxes of momentum $\boldsymbol{\tau}$, sensible heat $h$, and water vapor $r$ in terms of large-scale flow variables. Likewise, we must assume the forms of horizontal components $\mathbf{F}_H$, $Q_H$, and $E_H$. The art of parameterization is in choosing the specific forms for these terms (Section IV, F).

The vertical fluxes of momentum $\boldsymbol{\tau}_s$, sensible heat $h_s$, and water vapor $r_s$ in the surface boundary layer are empirically expressed by

$$\begin{aligned} \boldsymbol{\tau}_s &= \rho_s C_D V_s \mathbf{V}_s, \\ h_s &= -\rho_s C_p C_H V_s (\theta_s - \theta_g), \\ r_s &= -\rho_s C_E V_s (q_s - q_g), \end{aligned} \tag{4.14}$$

where subscript s refers to the values in the surface boundary layer, usually at the anemometer level about 10 meters above the earth's surface. Subscript $g$ refers to the values at the earth's surface; $V_s$ denotes the magnitude of $\mathbf{V}_s$; and $C_D$, $C_H$, and $C_E$ represent the *bulk transfer* (or exchange) *coefficient* for momentum, sensible heat, and water vapor, respectively.

Formulation (4.14) allows the linkage of atmospheric state—temperature and specific humidity—with surface temperature $T_g$, surface specific humidity $q_g$, and roughness of the ground. Surface temperature $T_g$ may be determined by applying a heat balance equation considering all factors influencing the heat balance at the ground. Incoming solar and terrestrial radiation fluxes are balanced by outgoing blackbody infrared radiation flux and sensible and latent heat transports to the atmosphere and into the soil. Similarly, surface specific humidity $q_g$ may be calculated by knowing the moisture budget at the earth's surface. Knowledge of the thermal characteristics of the lithosphere and ground hydrology related to soil moisture is required to solve heat and moisture budgets at the earth's surface.

## E. Influence of the Oceans

The oceans influence the atmosphere more than do the lands because they cover approximately 70% of the earth's surface and have large specific heat capacity. Just as with land, the important variable characterizing the thermal property for air–sea interactions is sea-surface temperature.

To determine sea-surface temperature, the same principle of surface energy balance holds for the oceans except that we now deal with the hydrosphere instead of the lithosphere. Knowledge of ocean circulations is needed to evaluate heat transport to the deeper oceans and the net gain or loss of heat transport by ocean currents.

In the polar regions, sea ice is the major factor controlling the exchange of heat, water vapor, and momentum between the ocean and the atmosphere. Ice cuts off the heat and water vapor transports from the ocean to the atmosphere and increases the albedo. Thus, similar to snow cover over land, sea ice contributes to cooling over the ice surface which, in turn, thickens the ice. In this sense, ice provides a positive climatic feedback because its presence causes a cold climate and produces more ice.

During the past 10 years, significant developments have evolved in the numerical modeling of ocean circulation (National Academy of Sciences, 1975). At the same time, many difficulties are brought to light in dealing with the physical and computational aspects of numerical modeling of oceans. Despite the many questions fundamental to the study of oceans, efforts have been made to couple an ocean model with a model of the atmosphere in investigating the circulations in the atmosphere and oceans as one interactive system rather than the ocean forced by the atmosphere or vice versa (Manabe *et al.*, 1975; Bryan *et al.*, 1975). Because of increased freedom in a joint atmosphere–ocean model by eliminating prescribed boundary conditions at the interface, stable solutions of such a joint system, if in existence, are not necessarily unique. This question has direct relevance to understanding the nature of climatic changes (Lorenz, 1970).

## F. Parameterization of Subgrid-Scale Processes

Any numerical model of the atmosphere must use a finite resolution in representing the continuous medium. Certain physical and dynamical phenomena, such as severe storms, cumulus convection, and gravity waves whose scales are smaller than the computational resolution, are not represented in large-scale models. We refer to such unrepresented motions as *subgrid-scale processes.* Here the term grid-scale refers to the physical scale corresponding to the finest computational resolution. Also, truncation errors arise in the integration of atmospheric equations. Owing to the nonlinearity of fluid dynamics equations, the use of finite resolution causes yet another misrepresentation—*aliasing errors.* Nonlinear interactions with the velocity field induce smaller scale structure with increasing complexity. This cascading process will be misrepresented in a model with a finite number of degrees of freedom and cause aliasing errors (Section V, H).

Contribution of subgrid-scale motions to larger scales may be judged first by the amount of energy contained in the subgrid-scale motions and second by the rate of energy transfer from small- to large-scale motions. It is customary to use Fourier representation to express the distribution of kinetic energy in longitudinal scales. Figure 2 is a collection of several observations of longitudinal spectra of eddy kinetic energy (Leith, 1971). The abscissa is the longitudinal wave number $k$ defined by $2\pi\, a \cos \phi/L$,

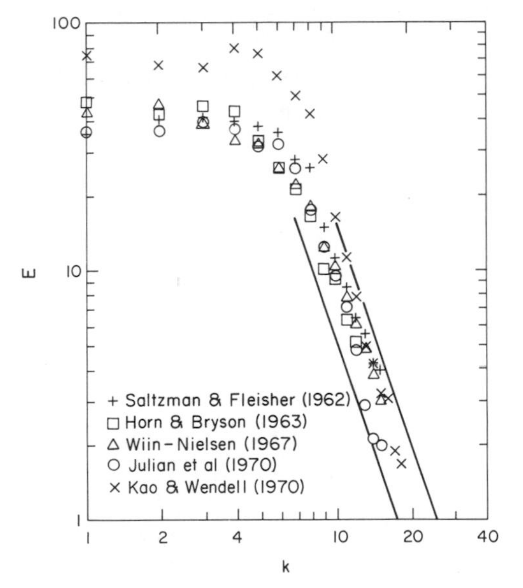


Fig. 2. Observations of longitudinal spectra of eddy kinetic energy $E$ in units of $10^{-4}$ rad$^2$ day$^{-2}$ as a function of longitudinal wave number $k$. From Leith (1971).

where $a$ is the earth's radius, $\phi$ the latitude, and $L$ the wavelength. Although these spectra are for a variety of seasons, latitudes, and pressure levels, all spectra at higher wave numbers show a tendency toward $k^{-3}$.

If we allow an extremely fine resolution in a numerical model, the ultimate energy dissipation takes place by molecular viscosity at very small scales. In practice, the smallest grid size is usually far larger than desired in an ideal situation. One objective in the parameterization of subgrid-scale motions is to simulate a physical mechanism to provide an energy sink withdrawing the energy comparable to cascading energy at the grid scale in an ideal situation.

Leith (1969) discussed the formulation of artificial or eddy viscosity coefficients designed to simulate the energy-cascading process based on three- and two-dimensional turbulence theories. Smagorinsky (1963) had proposed estimating the energy cascade rate from the deformation of flow assuming three-dimensional and quasi-nondivergent turbulence motions. Smagorinsky's form of eddy viscosity has been used to simulate the subgrid-scale dissipation terms $\mathbf{F}_s$ in (4.2) and $\mathbf{F}_H$ in (4.8), the horizontal diffusion of sensible heat $Q_H$ in (4.10), and the horizontal diffusion of water vapor $E_H$ in (4.12) (Smagorinsky *et al.*, 1965; Oliger *et al.*, 1970).

There are many subgrid-scale phenomena that have not yet been discussed. Bretherton (1972) described necessary steps for parameterization of subgrid-scale processes: (1) the phenomenon involved must be identified and its role meaningfully distinguished from other processes operating simultaneously in the atmosphere; (2) its importance for large-scale motion must be determined; (3) through case studies the relevant physics and dynamics must be understood; and (4) quantitative rules must be formulated for large-scale variables in a specific prediction model.

Although these steps are essential, the difficulty in carrying them out varies enormously depending on the physical phenomenon. In some cases, multiple physical processes such as thunderstorms operate simultaneously and a practical problem may arise in modeling the phenomenon. The importance of a particular physical process is even difficult to judge until at least a crude parameterization has actually been tested in a prediction model and experiments conducted with and without its parameterization.

For short-range forecasts, the influence of various subgrid-scale motions is relatively minor. For long-range prediction and climate simulations, the calculated evolution of the atmospheric circulation appears to be sensitive to the subgrid-scale vertical transport of momentum (e.g., Stone *et al.*, 1974) and, presumably, to the subgrid-scale vertical and horizontal transports of water vapor and energy as well as momentum. Better parameterizations of subgrid-scale motions including the ensemble of cumulus clouds have high priority in numerical modeling.

## G. Effects of Upper Boundary Conditions

The vertical extent of a prediction model is inevitably finite, and the question remains as to the effect of upper boundary conditions in the model. It has been suggested that the simple boundary condition of no mass flux at the model top (3.16) causes distortion of wave modes, and, therefore, the upper boundary condition may contribute to errors in weather prediction and climate simulation. An effect of this boundary condition may be seen by comparing two simulated Januaries, one by a 5° 12-layer model (Kasahara *et al.*, 1973) and the other by a 5° six-layer model (Kasahara and Washington, 1971). In the six-layer model, the top boundary condition $w = 0$ is applied at 18 km and in the 12-layer model at 36 km. The six lowest layers of the 12-layer model correspond exactly to the six-layer model of the troposphere. In comparing the tropospheric circulation, we did not find much improvement with the 12-layer model, except that the structure of the upper tropospheric jet streams improves. One conceivable reason for this unexpected result is that by not having the lid at 18 km in the 12-layer model, eddy kinetic energy in the troposphere is reduced by an amount transmitted to the lower stratosphere by the vertical propagation of planetary-scale waves. Although this mechanism is physically correct, it adversely affects the reduction in tropospheric eddy kinetic energy which is already smaller than observed in the six-layer model. The lid at 18 km in the six-layer model fortuitously helped contain eddy kinetic energy in the troposphere. This is an example that the introduction of realistic physics does not necessarily improve the result unless these particular physics are most sensitive in influencing the model performance. Incidentally, the simulation of the lower stratospheric circulation by the 12-layer model was realistic in many respects. The effect of the model lid appears to be confined to the two top layers in both the six- and 12-layer models (Williams, 1976).

For forced oscillations in a simple analytical model, the upper boundary condition may be selected by applying the so-called radiation condition. For free oscillations, however, the radiation condition depends on the frequency of wave motions. Therefore, it is difficult to implement the radiation condition as an upper boundary condition in numerical models. Since the radiation condition applied to inviscid and adiabatic motions is designed to simulate the absorption of upward-propagating waves in the upper atmosphere due to molecular viscosity and conductivity, one can introduce a small dissipation in the upper boundary conditions. Upward-propagating waves may also be absorbed at a critical level where the mean horizontal velocity is equal to the horizontal phase velocity and momentum is transferred to the mean flow. Thus, in a realistic numerical model where the mean horizontal velocity depends on height and effects of viscosity and conductivity are included at a sufficiently high level of the atmosphere, the selection of upper boundary conditions may not be as crucial as in a simple inviscid atmospheric model, except near the top boundary (GCM Steering Committee, 1975).

## V. Numerical Integration Methods

### A. Types of Differential Equations

The evolution of flow patterns may be determined by integrating relevant prediction equations with time starting from initial conditions at some moment $t = t_0$. Because differential equations are nonlinear and not easily solved analytically, we must resort to the numerical approach. The numerical methods vary according to the types of differential equations. There are two main types of differential equations—*ordinary* and *partial*. The former contains only one independent variable and its total derivatives, whereas the latter has several independent variables and partial derivatives.

Some essential numerical aspects of large-scale prediction equations are contained in the partial differential equations governing an incompressible, homogeneous, and hydrostatic fluid (such as water) in the form

$$\mathbf{V}_t + \mathbf{V} \cdot \nabla\mathbf{V} + f\mathbf{K} \times \mathbf{V} + g\,\nabla h = K\,\nabla^2\mathbf{V}, \tag{5.1a}$$

$$h_t + \nabla \cdot (h\mathbf{V}) = Q, \tag{5.1b}$$

where $h$ denotes the height of free surface, $\mathbf{V}$ the horizontal velocity, $K$ the eddy viscosity, and $Q$ the mass source. We use Cartesian coordinates $(x, y)$ as a reference, where $x$ and $y$ are directed eastward and northward, and $\mathbf{K}$ is a unit vector directed upward. The gradient operator on any scalar function $A$ is defined by

$$\nabla A = \mathbf{i}(\partial A/\partial x) + \mathbf{j}(\partial A/\partial y), \tag{5.2}$$

where $\mathbf{i}$ and $\mathbf{j}$ are unit vectors in $x$ and $y$ directions. Laplacian operator $\nabla^2 A$ is given by

$$\nabla^2 A = \partial^2 A/\partial x^2 + \partial^2 A/\partial y^2.$$

The right-hand side of (5.1a) denotes the viscous term contributing energy dissipation, and the right-hand side of (5.1b) represents the mass source corresponding to heating/cooling in the thermodynamic equation. Equations (5.1) without the viscous and mass source terms are called *shallow-water equations* or *barotropic primitive equations.*

If the right-hand side of (5.1a) dominates, the type is *parabolic*, but the viscous effect of large-scale motions is small. When the right-hand sides of (5.1) are ignored and we introduce a new variable

$$C_g = (gh)^{1/2}, \tag{5.3}$$

the shallow-water equations are written

$$w_t + \mathsf{A}w_x + \mathsf{B}w_y = \mathsf{R}w, \tag{5.4}$$

where $u$ and $v$ are the $x$ and $y$ components of velocity $\mathbf{V}$ and

$$w = \begin{pmatrix} u \\ v \\ 2C_g \end{pmatrix}, \qquad \mathsf{R} = \begin{pmatrix} 0 & f & 0 \\ -f & 0 & 0 \\ 0 & 0 & 0 \end{pmatrix},$$

$$\mathsf{A} = \begin{pmatrix} u & 0 & C_g \\ 0 & u & 0 \\ C_g & 0 & u \end{pmatrix}, \qquad \mathsf{B} = \begin{pmatrix} v & 0 & 0 \\ 0 & v & C_g \\ 0 & C_g & v \end{pmatrix}.$$

Because each matrix $\mathsf{A}$ and $\mathsf{B}$ has three distinct real eigenvalues, system (5.4) is *hyperbolic*. The eigenvalues correspond to the local characteristic velocities of the system. The right-hand side of (5.4) is a linear term on $w$; hence, the type of differential equation is not affected by the Coriolis term.

Applying $\mathbf{K} \cdot \nabla \times$ to (5.1a) without the viscous term and assuming that the flow is nondivergent $\nabla \cdot \mathbf{V} = 0$, we obtain the barotropic nondivergent vorticity equation:

$$\begin{aligned} \zeta_t &= J(\zeta + f, \psi) \\ &= (\zeta + f)_x \psi_y - (\zeta + f)_y \psi_x, \end{aligned} \tag{5.5}$$

where $\psi$ is stream function and $\zeta = \mathbf{K} \cdot \nabla \times \mathbf{V} = \nabla^2 \psi$ represents the vertical component of vorticity.

Equation (5.5) is a nonlinear partial differential equation of third order. To obtain stream function $\psi$ from vorticity $\zeta$, we need to solve Poisson's equation which is *elliptic*. Numerical methods to solve various quasi-geostrophic vorticity equations of type (5.5) are described, for example, in Thompson (1961) and Haltiner (1971).

Although we encounter various partial differential equations in atmospheric modeling, the hyperbolic type appears most often.

### B. Time-Differencing Schemes

If the space derivatives of a partial differential equation are expressed by finite differences (Section V, C) or spectral representation (Section V, F), the resulting equations become ordinary differential equations, involving time $t$ as the only independent variable, which may be expressed by

$$du/dt = F(u, t), \tag{5.6}$$

where the dependent variable $u = u(t)$ is a vector. An initial condition

$$u(0) = u_0 \tag{5.7}$$

is required to complete the system.

To integrate (5.6) numerically, we replace the time derivative with a finite-difference quotient. The resulting equation is a *difference equation.* The usefulness of a particular difference scheme for solving (5.6) depends on the functional form of $F(u, t)$.

Since we are concerned with wave motions, we consider an oscillatory motion described by

$$du/dt = i\omega u \tag{5.8}$$

with the same initial condition (5.7), where $\omega$ denotes frequency (constant). The solution then is given by

$$u(t) = u_0 \exp i\omega t. \tag{5.9}$$

To express a difference equation, we define grid points $t_n = n\,\Delta t$ with time step $\Delta t$ and $n = 0, 1, 2, \ldots$. We write $u_n$ to approximate $u(t_n)$.

By Taylor expansion, we obtain

$$\begin{aligned} u_{n+1} &= u_n + u_t\,\Delta t + \tfrac{1}{2}u_{tt}(\Delta t)^2 + \cdots \\ &= [1 + i\omega\,\Delta t - \tfrac{1}{2}(\omega\,\Delta t)^2 + \cdots]u_n \\ &= u_n \exp i\omega\,\Delta t, \end{aligned} \tag{5.10}$$

where $u_t$ and $u_{tt}$ are evaluated at $t = n\,\Delta t$. The above equation can be written as

$$u_{n+1} = Cu_n, \tag{5.11}$$

where $C$ is the *amplification factor.*

Repeating (5.11) from the initial condition (5.7) yields solution of the difference equation as

$$u_n = C^n u_0.$$

To have a meaningful solution $u_n$, $|C^n|$ must be bounded during the integration period irrespective of the choice of $\Delta t$ (Richtmyer and Morton, 1967). For stability, we must have $|C| \leqslant 1$.

For instance, we approximate $u_{n+1}$ by the first two terms in (5.10) as a simple extrapolation from $u_n$ with the gradient $u_t$ evaluated at $t = n\,\Delta t$. This is *Euler's scheme*, and it is unconditionally unstable because $C = 1 + i\omega\,\Delta t$ and the magnitude of $C$ is greater than unity unless $\omega = 0$. Note that the scheme is conditionally stable if it is applied to an exponentially damped problem (i.e., $\omega = i\nu$) where $\nu$ is real and positive). In this case, the stability condition is $\nu\,\Delta t \leqslant 2$. Also, the Euler scheme is explicit and first-order in accuracy as seen from the Taylor expansion (5.10).

Table I shows various time difference schemes, which appear often in atmospheric modeling, used to solve (5.6) and corresponding amplification factors $C$ applied specifically to (5.8). The magnitude of $C$ is a function of $\omega\,\Delta t$, and Fig. 3 shows the dependence of $|C|$ on $|\omega|\,\Delta t$ for various schemes.

TABLE I

VARIOUS TIME DIFFERENCE SCHEMES APPLIED TO EQ. (5.6)

Euler (explicit, first order, $E$)

$$U_{n+1} = U_n + F(U_n)\,\Delta t \qquad C = 1 + i\omega\,\Delta t$$

Backward (implicit, first order, $B$)

$$U_{n+1} = U_n + F(U_{n+1})\,\Delta t \qquad C = (1 - i\omega\,\Delta t)^{-1}$$

Crank–Nicholson (implicit, second order, $C$)

$$U_{n+1} = U_n + \tfrac{1}{2}[F(U_n) + F(U_{n+1})]\,\Delta t$$
$$C = (1 + \tfrac{1}{2}i\omega\,\Delta t)(1 - \tfrac{1}{2}i\omega\,\Delta t)^{-1}$$

Leap-frog (explicit, second order, $L_1$, $L_2$)

$$U_{n+1} = U_{n-1} + 2F(U_n)\,\Delta t$$
$$C_{1,2} = i\omega\,\Delta t \pm [1 - (\omega\,\Delta t)^2]^{1/2}$$

Iterative schemes (explicit)

$$U^*_{n+1} = U_n + F(U_n)\,\Delta t$$
$$U_{n+1} = U_n + [(1 - \alpha)F(U_n) + \alpha F(U^*_{n+1})]$$
$$C = 1 + i\omega\,\Delta t - \alpha(\omega\,\Delta t)^2$$

(a) $\alpha = 1$, Euler-backward (first order, $M$).

(b) $\alpha = 0.5$, Euler-midpoint (second order, $I$).

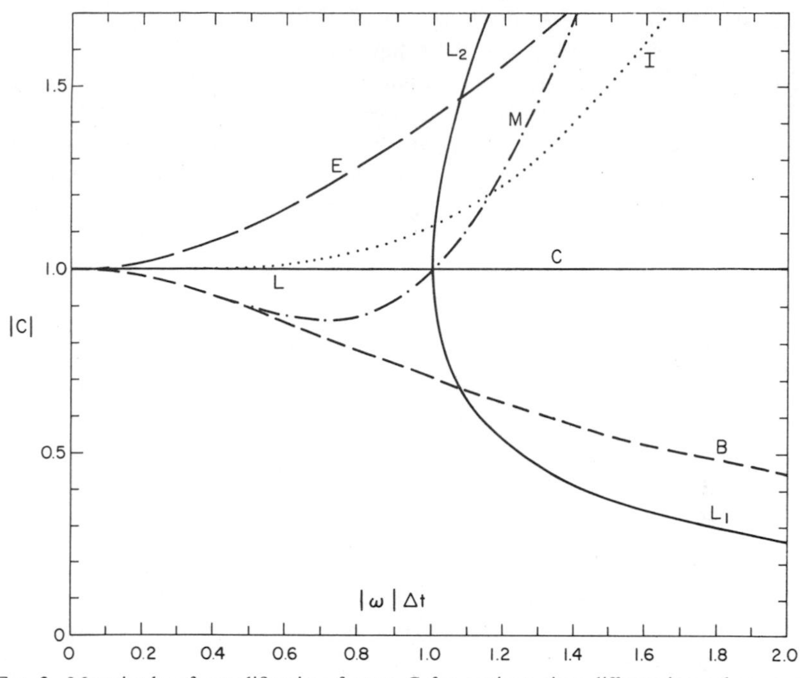


FIG. 3. Magnitude of amplification factor $C$ for various time-differencing schemes as a function of $|\omega|\Delta t$.

In Table I, within parentheses following the scheme, we indicate explicit or implicit and first or second order in accuracy. The letter $E$, $B$, $C$, etc., corresponds to the stability curve (Fig. 3). If $|C|$ is greater than unity, the scheme is unstable.

The Crank–Nicholson scheme is accurate and stable for a long time integration, but due to its implicit nature the integration can be time-consuming. Leap frog is the most common scheme in atmospheric modeling, but because three time levels are involved the scheme produces two independent solutions.

In the limit $\Delta t \to 0$, $C_1{}^n \to e^{i\omega t}$ and $C_2{}^n \to (-1)^n e^{-i\omega t}$, so that the first solution is in a physical mode and the second a computational mode (Platzman, 1954). The computational mode may oscillate every time step with a phase lag of approximately 180° relative to the physical mode. The computational mode must be controlled to avoid separation of the two modes.

To control mode separation with the leap-frog scheme, we may apply a smoother or a time filter. The Euler-backward scheme proposed by Kurihara (1965a) and Matsuno (1966a) is often used to eliminate high-frequency oscillations during numerical integrations of primitive equation models

(Arakawa, 1972). The rate of damping of Euler-backward scheme $M$ for $|\omega|\,\Delta t < 0.6$ is similar to backward scheme $B$.

The last scheme in Table I is second order in accuracy but is unstable and is not recommended. Matsuno (1966b) further formulated three-step schemes which are second and third order in accuracy and give damping for high-frequency motions. They have the desirable characteristics of time filters, since only high-frequency motions are eliminated while low-frequency motions are virtually unaffected. These iterative schemes are similar to the well-known Runge–Kutta method (e.g., Milne, 1953; Henrici, 1962). For more time-differencing schemes in numerical weather prediction, see Miyakoda (1962), Kurihara (1965a), Lilly (1965), and Young (1968).

## C. Finite-Difference Method

We consider a system of conservation equations

$$w_t + F_x = 0, \tag{5.12}$$

where $w$ is a vector whose components are dependent variables. We assume a quasi-linear system in that $F_x = Aw_x$, where $A$ is a matrix whose components depend on $w$ only. Equation (5.12) is hyperbolic if the eigenvalues of $A$ are all real.

To solve (5.12) for $t > 0$ with prescribed values at $t = 0$,

$$w(x, 0) = w_0(x)$$

in the domain $-\infty < x < \infty$ is referred to as the *initial value problem* (Cauchy problem). If an infinite space domain is physically unrealistic, we adopt instead a finite space domain with periodic boundary conditions. When periodic boundary conditions are not appropriate, we no longer have a pure initial value problem, but a *mixed initial and boundary value problem* with specified boundary conditions.

To express the difference equation, we divide the $(x, t)$ plane by the set of points given by $x_j = j\,\Delta x$ and $t_n = n\,\Delta t$, where $\Delta x$ and $\Delta t$ are increments of $x$ and $t$ and $j = \ldots, -2, -1, 0, 1, 2, \ldots$ and $n = 0, 1, 2, \ldots$ The grid points are given at $t_n = n\,\Delta t$ and $x_j = j\,\Delta x$. An approximation to $w(x_j, t_n)$ is denoted by $W_j^n$.

A simple but most important explicit second-order difference scheme is the leap-frog or step-over scheme originally used by Richardson (1922):

$$W_j^{n+1} = W_j^{n-1} - (\Delta t/\Delta x)(F_{j+1}^n - F_{j-1}^n). \tag{5.13}$$

To analyze the stability of this scheme, we further assume that matrix $A \equiv dF/dW$ is constant, so that $F^n_{j+1} - F^n_{j-1} = A(W^n_{j+1} - W^n_{j-1})$. Hence, the dependence of $W_j^n$ on $x$ can be expressed by

$$W^n_j = \hat{W}^n \exp(ikj\,\Delta x), \tag{5.14}$$

where $k$ is the wave number and $\hat{W}^n$ is the amplitude. The solution $W^{n+1}_j$ of (5.13) has the same form as (5.14), but the amplitude is given by

$$\hat{W}^{n+1} = G\hat{W}^n, \tag{5.15}$$

where $G$ is the *amplification matrix*. Substituting (5.14) and (5.15) into (5.13), we find that $G$ satisfies the equation

$$G^2 + 2i[A(\Delta t/\Delta x)\sin(k\,\Delta x)]G - 1 = 0. \tag{5.16}$$

To obtain bounded solutions of (5.13), as in the differential equation (5.12), the eigenvalues of $G$ should not exceed unity in absolute value. This is essentially the von Neumann necessary condition for stability of a finite-difference scheme (Richtmyer and Morton, 1967).

To satisfy the stability condition for the leap-frog scheme, we see from (5.16) that

$$|a|\,\Delta t/\Delta x \leqslant 1, \tag{5.17}$$

where $a$ is an eigenvalue of matrix $A$. In this case, we have $|G| = 1$ for all values of $k\,\Delta x$. The general solution of (5.12) is a function of $x - at$ and represents a wave moving with velocity $a$. In the difference system (5.13), a perturbation at a grid point can move only one mesh interval $\Delta x$ in one time step $\Delta t$. The velocity of influence is then $\Delta x/\Delta t$. The stability condition (5.17) requires that the wave speed $|a|$ must be less than or equal to the speed of grid influence. This condition is usually needed for explicit difference schemes for a hyperbolic system as pointed out by Courant *et al.* (1928).

Since the von Neumann stability analysis is based on the Fourier transform of a difference equation, the condition is applicable only to linear equations with constant coefficients and periodic boundary conditions. Yet the von Neumann condition is an important measure of stability and must be satisfied in the formulation of finite-difference equations.

The leap-frog scheme is very useful and perhaps the most important of explicit methods, but there is a well-known deficiency as already pointed out in Section V, B. Because the computational mesh can be divided into two parts and the difference calculation proceeds independently of the two parts, errors in the computation may cause out-of-phase values on the two meshes. The mesh separation is sometimes caused by starting from poor initial conditions which require two time levels $t = 0$ and $t = \Delta t$. A remedy such as smoothing often prevents separation of values on the two meshes.

Because of the large number of difference schemes to solve (5.12) and no space here to describe them, the reader may refer to Richtmyer (1963), Taylor *et al.* (1972), Anderson (1974), Strang (1968), Eilon *et al.* (1972), Turkel (1974), and Gottlieb and Turkel (1974). Most of these are variants of Lax–Wendroff (1960, 1964) for one- and two-dimensional space problems with and without viscous effects. Lax–Wendroff schemes are effective for high-speed flows, particularly the propagation of discontinuities (Houghton and Kasahara, 1968), but they are often too dissipative for low-speed flows such as large-scale atmospheric flows.

One aspect of solving a multidimensional problem by explicit difference methods is that the condition of stability often becomes more stringent than in the one-dimensional case. This difficulty may be avoided by application of *fractional steps* or *splitting-up*. The idea is to solve a multidimensional problem by dividing it into a series of steps, each containing difference operations of only one dimension or some parts of the differential equation involved. The procedure was originally developed for solution of parabolic and elliptic partial differential equations, known as the alternate direction method (Peaceman and Rachford, 1955; Douglas, 1955). A similar approach was extended to solution of multidimensional hyperbolic problems and used by Marchuk (1964) and Leith (1965) for numerical weather prediction. The advantage of the fractional step is that computational stability is subject to the difference scheme of each step rather than of the combined steps. Hence, it is often possible to take a larger time step than required without splitting-up. For more information, see Crowley (1967), Fromm (1968), Strang (1968), Gourlay and Morris (1970), Marchuk (1974), and Turkel (1974).

Difference schemes discussed so far are mostly of second-order accuracy in both time and space. To increase their accuracy, attempts have been made to formulate third- and fourth-order accuracy schemes. For example, Burstein and Mirin (1969) developed a third-order method based on a three-step approach in analogy to the two-step procedure by Richtmyer (1963). Zwas and Abarbanel (1971) formulated third- and fourth-order accuracy schemes using the same basic approach of Lax and Wendroff (1960). These examples show a great deal of complication in deriving higher order schemes, even for one-dimensional problems.

One compromise in reducing the arithmetics in higher order schemes is to use a fourth-order space difference with a second-order time differencing.

This approach does not give uniform fourth-order accuracy in space and time, but in numerical weather prediction truncation errors are usually more severe in space differencing than time differencing, since it is easier to take a shorter time increment than a smaller space increment because of computer storage limitations. Molenkamp (1968) compared the performance of several second- and fourth-order schemes with the advection equation and showed a gain in accuracy of the fourth-order scheme by Roberts and Weiss (1966) over second-order schemes. Kreiss and Oliger (1972) and Williamson and Browning (1973) tested the fourth-order leap-frog scheme for shallow-water equations and demonstrated an increase in accuracy over the second-order leap-frog scheme. Gerrity *et al.* (1972) made a similar comparison of second- and fourth-order accuracy difference schemes and confirmed the effectiveness of the fourth-order scheme.

### D. Quadratic Conserving Schemes



The von Neumann stability condition discussed in Section V, C is based on the Fourier transform. A different type of stability analysis—the *energy method*—is also useful, particularly for nonlinear problems. The essential idea is to examine the variance of solution of a difference equation; the difference scheme is stable if the total variance of solution within the integration domain does not increase faster than by $1 + 0(\Delta t)$. Such a difference equation is called *quadratic conserving scheme.* A theoretical discussion of the energy method is presented, for example, in Richtmyer and Morton (1967). Here, we consider the practical aspects of various conserving schemes developed for numerical weather prediction.

As an example of a conserving scheme, consider a simple nonlinear hyperbolic equation

$$\partial u/\partial t = -u(\partial u/\partial x), \tag{5.18}$$

where $u$ is the dependent variable defined in a cyclic continuous domain $x = 0$ to $x = L$. We observe that

$$\frac{\partial}{\partial t}\int_0^L u^\gamma \, dx = 0.$$

where $\gamma$ is a natural number.

The well-known space-centered scheme

$$\frac{\partial u_j}{\partial t} = -\frac{(u_{j+1})^2 - (u_{j-1})^2}{4\Delta x}, \tag{5.19}$$

where $u(x_j, t)$ is denoted by $u_j$, conserves the first moment $\gamma = 1$, or momentum, as expressed by

$$\frac{\partial}{\partial t}\sum_{j=1}^{J} u_j^{\gamma} = 0, \tag{5.20}$$

where $J = L/\Delta x$, a natural number.

Conservation of total momentum does not ensure computational stability since the magnitude of solution is not necessarily bounded in the domain. On the other hand, if the total sum of the second moment $\gamma = 2$, or kinetic energy, is conserved, the magnitude of solution must be bounded. Thus, a difference scheme may be stable if it conserves the sums of both $u_j$ and $u_j^2$. Inspection can verify that (5.19) does not conserve the sum of $u_j^2$.

We rewrite (5.18) in the form

$$\frac{\partial u}{\partial t} = -\frac{1}{3}\left[u\frac{\partial u}{\partial x} + \frac{\partial}{\partial x}(u^2)\right] \tag{5.21}$$

and approximate it by

$$\frac{\partial u_j}{\partial t} = -\frac{1}{3}\left[u_j\left(\frac{u_{j+1} - u_{j-1}}{2\Delta x}\right) + \frac{(u_{j+1})^2 - (u_{j-1})^2}{2\Delta x}\right]. \tag{5.22}$$

We can show that the above scheme conserves the sums of both $u_j$ and $u_j^2$ (Fornberg, 1973; Kreiss and Oliger, 1973); hence, (5.22) is stable. Actually, the scheme is only quasi-conserving if the time derivative is discretized by a leap-frog procedure. Nevertheless, quadratic conserving schemes are less prone to nonlinear instability (Section V, H).

The nondivergent barotropic vorticity equation (5.5) without Coriolis term $f = 0$ yields the conservation laws of total vorticity, squared vorticity, and kinetic energy. In a domain $S$ enclosed by a rigid boundary along which $\psi$ vanishes, the area integrals of the time rate of change of $\zeta$, $\zeta^2$, and $|\nabla\psi|^2$ vanish (Fjørtoft, 1953). Arakawa (1966) invented unique second- and fourth-order difference schemes to conserve the sums of vorticity, squared vorticity, and kinetic energy over the domain. Arakawa's conserving schemes have been used extensively in many fluid dynamics problems (e.g., Lilly, 1969; Orszag, 1971a; Bretherton and Karweit, 1975).

Arakawa's idea (1966) to develop a quadratic conserving scheme was extended by Lilly (1965) to formulate a total energy-conserving scheme for the shallow-water equations. Bryan (1966) also discussed formulation of a quadratic conserving scheme for the two-dimensional incompressible advection equation applied to irregularly spaced grid points.

We demonstrate the formulation of total energy-conserving schemes for shallow-water equations (5.1). Without dissipation and source terms, (5.1a, b) are rewritten as

$$\partial(h\mathbf{V})/\partial t + \nabla \cdot (\mathbf{V}h\mathbf{V}) + f\mathbf{K} \times \mathbf{V}h + \nabla(gh^2/2) = 0, \tag{5.23}$$

$$(\partial h/\partial t) + \nabla \cdot (h\mathbf{V}) = 0. \tag{5.24}$$

Combination of the above two equations yields the equation of total energy

$$\partial[\tfrac{1}{2}(\mathbf{V} \cdot \mathbf{V}h + gh^2)]/\partial t + \nabla \cdot [\mathbf{V}(\tfrac{1}{2}\mathbf{V} \cdot \mathbf{V}h + gh^2)] = 0. \tag{5.25}$$

The space derivative terms in (5.23)–(5.25) vanish over the closed domain $S$ enclosed by the boundary along which normal velocity vanishes. Thus, (5.23) without the Coriolis term, (5.24) and (5.25) express conservation of momentum, mass, and total energy, respectively.

We consider a two-dimensional rectangular grid with horizontal increments of $\Delta x$ and $\Delta y$. All dependent variables are given at grid points $x_j = j\,\Delta x$ and $y_l = l\,\Delta y$, where $j$ and $l$ are natural numbers, shown by dots in Fig. 4. We introduce supplementary points denoted by crosses. The dependent variables are not given at these supplementary points, but the values may be generated by some interpolation from neighboring points. At this stage we simply assume that some values are given at these supplementary points and discuss later how to obtain interpolated values.

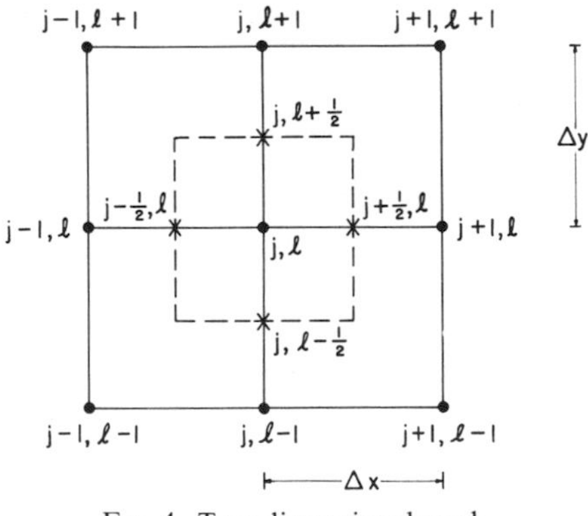


FIG. 4. Two-dimensional mesh.

We define the difference operators

$$\delta_x A = [A(x_j + (\Delta x/2), y_l) - A(x_j - (\Delta x/2), y_l)]/\Delta x,$$

$$\delta_y A = [A(x_j, y_l + (\Delta y/2)) - A(x_j, y_l - (\Delta y/2))]/\Delta y,$$

for any variable $A$. We combine the two operators to give

$$\begin{aligned} \boldsymbol{\delta} A &= \delta_x A\mathbf{i} + \delta_y A\mathbf{j}, \\ \boldsymbol{\delta} \cdot \mathbf{V} &= \delta_x u\mathbf{i} + \delta_y v\mathbf{j}, \end{aligned} \tag{5.26}$$

where **i** and **j** are unit vectors in the $x$ and $y$ directions.

Using notation (5.26), we write a mass-conserving difference form of (5.24) applicable at grid point $(j, l)$:

$$(\partial h/\partial t) + \boldsymbol{\delta} \cdot (h\mathbf{V}) = 0. \tag{5.27}$$

Differentiating $h\mathbf{V}$ in (5.23) individually with respect to time and utilizing (5.24), we have

$$h(\partial \mathbf{V}/\partial t) + \nabla \cdot (\mathbf{V}h\mathbf{V}) + f\mathbf{K} \times \mathbf{V}h + \nabla(gh^2/2) - \mathbf{V}\nabla \cdot (h\mathbf{V}) = 0.$$

We write a difference equation approximating the above:

$$h(\partial \mathbf{V}/\partial t) + \boldsymbol{\delta} \cdot (\mathbf{V}h\mathbf{V}) + f\mathbf{K} \times \mathbf{V}h + \boldsymbol{\delta}(gh^2/2) - \mathbf{V}\boldsymbol{\delta} \cdot (h\mathbf{V}) = 0. \tag{5.28}$$

This differencing scheme conserves momentum if the Coriolis term is neglected.

Using (5.27) and (5.28), we form the total energy equation:

$$\begin{aligned} h\frac{\partial}{\partial t}\left(\frac{\mathbf{V} \cdot \mathbf{V}}{2}\right) + \frac{\mathbf{V} \cdot \mathbf{V}}{2}\frac{\partial h}{\partial t} + gh\frac{\partial h}{\partial t} &= -\mathbf{V} \cdot [\boldsymbol{\delta} \cdot (\mathbf{V}h\mathbf{V})] - \mathbf{V} \cdot \boldsymbol{\delta}(gh^2/2) \\ &\quad + \tfrac{1}{2}(\mathbf{V} \cdot \mathbf{V})\boldsymbol{\delta} \cdot (h\mathbf{V}) - gh\, \boldsymbol{\delta} \cdot (h\mathbf{V}). \end{aligned}$$

The left-hand side is the time rate of change of total energy. To conserve total energy, the sum of the right side over all grid points, indicated by $\Sigma$, in the closed domain must vanish. This requires that

$$\sum \tfrac{1}{2}(\mathbf{V} \cdot \mathbf{V})\boldsymbol{\delta} \cdot (h\mathbf{V}) = \sum \mathbf{V} \cdot [\boldsymbol{\delta} \cdot (\mathbf{V}h\mathbf{V})], \tag{5.29}$$

$$\sum \mathbf{V} \cdot \boldsymbol{\delta}(gh^2/2) + \sum gh\, \boldsymbol{\delta} \cdot (h\mathbf{V}) = 0. \tag{5.30}$$

Also, to conserve the total mass summed over all grid points, we require that

$$\sum \boldsymbol{\delta} \cdot (h\mathbf{V}) = 0. \tag{5.31}$$

In applying operators $\delta_x$ and $\delta_y$, we must define the values of the dependent variables at the supplementary points shown in Fig. 4. The essence of formulating a scheme of conserving total energy as well as momentum and mass is to define the supplementary values so as to satisfy the requirements (5.29)–(5.31).

We can show that the following two schemes satisfy requirements (5.29)–(5.31):

*scheme I*:

$$\frac{\partial hu}{\partial t} + \delta_x(\overline{hu}^x \overline{u}^x) + \delta_y(\overline{hu}^y \overline{v}^y) - fhv + gh\,\delta_x(\overline{h}^x) = 0,$$

$$\frac{\partial hv}{\partial t} + \delta_x(\overline{hu}^x \overline{v}^x) + \delta_y(\overline{hv}^y \overline{v}^y) + fhu + gh\,\delta_y(\overline{h}^y) = 0,$$

$$\frac{\partial h}{\partial t} + \delta_x(\overline{hu}^x) + \delta_y(\overline{hv}^y) = 0. \tag{5.32}$$

*scheme II*:

$$\frac{\partial hu}{\partial t} + \delta_x(\overline{h}^x \overline{u}^x \overline{u}^x) + \delta_x(\overline{h}^y \overline{u}^y \overline{v}^y) - fhv + g\,\overline{\delta_x(h^2/2)}^x = 0,$$

$$\frac{\partial hv}{\partial t} + \delta_x(\overline{h}^x \overline{u}^x \overline{v}^x) + \delta_y(\overline{h}^y \overline{v}^y \overline{v}^y) + fhu + g\,\overline{\delta_y(h^2/2)}^y = 0,$$

$$\frac{\partial h}{\partial t} + \delta_x(\overline{h}^x \overline{u}^x) + \delta_y(\overline{h}^y \overline{v}^y) = 0. \tag{5.33}$$

where, for any scalar $A$,

$$\begin{aligned} \overline{A}^x &= \tfrac{1}{2}[A(x_j + (\Delta x/2), y_l) + A(x_j - (\Delta x/2), y_l)], \\ \overline{A}^y &= \tfrac{1}{2}[A(x_j, y_l + (\Delta y/2)) + A(x_j, y_l - (\Delta y/2))]. \end{aligned} \tag{5.34}$$

This averaging notation was used by Shuman (1962).

Scheme I is identical to scheme B of Grammeltvedt (1969), which is the Cartesian version of the conserving scheme formulated by Grimmer and Shaw (1967) for spherical coordinates. Modification of the present schemes for a spherical grid is straightforward. The spherical version of scheme II is identical to a conserving scheme used by Tiedtke (1972), who compared the results of numerical calculations with the spherical versions of schemes I and II and found no significant differences between the performance of two schemes for a 10-day period. Gary (1969) tested a scheme similar to II for a spherical grid in conjunction with a box-type grid developed by Kurihara (1965b). The spherical version of scheme II was used also by Holloway *et al.* (1973). Williamson (1971) and Sadourny (1972) formulated conserving schemes on a spherical geodesic grid (Section V, E).

In the present formulation, all dependent variables are given at the same grid points (Fig. 4), but this is not necessary to derive conserving schemes. Lilly (1965) and Grammeltvedt (1969) formulated total energy-conserving schemes for a staggered mesh in which $u$, $v$, and $h$ are placed at different locations.

As mentioned earlier, total energy is exactly conserved as long as we do not discretize the time derivative. If we use Crank–Nicholson implicit time differencing, total energy is conserved for the difference equations. On the other hand, if we use leap-frog time differencing, the difference equations are only quasi-conserving and nonlinear instabilities (Section V, H) may occur. Such instabilities, however, may be controlled by a very small amount of dissipation (Kreiss and Oliger, 1973).

The second-order energy-conserving scheme requiring averaging operators is generally more time consuming than a straightforward second-order leap-frog scheme on a staggered mesh such as used by Washington and Kasahara (1970). Gary (1969) compared two difference schemes—one energy-conserving and the other not—on two different global grid systems. Although the conserving scheme was much less prone to nonlinear instability, it was not entirely obvious which scheme gave more accurate results for the same computing time and storage.

Kreiss and Oliger (1972) showed that a centered fourth-order scheme with a small dissipation term satisfies the bound of first and second moments of solution. This approach seems more accurate than second-order energy-conserving schemes for the same amount of computing.

## E. Problems in Mapping Flows over a Sphere

To represent flows over a sphere with a quasi-homogeneous grid, a rectangular grid on a map projection is often used. For example, a stereographic projection may be suitable for hemispherical flows. To extend mapping to the entire globe, two stereographic grids may be used, one for each hemisphere with an overlapping of the two grids over the equator where interpolation is required to combine the two grids (Stoker and Isaacson, 1975). Actually, a Mercator projection has less map distortion in lower latitudes. Phillips (1959a) designed a combination grid with a polar stereographic projection over higher latitudes and a Mercator projection over lower latitudes. The two projection grids overlapped in the midlatitudes and an interpolation was used to transform values at grid points from one projection to the other. Although the combination grids provided a fairly uniform distribution of grid points over a sphere, the need for an interpolation procedure adds complexity in dealing with the combination grids.

To design a quasi-homogeneous grid over a sphere, Sadourny *et al.* (1968) and Williamson (1968) used a *spherical geodesic grid* based on the icosahedron. The icosahedron is the highest degree regular polyhedron consisting of 20 equilateral-triangular faces constructed within a sphere with its 12 vertices on the sphere. The vertices of the triangles can be connected by great circles to form 20 congruent major spherical triangles which are then divided into smaller grid triangles.

Sadourny *et al.* (1968) and Williamson (1968) developed and tested difference schemes for the barotropic nondivergent vorticity equation over spherical geodesic grids. Williamson (1970) also developed conserving difference schemes for the barotropic primitive equations and found that the grid resolution must be less than $2\frac{1}{2}^{\circ}$ in order that truncation errors not dominate the mass flux calculation. Williamson (1971) analyzed the source of truncation errors and found that the schemes are first order because of the nonuniform nature of the grid intervals. Although the major triangles are equilateral, the grid intervals of the divided small grids vary on the order of 10% of the mean value. This small variation produces truncation errors of first order that are significant for mass divergence calculations, but less significant for advection calculations.

It is natural to construct a spherical grid with constant increments in longitude and latitude. This grid system—the *longitude–latitude grid*—has a number of computational disadvantages. Poles are singular where wind components in spherical coordinates are undefined and a computational stability condition is stringent near the poles. The maximum allowable time step with an explicit time-difference scheme is reduced proportionally to the cosine of latitude because of the convergence of meridians near the poles. Hence, a small time step is required leading to an excessively long computation if applied to the entire grid.

To avoid this computational difficulty, a number of expediences are proposed. The reduction of grid points along parallels (full circles at constant latitude) near the poles is an obvious solution to the pole problem and many researchers have designed various quasi-homogeneous grids.

Kurihara (1965b) formulated a spherical grid in which the grid points were placed on parallels and the number of points reduced gradually in successive parallels at higher latitudes such that the area of each grid square enclosing the grid points was approximately constant over the entire sphere. However, the longitude of grid points over one parallel did not coincide with that on consecutive parallels. Kurihara grids were used by Kurihara and Holloway (1967) for general circulation experiments and by Miyakoda *et al.* (1971) for short-range forecasts.

Holloway and Manabe (1971) noted the problems when the Kurihara grid is used for a long-time integration of a global general circulation model. Excessively high pressure developed over polar regions and low- and high-pressure belts shifted toward the equator. These deficiencies seemed to be due to inaccuracy in the calculation of divergence caused by lack of sufficient resolution near the poles. Dey (1969) observed similar deficiencies in use of the equal-area grid for short-range forecasts with real data. Sankar-Rao and Umscheid (1969) investigated the accuracy of the equal-area grid applied to the barotropic primitive equations and found that it produces a large distortion of wave patterns near the poles unless the longitudinal grid spacing is sufficiently reduced near the poles. Shuman (1970) analyzed the error of divergence calculation associated with a coarse grid resolution near the pole and concluded that this type of error arises from inaccuracy in dealing with the curvilinear components of the velocity vector in spherical coordinates and may be eliminated by using a grid with a constant longitude increment. A similar conclusion was reached by Williamson and Browning (1973). Earlier, Gates and Riegel (1962) examined the phase error of non-divergent wave propagation in the grid in which the longitudinal grid resolution was decreased from lower to higher latitudes.

To limit the reduction of longitudinal grid resolution to the polar regions, Washington and Kasahara (1970) in a general circulation model used a spherical grid where points were spaced at 5° longitude and latitude intervals below 70° latitude and the longitudinal increment was increased to 10°, 10°, and 30° at 75°, 80°, and 85° latitudes, respectively. Thus, at 85° latitude there were 12 grid points compared with only 4 points for the equal-area grid with comparable resolution. Oliger *et al.* (1970) improved the method of skipping grid points in the polar regions to gradually reduce the grid points poleward from 60° latitude.

Gary (1969) compared two difference schemes formulated for two grid systems—the equal-area grid of Kurihara and the skipped grid near the poles. Although Gary's objective was to test the two difference schemes used for general circulation studies, he pointed out the need for sufficient grid resolution near the poles for the satisfactory performance of any difference scheme. Kida (1974) also tested two difference schemes with grid points skipped near the poles and reaffirmed the need for a sufficient grid resolution near the poles.

It is desirable to use constant latitude and longitude angular increments for difference schemes if the spherical coordinate system is employed. Grimmer and Shaw (1967) formulated a conserving difference scheme in a longitude–latitude grid and used time steps varied latitudinally to avoid computational instability. They also examined computational errors caused by reducing the longitudinal grid resolution near the poles. Tiedtke (1972) made a similar study.

The use of shorter time steps for the polar region of a constant latitude–longitude grid, practiced by Grimmer and Shaw (1967) and Corby *et al.* (1972), is perhaps best suited to calculate the flow over a sphere with an explicit difference scheme. However, it is computationally expensive to use short time steps even only over the polar regions. (Grimmer and Shaw pointed out that over 50% of the computing time was spent integrating the two northernmost rows, which represent only about 2% of the hemispheric area.)

Another solution to the pole problem is to use an implicit time difference scheme only in the longitudinal direction so that the same time step may be used throughout the entire globe. This approach has not yet been tested.

A simpler approach is to filter out short waves subject to linear computational instability near the poles. In this way, the same time increment can be used throughout the grid. In the Arakawa–Mintz general circulation model (Gates *et al.*, 1971), smoothing was performed on the zonal mass flux in the divergence terms and the zonal pressure gradient in the prediction equations. This was designed to eliminate fast gravity waves in the zonal direction. Vanderman (1972) applied weighting averaging in the east–west direction to the tendency values of dependent variables near the poles to eliminate short wave disturbances.

Umscheid and Sankar-Rao (1971) and Arakawa (1972) used Fourier filtering to eliminate short wave components in the zonal mass flux and the zonal pressure gradient near the poles. Williamson (1976) also applied longitudinal Fourier filtering to all prognostic variables to eliminate short wavelength, fast-moving waves near the poles. Holloway *et al.* (1973) applied space filtering to all prognostic variables at each time step. The filtering was performed by Fourier analysis and limited the east–west wavelength of the shortest wave at any latitude to approximately twice the grid increment at the equator. For wind velocity, Holloway *et al.* (1973) considered it desirable to transform velocity components onto a polar stereographic projection before Fourier filtering was performed. However, Merilees (1974a) pointed out that Fourier filtering for zonal and meridional velocity components can be performed without transformation onto a stereographic projection.

## F. Spectral Methods

In the numerical integration of prediction equations, so far we approximate any dependent variable in terms of its values at discrete points in physical space and time. The evolution of the flow field is determined by extrapolating in time its values at the grid points in space.

An alternative is to represent the field of any dependent variable in physical space as a finite series of smooth and, preferably, orthogonal functions. In this case, the prediction equations are expressed by ordinary differential equations for the expansion coefficients, which depend only on time. Since expansion coefficients are referred to as spectra, we call this approach the *spectral method.*

To demonstrate the basic idea of the spectral method, we consider the differential equation

$$\partial w/\partial t = F(w) \tag{5.35}$$

over a closed domain $S$ with a suitable boundary condition on $w$ along the domain boundary. The dependent variable $w$ is a function of independent variable $x$. We approximate $w$ by $W$ in terms of a finite series of linearly independent basis functions $\phi_j(x)$:

$$W(x, t) = \sum_{j=1}^{N} C_j(t)\phi_j(x), \tag{5.36}$$

where the coefficients $C_j(t)$ are functions of time $t$.

In deriving the prediction equations for expansion coefficients $C_j(t)$, we make use of the *Galerkin approximation.* Namely, the residue $\partial W/\partial t - F(W)$, obtained by substitution of (5.36) into (5.35), is forced to be zero in an averaged sense over domain $S$ with basis functions $\phi_j$ as weights:

$$\int_S \phi_k \left( \frac{\partial W}{\partial t} - F(W) \right) dx = 0 \qquad \text{for} \quad 1 \leqslant k \leqslant N. \tag{5.37}$$

From this integral, we obtain a system of $N$ equations for $N$ unknowns $dC_k/dt$. This system of $N$ equations can be simplified considerably if the basis functions are orthonormal, so that $\int \phi_j \phi_k \, dx = 1$ if $j = k$ and 0 if $j \neq k$ when the integration is performed over domain $S$.

The spectral form of the two-dimensional and nondivergent vorticity equation (5.5) over a sphere was formulated by Silberman (1954), Platzman (1960), and Baer and Platzman (1961) using spherical harmonics as basis functions. Results of extensive computations with simulated and observed data were presented by Kubota *et al.* (1961), Baer (1964), and Ellsaesser (1966).

The advantages of the spectral form of the vorticity equation over a sphere are as follows. (1) A mapping of the sphere is not necessary, so that the difficulties with finite-difference methods near the poles do not arise. (2) The truncated spectral equations possess the conservation properties of energy and enstrophy demonstrated by Lorenz (1960) for Cartesian coordinates. This property will help stabilize the time-dependent solutions of spectral equations. (3) The linear part of the spectral equations can be treated exactly in both space and time, because spherical harmonics are the eigensolutions of a linearized vorticity equation.

The principal difficulty in carrying out the integration of nonlinear spectral equations is that a large amount of arithmetic is needed to calculate the nonlinear interaction terms. Orszag (1970) and Eliasen *et al.* (1970) independently proposed a transform method to speed up calculation of the nonlinear interaction terms and eliminate the storage requirement of a large number of interaction coefficients in computers. The basic idea was to evaluate the nonlinear advection terms in the physical domain and then to expand them in spherical harmonics.

Eliasen *et al.* (1970), Machenhauer and Rasmussen (1972), and Bourke (1972) formulated spectral models with the transform method for barotropic primitive equations, and Machenhauer and Daley (1972), Bourke (1974), and Hoskins and Simmons (1975) for a baroclinic primitive equation model. Bourke (1972) detailed a timing comparison between the interaction coefficient method and the transform method and showed a significant gain in efficiency in solving barotropic primitive equations by the transform method. Doron *et al.* (1974) made a comparative study of the accuracy of Bourke's spectral model with two spectral resolutions against a second- order finite-difference model also with two different horizontal resolutions. The results favored the spectral model under comparable computing time and storage.

As basis functions used to formulate a spectral model of the primitive equations over a sphere, the choice of spherical harmonics is natural, but it is not at all clear that it is the most efficient. Since Hough harmonics are eigensolutions of linearized primitive equations over a sphere, Flattery (1971) and Kasahara (1976) suggested that Hough harmonics may offer advantages over spherical harmonics.

The application of simpler functions than spherical harmonics, such as special double Fourier functions by Robert (1966), may be a method of calculating accurate derivatives of the variables at discrete grid points. This approximate representation of variables analytically for evaluation of their derivatives is called *pseudospectral approximation.*

Let us consider the simple advection equation

$$\partial w/\partial t + a(x)(\partial w/\partial x) = 0, \tag{5.38}$$

where $w(x, t)$ and $a(x)$ are assumed functions of $x$ with period $2\pi$. The essence of pseudospectral approximation is to choose a suitable series expansion of the spatial dependence of $w(x, t)$ to evaluate accurately derivatives $\partial w/\partial x$ (Orszag, 1972).

Let $W_n \equiv w(x_n, t)$ denote the values of $w$ at $N = 2K + 1$ equally spaced points $x_n = 2\pi n/N$ $(n = 0, 1, \ldots, N)$. With the periodic boundary condition, $W_0 = W_N$, we express

$$W_n = \sum_{|k| \leq K} f(k) \exp(ikx_n), \qquad (n = 0, 1, \ldots, N - 1) \tag{5.39}$$

where

$$f(k) = \frac{1}{N} \sum_{n=0}^{N-1} W_n \exp(-ikx_n), \qquad (|k| \leq K).$$

A good approximation to $\partial w/\partial x$ at $x = x_n$ is

$$\left(\frac{\partial w}{\partial x}\right)_n = \sum_{|k| \leq K} ikf(k) \exp(ikx_n), \qquad (n = 0, 1, \ldots, N - 1).$$

Merilees (1973) formulated an algorithm for application of the pseudospectral method to the numerical integration of shallow-water equations over a sphere. Merilees (1974b) made a series of numerical experiments to test the accuracy of the pseudospectral method. The method seemed to be more accurate than a fourth-order finite-difference scheme for the same resolution, but the two methods were of comparable efficiency.

Orszag (1974) gave a comprehensive review of various spectral methods for numerical solution of flow problems in spherical geometries. On the face of it, pseudospectral methods are significantly more efficient than spectral methods, but there are serious problems with pseudospectral methods. For example, spherical harmonic expansions give uniform resolution over a sphere so that the spectral method is free from problems related to the poles. On the other hand, the computational stability condition of the pseudospectral method is similar to that of a finite-difference method and, therefore, we have the computational stability problem near the poles. Merilees (1974b) suggested Fourier filtering to eliminate computationally unstable modes. The application of Fourier filtering is simple in the pseudospectral method since all the variables are expanded into Fourier series for calculation of their derivatives.

Kreiss and Oliger (1973) discussed an error estimate of the pseudospectral method applied to (5.38). If the advection velocity $a$ is constant, the pseudospectral method is very accurate, that is, the error is never larger than the error committed by approximating the initial function by the truncated Fourier series (5.39). For equations with variable coefficients, it is difficult to estimate the accuracy of the pseudospectral method. Just as with computational instability for nonlinear finite-difference equations, the pseudospectral approximation may cause nonlinear instability because of aliasing errors in misrepresenting the products of two functions by the same number of discrete points used to expand the functions.

As a variant of the pseudospectral method, Price and MacPherson (1973) fit cubic spline polynomials to spatial variations of the dependent variables to calculate spatial derivatives in primitive equations on a variable-area telescoping grid.

A modified version of the spectral method is the *finite-element method* (e.g., Norrie and de Vries, 1973; Gary, 1974). The basis functions $\phi_j$ of (5.36) in the spectral method are defined over the entire domain, but the basis functions in the finite-element method are piecewise polynomials. These polynomials are local so that they are nonzero over only a small finite element. For example, piecewise linear "roof" functions have been used as basis functions. Coefficients $C_j(t)$ are then determined by the integral condition (5.37).

Cullen (1974) applied the finite-element method to solve the shallow-water equations over a sphere using a geodesic grid based on the icosahedron (Section V, E). A similar study was made by Hinsman (1975). Wang *et al.* (1972) applied piecewise Hermite cubic polynomials as basis functions to solve one-dimensional gravity wave equations.

The finite-element method has several advantages. Derivatives are approximated more accurately than by the usual finite differences. Because of the Galerkin procedure, the scheme can be quadratic-conserving. The method is applicable to arbitrarily shaped grid triangles and lateral boundaries. A disadvantage is that its computing algorithm may become very complicated and require storage of coefficients if they are different at each grid point. Although favorable results were reported with the finite-element method, no detailed study is available of the accuracy and efficiency of the finite-element method applied to prediction equations over a sphere in comparison with the finite-difference and spectral methods. Nevertheless, the finite-element method may offer an interesting alternative to difference and spectral methods.

### G. Implicit Methods

Explicit schemes are convenient to apply but time step $\Delta t$ is restricted by stability conditions, and the maximum allowable time step is usually determined by the fastest characteristic wave speed in the system. For primitive equation models, a short time step is dictated by fast-moving gravity waves with a speed of 300 m $\sec^{-1}$ rather than meteorologically significant waves with one order of magnitude less speed.

Although application of an implicit method such as the Crank–Nicholson scheme (Section V, B) is an obvious solution to avoid using a short time step, full treatment of the implicit method is time consuming (Gustafsson, 1971). One remedy to reduce the amount of computation is the application of fractional steps or splitting up (Section V, C). This approach is particularly popular in the USSR (Marchuk, 1974).

Another alternative is to apply an implicit procedure only to gravity wave components, whereas advective and Coriolis terms are calculated explicitly. This approach is called a *semi-implicit scheme.* Robert (1969) formulated a semi-implicit scheme for shallow-water equations over a sphere. Pressure gradient terms in the equations of horizontal motion and the mass divergence term in the continuity equation are averaged at time levels, $n + 1$ and $n - 1$, while the remaining terms are evaluated at time level $n$. Leap-frog time differencing is used. Eliminating the velocity components $u^{n+1}$ and $v^{n+1}$ from the difference equations, we obtain an elliptic-type equation for $h^{n+1}$. Once $h^{n+1}$ is determined, $u^{n+1}$ and $v^{n+1}$ are calculated explicitly.

Since gravity wave components are calculated implicitly, the time step is limited only by the characteristic speed of meteorologically significant waves. Therefore, it is possible to take a time step several times longer than one used with an explicit scheme. Of course, extra time is required to solve an elliptic-type equation, but the gains from a larger time step are more substantial here than the need for additional computing time.

Robert (1969) applied the semi-implicit scheme for a spectral model, and a similar approach was also used for baroclinic primitive equation spectral models by Bourke (1974) and Hoskins and Simmons (1975). Kwizak and Robert (1971) and McPherson (1971) applied the semi-implicit scheme to finite-difference models. Robert *et al.* (1972) extended the algorithm to a finite-difference baroclinic primitive equation model.

Elvius and Sundström (1973) compared the stability and accuracy of two semi-implicit schemes and leap-frog schemes applied to shallow-water equations. With a larger time step in the semi-implicit schemes, the propagation speed of gravity waves is reduced considerably and, therefore, gravity wave components are calculated inaccurately. This does not matter too much since the amplitudes of gravity waves are very small in reality when compared with those of meteorological waves. On the other hand, a larger time step actually reduces the phase velocity error of low-frequency motions compared to the case with a shorter time step for the leap-frog schemes.

### H. Nonlinear Instability

Von Neumann analysis based on Fourier transform gives only a necessary condition for stability of nonlinear difference equations. Even though the von Neumann condition is satisfied, instability may arise from incorrect specification of boundary conditions or nonlinear terms in difference equations. Instability caused by nonlinear effects is called *nonlinear instability,* first reported and analyzed by Phillips (1959b) who encountered this type of instability in a numerical solution of the nonlinear barotropic vorticity equation.

The following example by Richtmyer (1963) demonstrates the nature of nonlinear instability. Consider a leap-frog difference scheme applied to (5.18):

$$u_j^{n+1} = u_j^{n-1} - \tfrac{1}{2}\lambda[(u_{j+1}^n)^2 - (u_{j-1}^n)^2], \tag{5.40}$$

where $\lambda = \Delta t/\Delta x$. The von Neumann condition for the linearized equation is $\lambda|u_{\max}| < 1$. If a smooth initial condition and a time step satisfying the von Neumann condition are used, the calculation runs satisfactorily for some time and then suddenly explodes and results in instability.

Equation (5.40) has an exact solution in the form

$$u_j^n = C^n \cos(\pi/2)j + S^n \sin(\pi/2)j + U^n \cos \pi j + V, \tag{5.41}$$

where $C^n$, $S^n$, and $U^n$ are the coefficients of waves with wavelengths of $4\Delta x$, $4\Delta x$, and $2\Delta x$, respectively, and $V$ is a constant. Substitution of (5.41) into (5.40) gives the recurrence relations for coefficients

$$\begin{aligned} C^{n+1} - C^{n-1} &= 2\lambda S^n(U^n - V), \\ S^{n+1} - S^{n-1} &= 2\lambda C^n(U^n + V), \\ U^{n+1} - U^{n-1} &= 0. \end{aligned} \tag{5.42}$$

The last equation indicates that $U^n$ takes two constants, say $A$ and $B$, on an alternate cycle. The first two equations then are combined to give

$$C^{n+2} - 2\mu C^n + C^{n-2} = 0, \tag{5.43}$$

where

$$\mu = 1 + 2\lambda^2(A + V)(B - V). \tag{5.44}$$

We obtain a similar equation for $S^n$. If $|\mu| \leqslant 1$, coefficients $C^n$ and $S^n$ are bounded. If $|\mu| > 1$, the solutions grow exponentially with $n$, and the stability condition is

$$-1 < \lambda^2(A + V)(B - V) < 0.$$

As long as constant value $V$ is fairly large so that $|V| > |A|$ and $|V| > |B|$, the above condition is satisfied within the specification of the von Neumann condition. However, if constant $V$ becomes sufficiently small compared with the solutions of $2\Delta x$ so that $A + V$ and $B - V$ may have the same sign, then instability sets in.

Phillips (1959b) made a similar analysis, but the effect of the constant term $V$ was not taken into account. We see that the presence of the constant term plays an important role in the criterion of instability. Kreiss and Oliger (1972) and Fornberg (1973) made extensive experiments with (5.40) and pointed out that instability is observed when $u^n$ is nearly zero. In other words, nonlinear instabilities do not occur if $u^n$ does not come close to zero. A heuristic discussion of this type of observation was presented also by Robert *et al.* (1970).

According to Phillips (1959b), the growth of $2\Delta x$ waves occurs because the grid system is incapable of resolving wavelengths shorter than two grid intervals; when such wavelengths are formed by the nonlinear interaction of longer waves, the grid system interprets them incorrectly. The errors caused by this kind of misrepresentation are called *aliasing errors*. The existence of such errors, regarded as a general type of truncation error arising from nonlinearity of the equations, was pointed out by Platzman (1958, 1961) who proposed filtering wave components whose wavelengths are less than four grid intervals to eliminate such errors. A calculation scheme for the barotropic vorticity equation based on such a filter was designed by Platzman and Baer (1958; see also Baer, 1961) and the results reported by Baer (1958, 1961). Phillips (1959b) also suggested such a filter and showed that it eliminated computational instability due to aliasing errors.

Orszag (1971b) pointed out that aliasing errors can be eliminated by filtering out all waves with wavelengths between $2\Delta x$ and $3\Delta x$. If the number of grid points is $K$, then the cutoff wave number is $K/2$, since the smallest resoluble wavelength is $2\Delta x$. If modes $k_1$ and $k_2$ interact to give $k = k_1 + k_2$, then the aliasing modes are $k_1 + k_2 \pm K$, $k_1 + k_2 \pm 2K$, etc. However, if we eliminate higher modes than $k/3$ (wavelengths smaller than $3\Delta x$), then $|k_1| < K/3$ and $|k_2| < K/3$ so that the aliasing modes are always greater than $K/3$ and are filtered out.

As argued by Orszag (1971a), it is not correct to infer that aliasing errors always induce nonlinear instability because it is possible to construct a difference scheme in which leaving out the aliasing interactions induces instability. We have discussed in Section V, D that quadratic conserving schemes are not susceptible to nonlinear instability despite the presence of aliasing interactions. What aliasing errors do in nonlinear calculations is presented by Platzman (1958, 1961) who showed that the best mean-square approximation to the product of two discrete functions is not necessarily the product function at the grid points, but rather the alias-free representation of the product of two trigonometric interpolation polynomials through the grid points.

We mention that (5.22) is a quadratic conserving scheme and less prone to nonlinear instability. Equation (5.22) has an exact solution in the form

(Kikuchi, 1972)

$$u_j = C \cos(\pi/2)j + S \sin(\pi/2)j + U \cos \pi j + V \tag{5.45}$$

in analogy to (5.41). Note that no discretization is made in time and the coefficients of space functions are continuous functions of time.

Substitution of (5.45) into (5.22) gives a system of ordinary differential equations for the coefficients:

$$\begin{aligned} dC/dt &= rS(U - 3V), \\ dS/dt &= rC(U + 3V), \\ dU/dt &= -2rCS, \\ dV/dt &= 0, \end{aligned} \tag{5.46}$$

where $r = (3\Delta x)^{-1}$. The last equation implies that $V$ = constant. Multiplying the first to third equations of (5.46) by $C$, $S$, and $U$, respectively, we obtain

$$d(C^2 + S^2 + U^2)/dt = 0.$$

Hence, the total energy is invariant with time. If the Crank–Nicholson implicit scheme is used to solve (5.22), we can show that the total energy is also conserved. However, if we solve (5.22) by leap frog, then nonlinear instabilities may occur. These can be controlled by a very small amount of smoothing (Kreiss and Oliger, 1973).

## I. Discretization in the Vertical

Since we deal with three-dimensional space, we must be concerned with the discretization of variables in the vertical as well as in the horizontal. There are two ways to represent the vertical structure of the atmosphere. One is to use discrete points in the vertical with vertical derivatives approximated by finite differences. The other is to represent variables as a finite series of differentiable functions.

In the second approach, Francis (1972) proposed application of Laguerre polynomials in $\ln(p/p_H)$, where $p$ is pressure and $p_H$ the surface pressure. Machenhauer and Daley (1972) used Legendre polynomials in $p/p_H$. Simons (1968) suggested Bessel functions, which are eigensolutions of the vertical derivative operator in quasi-geostrophic forecast equations. The use of eigensolutions of linearized prediction equations as basis functions to expand variables in the vertical is an interesting approach, although further research is needed to demonstrate its merit. See Kasahara (1976) for the eigensolutions in the vertical direction of linearized primitive equations.

The grid point approach by far dominates the spectral approach in discretization in the vertical. The variables are usually staggered to take maximum advantage of computer storage while retaining second-order accuracy in the derivatives. Arakawa (1972) and Arakawa and Mintz (1974) designed a vertical difference scheme which calculates consistently energy conversion between kinetic energy and total potential energy, and vice versa, and satisfies various integral properties of dynamical and thermodynamical quantities. Arakawa's vertical differencing was applied by Somerville *et al.* (1974) to a nine-level global circulation model. Corby *et al.* (1972) also formulated a vertical difference scheme satisfying global conservation aspects. Phillips (1974) investigated Arakawa's energy-conserving vertical difference scheme in conjunction with the reduction of horizontal truncation errors of the pressure gradient force in the transformed pressure coordinate system (Section III, C).

A major problem in vertical differencing is treating the earth's orography. Large mountain barriers influence atmospheric motions of all scales, and their incorporation into numerical prediction models is extremely important. Poor numerical treatment of orography contributes to forecasting inaccuracy, not only through the time integration of prediction equations, but also through the analysis of meteorological data.

There are essentially two approaches to incorporate dynamical effects of orography into numerical prediction models. One is to use the transformed coordinate, such as the sigma system in Section III, C, in which the earth's surface coincides with a horizontal coordinate surface. The other is to block physically the integration domain in any coordinate system covered by mountains.

The computational procedure with a transformed coordinate is straightforward, since the earth's surface is a coordinate surface and, therefore, no special treatment is required to handle lower boundary conditions along the mountain surfaces (Section III, C). However, as seen from (3.11), the pressure gradient in the generalized $s$-coordinate system is expressed, not by one term, but by the sum of two:

$$\begin{aligned}\text{Pressure gradient} &= \rho^{-1}\,\nabla_s p + g\,\nabla_s z \\ &= RT\,\nabla_s(\ln p) + \nabla_s(gz).\end{aligned} \tag{5.47}$$

In regions with steep mountains, the individual terms become large but with opposite sign. The required pressure gradient is a small residual of their sum and this causes significant error unless careful difference approximations are used.

Gary (1973) analyzed the magnitude of numerical errors of the pressure gradient in the $s$ system and pointed out that considerable error can appear not only near the mountains, but also throughout the entire model atmosphere. In a $p$ or $z$ coordinate, the pressure gradient is reduced to $\nabla_p(gz)$ or $RT\,\nabla_z(\ln p)$, respectively, and there is no need to take the difference of two large terms as in the transformed system.

One remedy practiced by Smagorinsky *et al.* (1967) and Kurihara (1968) is to calculate the pressure gradient on pressure surfaces after vertically interpolating data from the sigma system to the pressure system. However, it is difficult to estimate the magnitude of truncation errors in this procedure because of the vertical interpolation. Corby *et al.* (1972) proposed a special finite-difference scheme to make consistent approximations to evaluate the two components of the pressure gradient in the case of temperature varying linearly as log $p$. This is a relatively simple approach, but it appears to be effective in reducing errors in the pressure gradient calculation in the sigma system.

One source of errors in the evaluation of the pressure gradient in a transformed coordinate is horizontal differencing on coordinate surfaces. Therefore, a finer horizontal mesh is required to reduce truncation errors. Another source of errors, as pointed out by Sundqvist (1975), is in the evaluation of temperature on the transformed coordinate surfaces. If the magnitude of individual terms in (5.47) is 10 times larger than a typical value of their sum, 1% (2°–3°C) error in temperature $T$ causes 10% error in the pressure gradient calculation. Such a temperature error arises from inconsistencies between $T$, $p$, and $z$ due to their variations along vertical grid points. This means that we must choose grid point values of $T$, $p$, and $z$ consistently to reduce horizontal truncation errors. Phillips (1974) investigated a consistent vertical discretization of variables $T$ and $z$ in transformed pressure coordinate systems under energy conservation constraints. Phillips suggested a reference atmosphere to reduce orographic truncation errors.

Another approach to handle orography in the models is to block the integration domain covered by mountains in the $z$ or $p$ system. Kasahara and Washington (1971) presented a computational procedure to deal with orographic blocking in the $z$ system. Katayama *et al.* (1974) described another blocking procedure in the $p$ system, but no results were reported. A major advantage of blocking is that the orographic truncation errors are confined mostly to the neighborhood of mountains rather than spreading vertically, as in transformed systems. Blocking can also easily incorporate steep-slope mountains. A major disadvantage is that actual coding of the blocking procedure may become complicated for higher order difference schemes, and the approach may not be applicable to spectral modeling. Blocking in the $p$ system is much more complicated than in the $z$ system, because the location of the mountain top indicated by surface pressure varies in time and space within the coordinate system. Egger (1974) applied blocking in the sigma system to deal with steep mountains. All of the computing procedures discussed here have some inherent shortcomings, and further research is needed to improve handling of orographic effects in the models.

Before we conclude this section, we should comment on the lateral boundary conditions and mesh refinements for a limited-area forecasting model. For prediction of weather over a relatively small area of the globe with an exceedingly fine computational mesh, we must either (1) limit the forecasting domain over a specific area, or (2) refine a global computational mesh over the area concerned. In (1), we must impose artificial conditions along the lateral boundary of the forecasting domain. In (2), the effects of nonuniform grids must be investigated. Because this volume deals mainly with prediction models of the global atmosphere, this problem is not discussed here. A review paper by Haltiner and Williams (1975) on recent developments in numerical weather prediction contains other materials not discussed in this article.

## VI. Initial Conditions

### A. Initialization

The numerical integration of a prediction model requires an initial condition, consisting of three-dimensional distributions of dependent variables. Initial conditions for short- to medium-range forecasts are based on observed data collected worldwide at a specific time and then processed automatically (or manually) to interpolate the grid values of pressure (or temperature), horizontal velocity, and specific humidity. The automatic process of synoptic analysis is called *objective analysis*.

The fields of pressure (or temperature) and horizontal wind velocity thus analyzed are usually not suitable as initial conditions for primitive equation models. When the fields of pressure and horizontal velocity are not adjusted initially, large-amplitude oscillations may develop during the numerical integration. The cause of the oscillations is explained by the appearance of gravity waves as the result of initial imbalance of pressure and wind. Although such oscillations are eventually smoothed out by the dispersion of gravity waves and the action of momentum dissipation, it is desirable to eliminate meteorological noise from the initial conditions. The process of suitably adjusting the fields of pressure and wind to set up the initial condition for a prediction model is called *data initialization*; and various methods have been proposed. Unless observed data are sufficiently dense and accurate, a good data initialization scheme is required for successful short-range forecasting. It remains a problem particularly in the tropics where it is difficult to separate noise from meteorologically significant motions. The application of normal mode analyses may offer a solution.

Short-range forecasts are largely influenced by initial conditions, including the data initialization procedure. When the period of integration extends beyond two weeks or so, the model gradually approaches its own statistical equilibrium state. The time average of such equilibrium state is called *model climate*. The long-term performance of a prediction model can be evaluated by examining how closely the model climate compares with the real climate.

## B. Four-Dimensional Data Assimilation

The future atmospheric observing system will be a combination of various observing subsystems consisting of satellites, constant-level balloons, ocean buoys, etc., as well as conventional upper-air and surface networks (Joint Organizing Committee, 1973). Among the different subsystems are enormous variations in the characteristics and capabilities of what to measure and how often to transmit the data. Thus, to analyze all meteorological observations by different instruments and at different times, we can use a global circulation model as an integrator (Fig. 5). This process is called *four-dimensional data assimilation* and is important in application of global circulation models.

The basic idea of four-dimensional data assimilation is to utilize any observational data, assuming they are sufficiently accurate, in place of predicted values during numerical integration of a global model. Thus, the predicted values are used if no observations are available. To be more specific, let us consider a marching problem of one space dimension $x$ and time $t$ as shown schematically at the bottom of Fig. 5. At $t = t_0$, we have observed data for all $x$ from which the time integration proceeds for $t \geqslant t_0$. In addition to observed data at regular times $t = t_1$ and $t_2$ for all $x$, we have a moving platform which observes data for the interval of $x_1(t)$ to $x_2(t)$ at every small time increment $\Delta t$. Therefore, except at $t_0$, $t_1$, $t_2$, etc., the state of this simple atmosphere at any time consists of the actually observed state for the part $x_1(t)$ to $x_2(t)$ and the predicted values for the rest. Because predicted values will contain errors as do observed data, the two states of observation and prediction must be adjusted to give a representative state of the model atmosphere. This process then provides a continuous mapping of the current atmospheric state. Numerical weather prediction can be performed on a different and preferably much faster computer starting from an initial condition provided by this process.

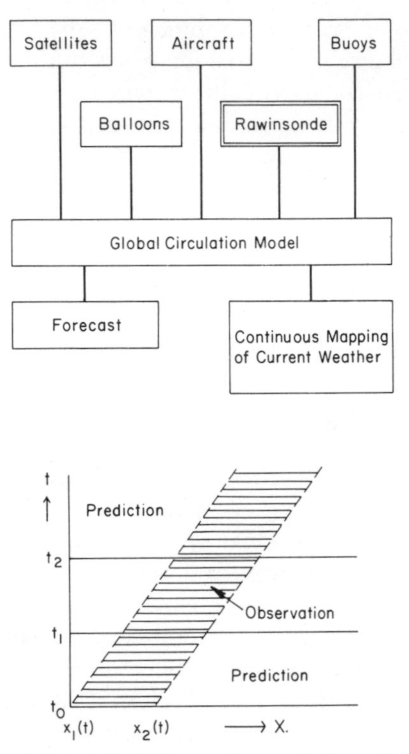


FIG. 5. Four-dimensional data assimilation and numerical weather prediction. The prediction model serves as an integrator of various observations. Bottom part illustrates how observed data from a moving platform will be mixed with prediction. From Kasahara (1972).

The success of four-dimensional data assimilation depends upon (1) the accuracy of a prediction model used as the integrator, (2) the method of blending observed and predicted data, and (3) the model's ability to infer the values of unobserved variables as the result of continuous feeding of incomplete, but not inaccurate, observed data. The four-dimensional assimilation is a promising approach to objective analysis of large-scale meteorological data and its active research is in progress (Bengtsson, 1975; McPherson, 1975).

## VII. Future Outlook

### A. Atmospheric Predictability

A notable application of global circulation models is to weather forecasting. It is, therefore, worthwhile to consider how predictable the atmosphere is by a numerical model. If we have no means for forecasting, all we

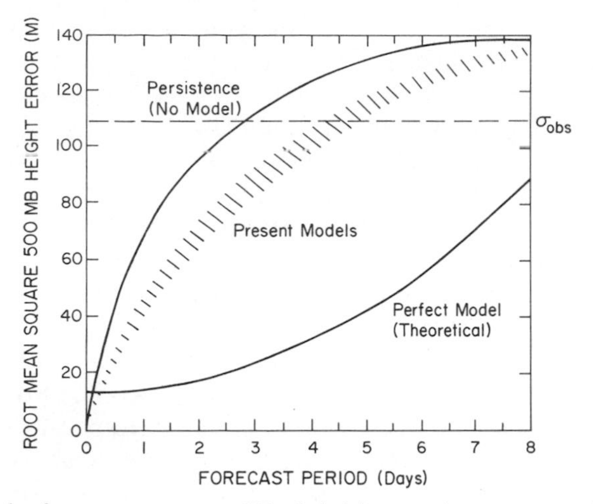


FIG. 6. Growth of root mean square 500 mb height error in meters as a function of the forecasting period.

can do is to use persistency, as discussed in the Introduction. Figure 6 shows the growth of root mean square (rms) 500-mb height error in meters as a function of the forecasting period. The upper curve represents error growth based on the persistency forecast. The lower curve indicates error growth when a "perfect" model is used. No numerical model is perfect when its performance is compared with the real atmosphere. On the other hand, any model can be made "perfect" by comparing its performance with the history of simulated data generated by the same model.

Even with a perfect model, the atmosphere cannot be predicted indefinitely. No matter how accurately instruments are designed to measure the atmosphere, the turbulent character of atmospheric motion limits accuracy of meteorological measurements. This uncertainty in the initial state grows during the numerical forecast. The magnitude of initial rms error for the perfect model experiment in Fig. 6 is expected from an operational satellite observing system for temperature. Error growth is calculated from the difference of two integrations with and without initial error using a global circulation model (Baumhefner and Julian, 1975). Predictability is lost if the growth of initial uncertainty has reached the level of rms difference of two randomly chosen states.

This inherent limit of predictability of the atmosphere was pointed out by Thompson (1957) and further by Lorenz (1969) based on relatively simple atmospheric models. Later, extensive numerical experiments were performed using various global circulation models. These results indicate that small errors double in two or three days and the limit of predictability is about two weeks. Recent studies on atmospheric predictability were summarized by Leith (1975).

The skill of extended-range forecasts with current operational and experimental prediction models falls within the hatched region in Fig. 6. Data are from various sources: Shuman (1972), Miyakoda *et al.* (1972), Druyan *et al.* (1975), and Baumhefner and Downey (1976). By comparing the performance of current prediction models with that of a perfect model, there is considerable room for improvement, although the error growth curve of the perfect model can be debated as an attainable target of a very good prediction model.

Robert (1974) analyzed the source of prediction error by another simulation study and suggested that 500-mb height errors for 36-hr forecasts over North America are due to inaccuracies in initial data (18%), to shortcomings in physical processes (35%), and numerical treatment (47%). Robert considers that the coarseness of grid resolutions is primarily responsible for errors in numerical treatment. We need extensive diagnostic analysis of short- to medium-range forecasts to find the source of forecasting biases.

### B. Climate Sensitivity

The predictability limit referred to so far concerns the model's ability to forecast the day-to-day evolution of the atmosphere in a deterministic sense. Presence of the predictability limit does not imply that the models are incapable of long-range forecasts of the average state of the atmosphere, namely climate. However, before testing a global model for long-range forecasts, there are a number of problems to solve. (1) The realistic energy balance in a prediction model must be maintained as in the real atmosphere; otherwise the model eventually seeks its own equilibrium climate different from the real one. (2) Because of the growth of initial errors during forecasting, single numerical forecasts do not provide useful information on forecasting relatively slow changes in climate. The level of uncertainty in single forecasts due to initial errors must be reduced by averaging many forecasts starting from slightly different initial conditions for the same case. (3) Slight changes in a prediction model such as using slightly different values of model parameters sometimes cause significant deviations in the forecast for an extended period. Thus, the uncertainties in model parameters contribute to predictability error which again can mask the truth of slow changes in climate.

Extensive numerical experiments are under way using various global circulation models to investigate model sensitivities to changes in initial conditions, model parameters, and external boundary conditions. Since atmospheric circulation models in these experiments specify ocean surface temperature and other factors, the model cannot describe the long-term behavior of the atmosphere in interaction with the oceans and other factors not included in the models. Nevertheless, when experiments are properly formulated, they are useful, within model limitations, in gauging climate sensitivity to advertent and inadvertent changes in environmental forcing. Results of such experiments, of course, must be examined carefully for statistical significance and viewed in terms of physical and mathematical shortcomings in modeling. See Chervin *et al.* (1976) as a most recent reference on climate sensitivity experiments with a global circulation model.

## C. Conclusions

We have reviewed in this article the mathematical and computational aspects of global circulation modeling with brief discussions of the physical processes included in the models.

A global circulation model is a principal tool for understanding:

- the transient behavior of the atmosphere as manifested in the large-scale fluctuations which control changes of the weather; this would lead to increasing the accuracy of short- to medium-range forecasting
- the factors that determine the statistical properties of the general circulation of the atmosphere which would lead to better understanding of the physical basis of climate.

These two objectives were set forth by the Global Atmospheric Research Program (GARP), an international cooperative research effort. There have been accelerated efforts in many countries to develop numerical models for weather prediction and climate research (Joint Organizing Committee, 1974).

There are a number of deficiencies in all of the current numerical models which must undergo improvement. Continued research is required to solve satisfactorily the numerical problems arising from applying finite-difference schemes near the poles. In this respect, the spectral approach is advantageous over the difference approach. It is also important to develop accurate methods of dealing with the earth's orography. Although evidence that the rigid top boundary significantly influences forecasts is not conclusive, it is desirable to investigate alternative boundary conditions in the models.

It is now apparent that to simulate eddy activity comparable to that of the real atmosphere, the horizontal resolution of the models should be less than 2.5° in longitude and latitude with a second-order difference scheme. More research is necessary to shed light on the trade-off between increasing the accuracy of a difference scheme and reducing the horizontal resolution of the computational mesh. A similar effort should be made to determine the effect of increased vertical resolution. Lack of adequate computational resolution in the models in both the horizontal and vertical is a prime suspect in forecasting failures.

Even in short- to medium-range forecasts, and particularly in climate simulation, heating and frictional terms in prediction models should be accurately formulated. Unfortunately, the details of physical processes in the atmosphere are often difficult to quantify or to treat within limited model calculations. Nevertheless, extensive efforts should be made to improve the parameterization of physical processes in global circulation models. Since the model responds collectively to physical processes, it is essential to improve the overall quality of physical parameterizations. We must also be aware of feedback between computational methods (for dynamical equations) and physical processes. For example, the parameterization of subgrid-scale processes may strongly depend on horizontal resolution and accuracy of finite-difference approximations.

The performance of prediction models should be examined for short-range forecasting and investigating major regional climatic features compared with real atmospheric data. Care must be taken in data initialization for short-range forecasts, since forecasting skills may depend on initial conditions.

Improvements in computational and physical aspects of global circulation models inevitably require a substantial increase in computing speed and storage. For example, a six-layer, 2.5° longitude–latitude grid version of the NCAR global circulation model takes about 2 hr on the Control Data Corporation 7600 computer for 1 model-day calculation. Doubling the horizontal resolution alone requires 8-fold more computing time. Considering the computing time needed to perform extended-range weather predictions and climate sensitivity studies, we urgently need much faster computers—at least one order of magnitude faster than what is available today.

## Acknowledgments

I thank Cecil Leith for permission to use Figs. 2 and 6 and David Baumhefner who recalculated skill scores in Fig. 1 based on data by Shuman (1972). Thanks are also due to George Platzman, Warren Washington, and David Williamson for their helpful comments on the manuscript and Ann Modahl for editorial assistance and typing. A part of this work was carried out while the author was a visitor at the Geophysical Institute, Tokyo University, during the fall quarter of 1975 under a travel grant from the Japan Society for the Promotion of Science, sponsored by the Japanese Ministry of Education. I had the benefit of stimulating discussions related to this subject with Professors K. Gambo and T. Matsuno.

## References

Anderson, P. A. (1974). *J. Comput. Phys.* **15**, 1.

Arakawa, A. (1966). *J. Comput. Phys.* **1**, 119.

Arakawa, A. (1972). "Design of the UCLA General Circulation Model," Tech. Rep. No. 7. Dept. Meteorol., University of California, Los Angeles.

Arakawa, A., and Mintz, Y. (1974). "The UCLA Atmospheric General Circulation Model" (notes distributed at the Workshop 25 Mar.–4 Apr. 1974). Dept. Meteorol., University of California, Los Angeles.

Baer, F. (1958). "The Extended Numerical Integration of a Simple Barotropic Model. Part II: Results," Tech. Rep. No. 3 (Grant NSF-G2159). Dept. Meteorol., University of Chicago, Chicago, Illinois.

Baer, F. (1961). *J. Meteorol.* **18**, 319.

Baer, F. (1964). *J. Atmos. Sci.* **21**, 260.

Baer, F., and Platzman, G. W. (1961). *J. Meteorol.* **18**, 393.

Baumhefner, D., and Downey, P. (1976). *Ann. Meteorol.* [N.S.] **11**, 205.

Baumhefner, D. P., and Julian, P. R. (1975). *Mon. Weather Rev.* **103**, 273.

Bengtsson, L. (1975). "4-Dimensional Assimilation of Meteorological Observations," GARP Publ. Ser. No. 15. Global Atmospheric Research Programme, WMO-ICSU Joint Organizing Committee, Geneva, Switzerland.

Bleck, R. (1974). *Mon. Weather Rev.* **102**, 813.

Bourke, W. (1972). *Mon. Weather Rev.* **9**, 683.

Bourke, W. (1974). *Mon. Weather Rev.* **102**, 687.

Bretherton, F. P. (1972). "Large-Scale Vertical Exchange of Mass, Momentum, Heat or Moisture-Gravity Wave Propagation and the Upper Boundary Condition." A report submitted to JOC Study Group Conference on the Parameterization of Sub-Grid Scale Processes, Leningrad, 1972.

Bretherton, F. P., and Karweit, M. (1975). *Numer. Models Ocean Circ., Proc. Symp., 1972* pp. 237–249.

Bryan, K. (1966). *Mon. Weather Rev.* **94**, 39.

Bryan, K., Manabe, S., and Pacanowski, R. C. (1975). *J. Phys. Oceanog.* **5**, 30.

Burstein, S. Z., and Mirin, A. (1969). "Third-Order Difference Methods for Hyperbolic Equations," Rep. NYO-1480-136. Courant Institute of Mathematical Sciences, New York University, New York.

Charney, J. G. (1948). *Geophys. Norv.* **17**, No. 2.

Charney, J. G. (1951). *In* "Compendium of Meteorology" (T. F. Malone, ed.), pp. 470–482. Am. Meteorol. Soc., Boston, Massachusetts.

Charney, J. G., Fjørtoft, R., and von Neumann, J. (1950). *Tellus* **2**, 237.

Chervin, R. M., Washington, W. M., and Schneider, S. H. (1976). *J. Atmos. Sci.* **33**, 413.

Corby, G. A., Gilchrist, A., and Newson, R. L. (1972). *Q. J. R. Meteorol. Soc.* **98**, 809.

Courant, R., Friedrichs, K. O., and Lewy, H. (1928). *Math. Ann.* **100**, 32.

Crowley, W. P. (1967). *J. Comput. Phys.* **1**, 471.

Cullen, M. J. P. (1974). *Q. J. R. Meteorol. Soc.* **100**, 555.

Dey, C. H. (1969). *Mon. Weather Rev.* **97**, 597.

Doron, E., Hollingsworth, A., Hoskins, B. J., and Simmons, A. J. (1974). *Q. J. R. Meteorol. Soc.* **100**, 371.

Douglas, J. (1955). *J. Soc. Ind. Appl. Math.* **3**, 42.

Druyan, L. M., Somerville, R. C. J., and Quirk, W. J. (1975). *Mon. Weather Rev.* **103**, 779.

Egger, J. (1974) *Mon. Weather Rev.* **102**, 847.

Eilon, B., Gottlieb, D., and Zwas, G. (1972). *J. Comput. Phys.* **9**, 387.

Eliasen, E., Machenhauer, B., and Rasmussen, E. (1970). "On a Numerical Method for Integration of the Hydrodynamical Equations with a Spectral Representation of the Horizontal Fields," Rep. No. 2. Dept. Meteorol., Copenhagen University, Denmark.

Eliassen, A. (1949). *Geophys. Norv.* **17**, No. 3.

Eliassen, A., and Rekustad, J.-E. (1971). *Geophys. Norv.* **28**, No. 3.

Ellsaesser, H. W. (1966). *J. Appl. Meteorol.* **5**, 246.

Elvius T., and Sundström, A. (1973). *Tellus* **25**, 132.

Ertel, H. (1938). "Methoden und Probleme der dynamischen Meteorologie." Springer-Verlag, Berlin and New York.
Fjørtoft, R. (1953). *Tellus* **5**, 225.
Flattery, T. W. (1971). *Proc. AWS Tech. Exchange Conf., 6th, 1970* Air Weather Serv. Tech. Rep. No. 242, pp. 42–53.
Fornberg, B. (1973). *Math. Comput.* **27**, 45.
Francis, P. E. (1972). *Q. J. R. Meteorol. Soc.* **98**, 662.
Fromm, J. E. (1968). *J. Comput. Phys.* **3**, 176.
Gary, J. M. (1969). *J. Comput. Phys.* **4**, 279.
Gary, J. M. (1973). *J. Atmos. Sci.* **30**, 223.
Gary, J. M. (1974). "The Numerical Solution of Partial Differential Equations" (course notes). Dept. Computer Sci., University of Colorado, Boulder.
Gates, W. L., and Riegel, C. A. (1962). *J. Geophys. Res.* **67**, 773.
Gates, W. L., Batten, E. S., Kahle, A. B., and Nelson, A. B. (1971). "A Documentation of the Mintz-Arakawa Two-Level Atmospheric General Circulation Model," Advanced Research Projects Agency Rep. R-877-ARPA. Rand Corporation, Santa Monica, California.
GCM Steering Committee, (1975). "Development and Use of the NCAR GCM," Tech. Note TN/STR-101. National Center for Atmospheric Research, Boulder, Colorado.
Gerrity, J. P., Jr., McPherson, R. D., and Polger, P. D. (1972). *Mon. Weather Rev.* **100**, 637.
Gottlieb, D., and Turkel, E. (1974). *J. Comput. Phys.* **15**, 251.
Gourlay, A. R., and Morris, J. L. (1970). *J. Comput. Phys.* **5**, 229.
Grammeltvedt, A. (1969). *Mon. Weather Rev.* **97**, 384.
Grimmer, M., and Shaw, D. B. (1967). *Q. J. R. Meteorol. Soc.* **93**, 337.
Gustafsson, B. (1971). *J. Comput. Phys.* **7**, 239.
Haltiner, G. J. (1971). "Numerical Weather Prediction." Wiley, New York.
Haltiner, G. J., and Williams, R. T. (1975). *Mon. Weather Rev.* **103**, 571.
Henrici, P. (1962). "Discrete Variable Methods in Ordinary Differential Equations." Wiley, New York.
Hinkelmann, K. H. (1969). *In* "Lectures on Numerical Short-Range Weather Prediction," (World Meteorol. Org. Regional Training Seminar) pp. 306–375. Hydrometeoizdat, Leningrad.
Hinsman, D. E. (1975). Master's Thesis, Naval Postgraduate School, Monterey, California.
Holloway, J. L., Jr., and Manabe, S. (1971). *Mon. Weather Rev.* **99**, 335.
Holloway, J. L., Jr., Spelman, M. J., and Manabe, S. (1973). *Mon. Weather Rev.* **101**, 69.
Horn, L. H., and Bryson, R. A. (1963). *J. Geophys. Res.* **68**, 1059.
Hoskins, B. J., and Simmons, A. J. (1975). *Q. J. R. Meteorol. Soc.* **101**, 637.
Houghton, D. D., and Kasahara, A. (1968). *Commun. Pure Appl. Math.* **21**, 1.
Joint Organizing Committee. (1972). "Parameterization of Sub-Grid Scale Processes," GARP Publ. Ser. No. 8. Global Atmospheric Research Programme, WMO-ICSU Joint Organizing Committee. Geneva, Switzerland.
Joint Organizing Committee. (1973). "The First GARP Global Experiment—Objectives and Plans," GARP Publ. Ser. No. 11. Global Atmospheric Research Programme, WMO-ICSU Joint Organizing Committee, Geneva, Switzerland.
Joint Organizing Committee. (1974). "Modelling for the First GARP Global Experiment," GARP Publ. Ser. No. 14. Global Atmospheric Research Programme, WMO-ICSU Joint Organizing Committee. Geneva, Switzerland.
Julian, P. R., Washington, W. M., Hembree, L., and Ridley, C. (1970). *J. Atmos. Sci.* **27**, 376.
Kao, S.-K., and Wendell, L. L. (1970). *J. Atmos. Sci.* **27**, 359.
Kasahara, A. (1972). *Bull. Am. Meteorol. Soc.* **53**, 252.

Kasahara, A. (1974). *Mon. Weather Rev.* **102**, 509.
Kasahara, A. (1976). *Mon. Weather Rev.* **104**, 669.
Kasahara, A., and Washington, W. M. (1971). *J. Atmos. Sci.* **28**, 657.
Kasahara, A., Sasamori, T., and Washington, W. M. (1973). *J. Atmos. Sci.* **30**, 1229.
Katayama, A., Kikuchi, Y., and Takigawa, Y. (1974). *In* "Modelling for the First GARP Global Experiment," GARP Publ. Ser. No. 14, pp. 174–188. Global Atmospheric Research Programme, WMO-ICSU Joint Organizing Committee. Geneva, Switzerland.
Kida, H. (1974). *J. Meteorol. Soc. Jpn.* [2] **52**, 1.
Kikuchi, Y. (1972). *Meteorol. Res. Notes, Meteorol. Soc. Jpn.* **110**, 83.
Kreiss, H.-O., and Oliger, J. (1972). *Tellus* **24**, 199.
Kreiss, H.-O., and Oliger, J. (1973). "Methods for the Approximate Solution of Time-Dependent Problems," GARP Publ. Ser. No. 10. Global Atmospheric Research Programme, WMO-ICSU, Joint Organizing Committee. Geneva, Switzerland.
Kubota, S., Hirose, M., Kikuchi, Y., and Kurihara, Y. (1961). *Pap. Meteorol. Geophys.* **12**, 199.
Kurihara, Y. (1965a). *Mon. Weather Rev.* **93**, 33.
Kurihara, Y. (1965b). *Mon. Weather Rev.* **93**, 399.
Kurihara, Y. (1968). *Mon. Weather Rev.* **96**, 654.
Kurihara, Y., and Holloway, J. L., Jr. (1967). *Mon. Weather Rev.* **95**, 509.
Kwizak, M., and Robert, A. J. (1971). *Mon. Weather Rev.* **99**, 32.
Lamb, H. (1932). "Hydrodynamics," 6th ed. Dover, New York.
Lax, P. D., and Wendroff, B. (1960). *Commun. Pure Appl. Math.* **13**, 217.
Lax, P. D., and Wendroff, B. (1964). *Commun. Pure Appl. Math.* **17**, 381.
Leith, C. E. (1965). *Methods Comput. Phys.* **4**, 1.
Leith, C. E. (1969). "Two-Dimensional Eddy Viscosity Coefficients." *Proc. WMO/IUGG Symp. Numer. Weather Prediction, 1968* I-41-44.
Leith, C. E. (1971). *J. Atmos. Sci.* **28**, 145.
Leith, C. E. (1975). *Rev. Geophys. Space Phys.* **13**, 681.
Lilly, D. K. (1965). *Mon. Weather Rev.* **93**, 11.
Lilly, D. K. (1969). *Phys. Fluids, Suppl.* **2**, 240.
Lorenz, E. N. (1960). *Tellus* **12**, 364.
Lorenz, E. N. (1969). *Tellus* **21**, 289.
Lorenz, E. N. (1970). *J. Appl. Meteorol.* **9**, 325.
Machenhauer, B., and Daley, R. (1972). "A Baroclinic Primitive Equation Model with a Spectral Representation in Three Dimensions," Rep. No. 4. Institute for Theoretical Meteorol., Copenhagen University, Denmark.
Machenhauer, B., and Rasmussen, E. (1972). "On the Integration of the Spectral Hydrodynamical Equations by a Transform Method," Rep. No. 3. Institute for Theoretical Meteorol., Copenhagen University, Denmark.
McPherson, R. D. (1971). *Mon. Weather Rev.* **99**, 242.
McPherson, R. D. (1975). *Bull. Am. Meteorol. Soc.* **56**, 1154.
Manabe, S., Bryan, K., and Spelman, M. J. (1975). *J. Phys. Oceanogr.* **5**, 3.
Marchuk, G. I. (1964). *Dokl. Acad. Sci. USSR, Earth Sci. Sect.* (*Engl. Transl.*) **156**, 5.
Marchuk, G. I. (1974). *In* "Numerical Methods in Weather Prediction" (A. Arakawa and Y. Mintz, eds.), p. 277, Academic Press, New York.
Matsuno, T. (1966a). *J. Meteorol. Soc. Jpn.* [2] **44**, 76.
Matsuno, T. (1966b). *J. Meteorol. Soc. Jpn.* [2] **44**, 85.
Merilees, P. E. (1973). *Atmosphere* **11**, 13.
Merilees, P. E. (1974a). *Mon. Weather Rev.* **102**, 82.
Merilees, P. E. (1974b). *Atmosphere* **12**, 77.

Milne, W. E. (1953). "Numerical Solution of Differential Equations." Wiley, New York.
Miyakoda, K. (1962). *Jpn. J. Geophys.* **3**, 75.
Miyakoda, K., Moyer, R. W., Stambler, H., Clarke, R. H., and Strickler, R. F. (1971). *J. Meteorol. Soc. Jpn.* [2] **49**, 521.
Miyakoda, K., Hembree, G. D., Strickler, R. G., and Shulman, I. (1972). *Mon. Weather Rev.* **100**, 836.
Molenkamp, C. R. (1968). *J Appl. Meteorol.* **7**, 160.
Monin, A. S., and Obukhov, A. M. (1959). *Tellus* **11**, 159.
National Academy of Sciences. (1975). *Numer. Models Ocean Circ., Proc. Symp., 1972* pp. 1–364.
Norrie, D. H., and de Vries, G. (1973). "The Finite Element Method." Academic Press, New York.
Oliger, J. E., Wellck, R. E., Kasahara, A., and Washington, W. M. (1970). Tech. Note TN-56. National Center for Atmospheric Research, Boulder, Colorado.
Orszag, S. A. (1970). *J. Atmos. Sci.* **27**, 890.
Orszag, S. A. (1971a). *J. Fluid Mech.* **49**, 75.
Orszag, S. A. (1971b). *J. Atmos. Sci.* **28**, 1074.
Orszag, S. A. (1972). *Stud. Appl. Math.* **51**, 253.
Orszag, S. A. (1974). *Mon. Weather Rev.* **102**, 56.
Peaceman, D. W., and Rachford, H. H., Jr. (1955). *J. Soc. Ind. Appl. Math.* **3**, 28.
Phillips, N. A. (1956). *Q. J. R. Meteorol. Soc.* **82**, 123.
Phillips, N. A. (1957). *J. Meteorol.* **14**, 184.
Phillips, N. A. (1959a). *Mon. Weather Rev.* **87**, 333.
Phillips, N. A. (1959b). *In* "The Atmosphere and the Sea in Motion" (B. Bolin, ed.), pp. 501–504. Rockefeller Inst. Press, New York.
Phillips, N. A. (1966). *J. Atmos. Sci.* **23**, 626.
Phillips, N. A. (1974). "Application of Arakawa's Energy Conserving Layer Model to Operational Numerical Weather Prediction," Office Note 104. U.S. Dept. Commerce, National Oceanic and Atmospheric Administration, National Weather Service, National Meteorol. Center, Washington, D. C.
Platzman, G. W. (1954). *Arch. Meteorol., Geophys. Bioklimatol., Ser. A* **7**, 29.
Platzman, G. W. (1958). "An Approximation to the Product of Discrete Functions," Tech. Rep. No. 2 (Grant NSF-G2159). Dept. Meteorol., University of Chicago, Chicago, Illinois.
Platzman, G. W. (1960). *J. Meteorol.* **17**, 635.
Platzman, G. W. (1961). *J. Meteorol.* **18**, 31.
Platzman, G. W. (1968). *Bull. Am. Meteorol. Soc.* **49**, 496.
Platzman, G. W., and Baer, F. (1958). "The Extended Numerical Integration of a Simple Barotropic Model. Part I. Program," Tech. Rep. No. 1 (Grant NSF-G2159). Dept. Meteorol., University of Chicago, Chicago, Illinois.
Price, G. V., and MacPherson, A. K. (1973). *J. Appl. Meteorol.* **12**, 1102.
Richardson, L. F. (1922). "Weather Prediction by Numerical Process." Cambridge Univ. Press, London and New York (Reprint, with a new introduction by Sydney Chapman, Dover, New York, 1965).
Richtmyer, R. D. (1963). "A Survey of Difference Methods for Non-Steady Fluid Dynamics," Tech. Note 63-2. National Center for Atmospheric Research, Boulder, Colorado.
Richtmyer, R. D., and Morton, K. W. (1967). "Difference Methods for Initial-Value Problems." Wiley (Interscience), New York.
Robert, A. J. (1966). *J. Meteorol. Soc. Jpn.* [2] **44**, 237.
Robert, A. J. (1969). "The Integration of a Spectral Model of the Atmosphere by the Implicit Method." *Proc. WMO/IUGG Symp. Numer. Weather Prediction, 1968* VII-19-24.

Robert, A. J. (1974). "Computational Resolution Requirements for Accurate Medium-Range Numerical Predictions." *Proc. Symp. Difference and Spectral Methods for Atmosphere and Ocean Dynamics Problems, 1973*, Part I, 83–102.
Robert, A. J., Shuman, F. G., and Gerrity, J. P., Jr. (1970). *Mon. Weather Rev.* **98**, 1.
Robert, A. J., Henderson, J., and Turnbull, C. (1972). *Mon. Weather Rev.* **100**, 329.
Roberts, K. V., and Weiss, N. O. (1966). *Math. Comput.* **20**, 272.
Sadourny, R. (1972). *Mon. Weather Rev.* **100**, 136.
Sadourny, R., Arakawa, A., and Mintz, Y. (1968). *Mon. Weather Rev.* **96**, 351.
Saltzman, B., and Fleisher, A. (1962). *J. Atmos. Sci.* **19**, 195.
Sankar-Rao, M., and Umscheid, L., Jr. (1969). *Mon. Weather Rev.* **97**, 659.
Shapiro, M. A. (1975). *Mon. Weather Rev.* **103**, 591.
Shuman, F. G. (1962). "Numerical Experiments with the Primitive Equations." *Proc. Int. Symp. Numer. Weather Prediction, 1960*, pp. 85–107.
Shuman, F. G. (1970). *J. Appl. Meteorol.* **9**, 564.
Shuman, F. G. (1972). "The Research and Development Program at the National Meteorological Center," Office Note 72. U.S. Dept. Commerce, National Oceanic and Atmospheric Administration, National Meteorol. Center, Washington, D. C.
Shuman, F. G., and Hovermale, J. B. (1968). *J. Appl. Meteorol.* **7**, 525.
Silberman, I. (1954). *J. Meteorol.* **11**, 27.
Simons, T. J. (1968). "A Three-Dimensional Spectral Prediction Equation," Atmos. Sci. Pap. No. 127. Dept. Atmos. Sci., Colorado State University, Fort Collins.
Smagorinsky, J. (1958). *Mon. Weather Rev.* **86**, 457.
Smagorinsky, J. (1963). *Mon. Weather Rev.* **91**, 99.
Smagorinsky, J., Manabe, S., and Holloway, J. L., Jr. (1965). *Mon. Weather Rev.* **93**, 727.
Smagorinsky, J., Strickler, R. F., Sangster, W. E., Manabe, S., Holloway, J. L., Jr., and Hembree, G. D. (1967). *Proc. Int. Symp. Dyn. Large-Scale Atmos. Processes, 1965* pp. 70–134.
Somerville, R. C. J., Stone, P. H., Halem, M., Hansen, J. E., Hogan, J. S., Druyan, L. M., Russell, G., Lacis, A. A., Quirk, W. J., and Tenenbaum, J. (1974). *J. Atmos. Sci.* **31**, 84.
Starr, V. P. (1945). *J. Meteorol.* **2**, 227.
Stoker, J. J., and Isaacson, E. (1975). "Final Report I," IMM 407. Courant Institute of Mathematical Sciences, New York University, New York.
Stone, P. H., Quirk, W. J., and Somerville, R. C. J. (1974). *Mon. Weather Rev.* **102**, 765.
Strang, G. W. (1968). *SIAM J. Numer. Anal.* **5**, 506.
Sundqvist, H. (1975). *Atmosphere* **13**, 81.
Sutcliffe, R. C. (1947). *Q. J. R. Meteorol. Soc.* **73**, 370.
Taylor, T. D., Ndefo, E., and Masson, B. W. (1972). *J. Comput. Phys.* **9**, 99.
Thompson, P. D. (1957). *Tellus* **9**, 69.
Thompson, P. D. (1961). "Numerical Weather Analysis and Prediction." Macmillan, New York.
Thompson, P. D. (1976). "The History of Numerical Weather Prediction in the United States," NCAR Manuscript 0301/76-03. National Center for Atmospheric Research, Boulder, Colorado.
Tiedtke, M. (1972). *Contrib. Atmos. Phys.* **45**, 43.
Turkel, E. (1974). *J. Comput. Phys.* **15**, 226.
Umscheid, L., Jr., and Sankar-Rao, M. (1971). *Mon. Weather Rev.* **99**, 686.
Vanderman, L. W. (1972). *Mon. Weather Rev.* **100**, 856.
Wang, H.-H., Halpern, P., Douglas, J., Jr., and Dupont, T. (1972). *Mon. Weather Rev.* **10**, 738.
Washington, W. M., and Kasahara, A. (1970). *Mon. Weather Rev.* **98**, 559.
Wiin-Nielsen, A. (1967). *Tellus* **19**, 540.
Williams, J. (1976). *Mon. Weather Rev.* **104**, 249.
Williamson, D. L. (1968). *Tellus* **20**, 622.

Williamson, D. L. (1970). *Mon. Weather Rev.* **98**, 512.
Williamson, D. L. (1971). *J. Comput. Phys.* **7**, 301.
Williamson, D. L. (1976). *Mon. Weather Rev.* **104**, 31.
Williamson, D. L., and Browning, G. L. (1973). *J. Appl. Meteorol.* **12**, 264.
Young, J. A. (1968). *Mon. Weather Rev.* **96**, 357.
Zwas, G., and Abarbanel, S. (1971). *Math. Comput.* **25**, 229.

# United Kingdom Meteorological Office Five-Level General Circulation Model

G. A. CORBY, A. GILCHRIST, AND P. R. ROWNTREE

METEOROLOGICAL OFFICE
BRACKNELL, ENGLAND

I. Introduction . . . 67
II. Coordinate System and Grid . . . 69
III. Basic Equations and Finite Difference Approximations . . . 70
IV. Dissipation Terms (Lateral Eddy Viscosity) . . . 74
V. Large-Scale Precipitation and Latent Heating . . . 75
VI. Simple Boundary Layer Parameterization . . . 76
VII. Representation of Surface Exchange Processes . . . 82
A. Momentum Exchange . . . 82
B. Sensible and Latent Heat Fluxes . . . 84
VIII. Treatment of Land and Ice Surfaces . . . 86
IX. Convective Interchange . . . 87
X. Radiation Scheme . . . 92
XI. Model's January and July Simulations . . . 96
List of Symbols . . . 107
References . . . 108

## I. Introduction

THE UNITED KINGDOM METEOROLOGICAL OFFICE general circulation model has been developed over a number of years. The first experiments to determine a suitable grid and finite difference systems were reported by Grimmer and Shaw (1967). The first version of the model was described by Corby *et al.* (1972), and the first long integration by Gilchrist *et al.* (1973). At that time, the grid, which was hemispheric only, was of regular latitude–longitude form with a spacing of 3° × 5°. This is approximately square in middle latitudes. To avoid very short time steps, the area from the pole to 81°N was treated as a polar cap so that only mean values of variables over the cap had to be carried. From 81°N to the latitude at which the longitudinal and latitudinal grid lengths were equal, a spatially variable time step, chosen so that the Courant–Friedrichs–Lewy condition for linear computational stability was satisfied locally, was used. When the model was reprogrammed for another computer, the grid and finite differences were changed to the second system tested by Grimmer and Shaw—namely, one in which the zonal grid length was approximately constant. In the particular test integrations they carried out it performed less well than the regular grid, but their test was one to which the latter was ideally suited. In real atmospheric situations the second grid was at a smaller disadvantage than the original tests suggested, and it had the advantage of being conceptually simpler. Since the original publication the model has been extended to the global domain, and development of the physical parameterizations has continued. The radiation, surface exchange, and convection schemes described here are all different from the originals, though clearly they have evolved from them.

The primary aim in developing the five-layer model has been to make it suitable in the first instance for investigating anomalies of the presently observed global climate. Thus, it has been extensively used in sea–surface temperature anomaly experiments (Rowntree, 1976a, b; Gilchrist, 1975a). This seems a logical way to proceed since testing against observable and reasonably well-documented atmospheric phenomena enables model characteristics and possible shortcomings to be identified. It can obviously serve to build up confidence in the performance of the model in circumstances when the simulations are less easily verifiable. On the whole, therefore, the model has been kept relatively simple; in particular a highly parameterized radiation scheme has generally been used (Section X) and the surface exchange scheme has so far been designed to cope with only a limited number of surface types (Section VII). It has nevertheless seemed important for our purposes to use a model that is capable of sufficiently realistic simulations for different times of year and is not confined in its applicability to circumstances of strong baroclinic instability or the annual mean circulation. In the present contribution we have illustrated the performance of the model for both January and July; at least as far as the Northern Hemisphere is concerned, the range of conditions represented by these two months is very wide indeed, and the fact that the simulations are reasonable gives one confidence that the model is capable of being used in a variety of atmospheric experiments. By virtue of its comparative simplicity, the model executes more quickly than some general circulation models. The version described here progresses at approximately six simulated days per 1 hr central processor time on an IBM 360/195 computer.

Simultaneously with the main line of development just sketched, other versions of the model have been written and applied in investigations beyond the scope of the simple five-layer model. For example, a 13-layer version with eight levels in the stratosphere was written and used in experiments concerning possible changes in the ozone layer (Newson, 1974a, b), and a version using a fully interactive radiation scheme with model-dependent cloud is being used to test the sensitivity of climate simulation to cloudiness parameters (Hunt, 1976).

## II. Coordinate System and Grid

The horizontal grid is designed to give a quasi-uniform resolution over the sphere; it is similar to the grid proposed by Kurihara (1965). The points lie on lines of latitude with the number per line decreasing toward the poles. It is thus an irregular grid and the finite differences have to be formulated in terms of fluxes across the boundaries of the volume that the variables at the grid points represent. In the horizontal, the "grid areas" surrounding the points are bounded by lines of latitude and longitude which are, respectively, midway between the lines of points and between pairs of points on a line. The formulation of the finite differences for a grid of this kind in conservation form has been discussed by Bryan (1966).

If $\Delta\theta$ is the latitudinal spacing, so that $N = \pi/2\Delta\theta$ is the number of rows between the equator and pole, the points are positioned at

$$\theta = \pm\Delta\theta/2, \pm 3\Delta\theta/2 \ldots \pm (\pi - \Delta\theta)/2.$$

Each row has the integral part of $4N \cos\theta$ (rounded) points, which we denote by $M$; this corresponds to $4N$ at the equator. It is known that, strictly applied, this form of grid does not provide an adequate resolution near the pole, because the truncation error rises steeply in high latitudes (Shuman, 1970).A minimum value of 16 has therefore been placed on $M$. The numbers at the two rows nearest the pole [which should be $4N \sin \Delta\theta/2$ $(=3)$ and $4N \sin 3\Delta\theta/2$ $(=9)$] are set to this value. It remains to decide the disposition of grid points around the lines of latitude. The choice that has been made is that, for the northern hemisphere, the "first" point at a line of latitude has been placed due south of the western boundary of the first grid area to the north. This has the property that the grid never becomes regular or almost regular over any part of the sphere; it seemed desirable in constructing the grid to avoid such special regions. The horizontal grid in the Southern Hemisphere is a mirror image of that in the Northern Hemisphere. The grid that has normally been used in the model and that is used in the integrations illustrated in this paper, has $N = 30$ corresponding to a grid length of approximately 330 km. This has a total of 4626 points on the globe, the numbers at selected latitudes being

$$1\tfrac{1}{2}^\circ(120);\ 4\tfrac{1}{2}^\circ(120) \ldots 19\tfrac{1}{2}^\circ(113) \ldots 31\tfrac{1}{2}^\circ(102) \ldots 40\tfrac{1}{2}^\circ(91)$$
$$49\tfrac{1}{2}^\circ(78) \ldots 58\tfrac{1}{2}^\circ(63) \ldots 70\tfrac{1}{2}^\circ(40) \ldots 79\tfrac{1}{2}^\circ(22) \ldots 88\tfrac{1}{2}^\circ(16)$$

Because the grid is quasi-uniform it is possible to use the same time step over almost all of it. For $N = 30$, a value of 10 min satisfies the linear stability criterion and has usually been used. However for the two lines closest to the poles where the number of points is greater than required for an even distribution, special action is required to avoid instability. At latitude $85\frac{1}{2}^\circ$ this consists of truncating the variables $u$, $v$, $T$, and $p_*$ at wave number 3 and doubling the time-meaning coefficient (see Section III), while retaining the same time step as elsewhere. At $88\frac{1}{2}^\circ$, however, the time step is halved, Fourier truncation is effected at wave number 1, and the time-meaning coefficient is double the value for $85\frac{1}{2}^\circ$. It was also found necessary to truncate the topographic heights consistently with the model variables.

In the vertical, the model uses the sigma-coordinate system (Phillips, 1957), since this allows the effects of mountains on the general circulation to be represented realistically without resort to internal boundaries. It has five layers of equal mass with the layer boundaries at $\sigma = 1.0, 0.8, 0.6, 0.4, 0.2$, and 0; the mean properties of the layers are represented by variables held at $\sigma = 0.9, 0.7, 0.5, 0.3$, and 0.1.

## III. Basic Equations and Finite Difference Approximations

The equations are used in so-called momentum or flux form since this facilitates the use of a finite difference formulation based on the energy conserving box method developed by Lilly (1965) and Arakawa (1966). They can be written in the form

$$\partial \mathbf{X}/\partial t + \nabla_3 \cdot \mathbf{V}_3 \mathbf{X} + \mathbf{Y} + \mathbf{Z} = \mathbf{N} \tag{1}$$

$$\partial \phi/\partial \sigma + RT/\sigma = 0 \tag{2}$$

$$\mathbf{X} = p_* \mathbf{x} = p_* \begin{bmatrix} \mathbf{V} \\ T \\ q \\ 1 \end{bmatrix} \quad \mathbf{Y} = \begin{bmatrix} p_* \nabla_p \phi \\ -\kappa T \omega/\sigma \\ 0 \\ 0 \end{bmatrix} \quad \mathbf{Z} = \begin{bmatrix} f' p_* \mathbf{l} \times \mathbf{V} \\ 0 \\ 0 \\ 0 \end{bmatrix}$$

$$\mathbf{N} = \begin{bmatrix} D_\mathbf{V} + g\, \partial \boldsymbol{\tau}/\partial \sigma \\ D_T + (g/c_p)(\partial H/\partial \sigma) + (p_*/c_p)(\dot{Q}_R + L\dot{P}) \\ D_q + g\, \partial M/\partial \sigma - p_* \dot{P} \\ 0 \end{bmatrix} \tag{3}$$

Certain approximations which tend to destroy the formal simplicity of the equations have already been made. It may be noted, however, that they are a consistent set with integral constraints similar to those of the full equations.

The most significant approximation which in effect transforms the equations from the full three-dimensional to a quasi-two-dimensional set is that for the large-scale motions that the model reproduces explicitly the atmosphere can be considered in hydrostatic equilibrium; scale analysis demonstrates the validity of this assumption. The vertical accelerations are ignored against the accelerations due to gravity. The vertical motion ($\dot{\sigma}$) occurring in $\nabla_3 \cdot$ is then obtained diagnostically from the convergence of the horizontal wind.

There are six equations (two for horizontal momentum, the thermodynamic, water vapor, and mass continuity equations plus the hydrostatic equation) in the six unknowns $p_*$, $u$, $v$, $T$, $q$, $\phi$. $Y$ contains the energy transformation terms that convert potential to kinetic energy and vice versa. $Z$ is the Coriolis term. Grouped in $N$ are the nonadiabatic terms that arise mainly from subgrid-scale motions and therefore have to be parameterized in terms of the model's explicit variables. They are left here largely in symbolic form; their treatment will be described later.

The approximation of the terms in finite differences is now considered.

Time differencing uses a centered or leap-frog time step, i.e., $\delta_t \overline{\mathbf{X}}^t$. This has the virtue that it does not damp the motions, and any damping that is required can then be applied explicitly by terms whose magnitude can be accurately controlled. The difficulty that tends to arise with this time step scheme is the development of a weak instability in which the difference between values of variables a time step apart grows progressively; it is overcome by using the time filter described by Asselin (1972).

$$\mathbf{x}^t = (1 - \alpha)(\mathbf{x}^t)' + (\alpha/2)[(\mathbf{x}^{t+\Delta t})' + \mathbf{x}^{t-\Delta t}] \tag{4}$$

where $(\ )'$ indicates a provisional value and the superscripts indicate the time level. $\alpha$ is given the value 0.02 over most of the grid but is increased near the pole (see Section II).

It may be noted that the scheme used in the model ensures conservation in space only and not in time. It may be shown that conservation in time is good when there is a linear change from time step to time step; it becomes poor in the presence of the weak time instability mentioned above.

Considering now the flux divergences, $\nabla_3 \cdot \mathbf{V}_3\mathbf{X}$, we omit for the moment this term as it arises in the thermodynamic equation and consider only the form it takes in the momentum, water vapor and mass equations. For these, the estimations follow the energy conserving system with the modifications to allow for an irregular grid suggested by Bryan (1966); further details are given in Grimmer and Shaw (1967). Thus we may write the finite difference approximation for the term as

$$(1/a \cos\theta)[\delta_\lambda \bar{U}^\lambda \bar{\mathbf{x}}^\lambda + \delta_\theta \bar{V}^\theta \bar{\mathbf{x}}^\theta \cos\theta] + \delta_\sigma \dot{\sigma} \bar{\mathbf{X}}^\sigma \tag{5}$$

It may be demonstrated that the form possesses linear and quadratic conservation properties both in the horizontal and in the vertical.

The estimates of $\dot{\sigma}$ and $\partial p_*/\partial t$ are obtained from vertical summation of the mass continuity equation which gives

$$(p_*\dot{\sigma})_{k+1/2} = -\sigma_{k+1/2}\,\delta_t\overline{p}_*{}^t - (1/a\cos\theta)\sum_{s=1}^{k}[\delta_\lambda\overline{U}_s{}^\lambda + \delta_\theta\overline{V}_s^\theta\cos\theta]\,\Delta\sigma_s \quad (6)$$

where $(k + \frac{1}{2})$ indicates that the value applies to the boundary between the $k$th and $(k + 1)$th layers. Substitution of $k = n$, for the lowest layer, gives $\delta_t\overline{p}_*{}^t$ since $\sigma = 1$ and $\dot{\sigma} = 0$ at the earth's surface.

The Coriolis terms are estimated in the most obvious way.

Values of $\phi$ are obtained by vertical integration of the hydrostatic relation (2) as follows:

$$\phi_k = \phi_* + RT_n\ln(1/\sigma_n) + \sum_{s=k}^{n-1}(R/2)(T_s + T_{s+1})\ln(\sigma_{s+1}/\sigma_s) \quad (7)$$

The logarithmic form of the equation is used for accuracy so that model values of $\phi$ correspond approximately with those calculated for the real atmosphere for the same vertical temperature structure. It will be noted that in deriving the thickness of the lowest half-layer in the model, $T_*$ is not used since it is determined outside the main equations, while in estimating thicknesses at higher levels the mean temperature of a layer is taken as the average of the temperatures at the top and bottom.

A problem arises in estimating the pressure gradient $\nabla_p\phi$ in sigma coordinates. It is

$$\nabla_\sigma\phi + (RT/p)\,\nabla_\sigma p \quad (8)$$

and, over mountains particularly, the required value is a small residual remaining after differencing two large terms. Unless they are estimated consistently, the overall truncation error for the term may be very severe. The forms used in the model are

$$(1/a\cos\theta)[\delta_\lambda\overline{\phi}^\lambda + R\overline{\overline{T}^\lambda\,\delta_\lambda\ln p_*}^\lambda]$$

and (9)

$$(1/a\cos\theta)[\overline{\cos\theta\,\delta_\theta\phi}^\theta + R\,\overline{\cos\theta\,\overline{T}^\theta\,\delta_\theta\ln p_*}^\theta]$$

in the first and second momentum equations, respectively. They have the property that, if applied to a horizontally homogeneous atmosphere, in which the pressure gradient term is zero, then, on replacing the horizontal differences by vertical differences, they reduce to the form adopted for the estimation of thicknesses, i.e.,

$$(\phi_1 - \phi_2) = (R/2)(T_1 + T_2) \ln (p_2/p_1) \tag{10}$$

where 1 and 2 refer to two successive grid points on a sigma surface. In this sense, therefore, the two parts of the term are estimated consistently, and the truncation error should be small. Experience with the formulation tends to bear out this expectation. Even when applied across the Andes where successive grid points may have violent changes of $p_*$, sometimes by as much as 300 mb, no computational difficulties have been found.

The estimation of the $\omega$ term in the thermodynamic equation is also known to require care since computational difficulties can occur, particularly in the vicinity of mountains (Holloway and Manabe, 1971). The analytical form is

$$-\kappa\omega T/\sigma = -(\kappa T/\sigma)(p_*\dot{\sigma} + \sigma\, \partial p_*/\partial t + \sigma \mathbf{V} \cdot \nabla p_*) \tag{11}$$

There is no obvious "best" finite difference estimate, and there are many possibilities. The solution adopted here is to use a formulation that reproduces the analytical constraint that the energy conversion terms vanish when integrated over the global domain. Specifically, the requirement is

$$\int\int\int_\Omega [p_* \mathbf{V} \cdot \nabla_p \phi - (RT\omega/\sigma) - \phi_*(\partial p_*/\partial t)]\, d\Omega = 0 \tag{12}$$

Inserting the finite difference estimates for the first and third terms and replacing the integration by a summation over all grid points, the following estimate of the $\omega$ term is deduced:

$$\begin{aligned}(\kappa T_k/2)\left[(\ln \sigma_{k+1}/\ln \sigma_k) \sum_{s=1}^{k} + (\ln \sigma_k/\ln \sigma_{k-1}) \sum_{s=1}^{k-1}\right](1/a\cos\theta)[\delta_\lambda \bar{U}_s{}^\lambda + \delta_\theta \bar{V}_s{}^\theta \cos\theta] \\ -(\kappa/a\cos\theta)[U_k \overline{\bar{T}_k^\lambda \delta_\lambda \ln p_*}^\lambda + V_k \overline{\cos\theta \bar{T}_k^\theta \delta_\theta \ln p_*}^\theta]\end{aligned} \tag{13}$$

A more detailed derivation of this result, including the modification required when the model has variable depth layers in the vertical is given in Gilchrist (1975c).

We return now to consider the estimation of $\nabla_3 \cdot \mathbf{V}_3 p_* T$ in the thermodynamic equation. The problem here is akin to that involved in estimating the pressure gradient term when $p_*$ changes sharply from one grid point to the next. In the simplest circumstances of a horizontally homogeneous atmosphere in which the temperature lapse rate is dry adiabatic, air rising over topography may be regarded as being subject to increments of temperature from two sources, *viz.*, advection along the sigma surface and adiabatic expansion, since there is a change of both temperature and pressure along the surface. At a particular grid point these two effects may be large individually, but they should cancel exactly in the simple circumstances just postulated. The thermodynamic equation in one dimension in this case reduces to

$$(\partial T/\partial t) + u(\partial T/\partial x) - u(\kappa T/p)(\partial p/\partial x_\sigma) = 0 \tag{14}$$

The finite difference formulation of the last term has already been fixed since it comes from the $\omega$ term. It can be written

$$u(1/c_p)(\overline{R\overline{T}^x\,\delta_x \ln p_*^x}) \tag{15}$$

Now observe that since $(1/c_p)$ is the adiabatic lapse rate per unit of geopotential and $R\overline{T}^x\,\delta_x \ln p_*$ is the consistent estimate of the geopotential difference between two successive grid points $i$ and $i+1$ on the same sigma surface

$$(1/c_p)R\overline{T}^x_{i+1/2}\,\delta_x \ln p_* = (T_{i+1} - T_i) = \Delta x\,\delta_x\,T_{i+1/2} \tag{16}$$

Thus, for exact cancellation of the terms to give zero temperature change at grid point, the thermodynamic equation should reduce to

$$\delta_t\overline{T}^t + u\overline{[\delta_x T - \kappa\overline{T}^x\,\delta_x \ln p_*^x]} = 0 \tag{17}$$

in the special circumstances considered. This is achieved by using for $\nabla_3 \cdot \mathbf{V}_3 p_* T$ the finite difference estimate

$$(1/a\cos\theta)[\delta_\lambda \widetilde{UT}^\lambda + \delta_\theta \widetilde{VT}^\theta \cos\theta] + \delta_\sigma \dot{\sigma}\overline{p_* T^\sigma} \tag{18}$$

since

$$\widetilde{AB}^x = A\overline{B}^x + B\overline{A}^x \tag{19}$$

## IV. Dissipation Terms (Lateral Eddy Viscosity)

Since the dynamical equations are nonlinear there is a continuous transfer of properties to higher and higher wave numbers. There are physical grounds, therefore, for adding a dissipative term to the finite difference system in order to simulate the transfer from scales of motion that can be represented explicitly on the grid to the smaller scales that cannot be so represented.

The problem can be approached in a number of ways, and various proposals have been made for the most appropriate form of the term. Thus, Shapiro (1971) suggests the use of a linear filter to eliminate the shortest waves responsible for aliasing; Lilly (1967) and Leith (1969) derive nonlinear forms which follow from the assumption that a uniform turbulence structure connects the explicit and implicit regimes; and Richtmyer and Morton (1967) design the term to prevent the occurrence of "shock-wave" phenomena in a system that is incapable of describing them. The practical constraints on the dissipation are that it should enable stable integrations to be carried out, and that its effect on the meteorologically important scales of motion should be as small as possible. The simplest form of dissipative term that has been extensively used is a linear viscosity, but it is not sufficiently selective with respect to the damping it produces at different wavelengths. In particular it leads to excessive damping of the baroclinic waves, which is both unrealistic and damaging. To overcome this difficulty some versions of the model have used Shapiro's linear filtering technique (Francis, 1975), but mostly simple nonlinear forms that are economical in computing have been preferred. They are

$$\begin{aligned} D_u &= (K'_u/p_*)[\nabla \cdot p_* \nabla u] \\ D_v &= (K'_v/p_*)[\nabla \cdot p_* \nabla v] \\ D_T &= (K'_T/p_*)[\nabla \cdot p_*^{k+1} \nabla(T/p_*^k)] \\ D_q &= (K'_q/p_*)[\nabla \cdot p_* \nabla q] \end{aligned} \tag{20}$$

where the effective nonlinear eddy viscosity coefficients $K'_u$, etc., are taken as proportional to the modulus of the respective terms within square brackets. The constant of proportionality is $10^{15}$ m$^3$ in the momentum equations; in the other equations, the value is such that on average, $K'_u \fallingdotseq K'_T \fallingdotseq K'_q$. These terms were evaluated using the obvious finite difference approximations; to maintain numerical stability a forward step was used in the time differencing. There are theoretical objections to a smoothing of this form in that it does not conserve first-order quantities, but, in practice, it has not been found that other forms that have a more logical basis but generally require considerably more computing have advantages.

## V. Large-Scale Precipitation and Latent Heating

Temperature and moisture are first incremented, treating the advection of temperature on a dry adiabatic basis and the advection of mixing ratio as though this were conserved, and then appropriate adjustments are made if an excess beyond saturation occurs. Thus if after completing a time step on the basis of dry conservative advection the temperature and mixing ratio at a point in the model are $T$ and $q$, and if in addition $q$ exceeds the saturation value $q^{sat}$ at temperature $T$, then we augment the temperature by $\Delta T$ where

$$\Delta T = (L/c_p)\{q - (q^{sat})'\} \tag{21}$$

$(q^{sat})'$ being the saturation value at the new temperature $T + \Delta T$.
We have approximately

$$\begin{aligned}(q^{sat})' &= q^{sat} + \{(\partial/\partial T)q^{sat}\}\,\Delta T \\ &= q^{sat} + Lq^{sat}\,\Delta T/(R_v T^2)\end{aligned} \tag{22}$$

$R_v$ being the gas constant for water vapor. We then have $(q - (q^{sat})')/\Delta t =$ rate of condensation per unit mass $= \dot{P}_L$

$$= (q - q^{sat})/\{\Delta t\,(1 + L^2 q^{sat}/c_p R_v T^2)\} \tag{23}$$

while $L/c_p$ times this expression gives the rate of change of temperature due to latent heating. For the derivation of $q^{sat}$ we use

$$q^{sat} = 0.622(e_s/p) \tag{24}$$

where $e_s$ is the saturation vapor pressure. The condensation threshold is taken as 100% relative humidity and condensed moisture is immediately precipitated without any provision for evaporation into layers below the level of condensation. (This procedure is not, however, followed in the convective scheme, where evaporation into lower layers is allowed.)

## VI. Simple Boundary Layer Parameterization

The convectively unstable boundary layer capped by an inversion appears to be the most frequently occurring vertical structure of the lowest part of the atmosphere over the sea (Charnock and Ellison, 1967); it is also common over land during the day. Analytical models that attempt to describe its essential features and its evolution have been presented in a number of papers (Ball, 1960; Betts, 1973; Carson, 1973; Tennekes, 1973). In modeling the general circulation, it is important to simulate the fluxes of heat and moisture into the free atmosphere correctly. They are controlled to a considerable degree by the boundary layer structure, being large when the capping inversion is sufficiently weak to be broken down by ascending buoyant air or when it effectively disappears as it does ahead of fronts or near cumulonimbus clouds. Conversely, when the inversion is strong, there is little exchange across it and fluxes into the free atmosphere are small. It seems reasonable to attempt to model these effects explicitly in a general circulation simulation, so following the approach of Charnock and Ellison (1967) in spirit if not in detail.

The simple boundary layer parameterization used in the model starts from the following idealizations.

(i) There is a unique boundary layer height which is thought of as the base of an inversion limiting interchange of air between the boundary layer and the free atmosphere.

(ii) The boundary layer top always lies within the bottom slab of the model; that is, it can rise only to the pressure level $0.8p_*$ (approximately 2000 m). By implication the strength of the capping inversion is related to its height, being more easily broken down as it approaches its upper limit.

(iii) The boundary layer is well mixed so that the potential temperature and mixing ratio are constant in the vertical down to the top of a thin surface layer. Surface exchanges of heat and moisture depend on differences of temperature and humidity across this surface layer.

(iv) If there is convection proceeding between the two lowest model layers the ascending air originates in the boundary layer, that is, the base of the cloud coincides with the base of the inversion.

It is a matter of observation that the convectively unstable boundary layer is maintained by buoyant convection, requiring an upward flux of heat at the earth's surface. If there is a downward flux a stable layer forms close to the ground, the most obvious example being the nocturnal inversion that forms and intensifies over land during the hours of darkness; commonly this happens under an existing daytime inversion and during the day the surface inversion erodes rapidly to allow convection up to the previous level. A realistic simulation of this sequence of events cannot be accommodated within a simple scheme with a single boundary layer height. Stable boundary layers also form over relatively cold surfaces, for example, in warm moist air moving poleward ahead of depressions or in initially unstable airstreams when they flow from the sea over a cold land surface in winter. A satisfactory scientific account of the evolution of a stable boundary layer in such circumstances is still lacking. The question of how to deal with stable boundary layers in a general circulation model is obviously very difficult, but it is to be hoped that it is not crucial since the important circumstances are those in which there is a substantial flux of heat and moisture from the boundary layer into the free atmosphere. In conformity with the general approach adopted in the model, the stable boundary layer is dealt with in a simple way. It is assumed that a downward heat flux at the surface leads to a sinking of the boundary layer top until it reaches a minimum value determined by mechanical mixing; that is, at this stage the neutrally buoyant layer is maintained by the turbulence of the wind. Such an approach cannot of course produce stable surface inversions, and this must lead to poor simulations in some circumstances. However, considering the diurnal effect (the most satisfactorily documented case) perhaps what we seek for a general circulation model is a minimum value such that significant convection (and therefore significant interaction with the free atmosphere) is possible following receipt of an appropriate amount of solar radiation at the surface and, in this at least, it is not evident *a priori* that the formulation is incapable of achieving a worthwhile simulation.

We turn now to the specific mechanisms by which the boundary layer as just described is inserted into the model. The position of the top is specified by means of a parameter $b$, the only additional variable. It is the fraction of the mass of the bottom slab of the model which lies within the boundary layer, so that if $b = 0$ the layer is vanishingly thin and if $b = 1$ it has its maximum depth.

In order to increment $b$, values of variables are required both at the top of, and representative of the mean within, the boundary layer. They are derived by linear extrapolation with respect to height from the two lowest model levels. Thus, we take

$$X_T = X_{0.9} - (\Delta X/\Delta\phi)\phi_9'(1 - 2b) \tag{25}$$

where $X$ denotes the variable, suffixes $T$ and 0.9 denote values at the top of the boundary layer and at the lowest model level, respectively, and $\Delta$ denotes a difference between variables at the lowest two levels; $\phi_9'$ is the geopotential thickness from the surface up to pressure $0.9p_*$ so that $|\phi_9'(1 - 2b)|$ is an estimate of the geopotential difference between the top of the boundary layer and the location of the lowest level variables. The formula is used for the wind components and also for the potential temperature $\Theta_T$ and mixing ratio $q_T$ when $0 \leqslant b \leqslant 0.5$. If $0.5 < b \leqslant 1$ the lowest level lies within the boundary layer, and it is then consistent with the initial postulates to take $\Theta_T = \Theta_{0.9}$ and $q_T = q_{0.9}$. In calculating the potential temperature, the reference pressure is $p_*$, the pressure at the earth's surface.

Deviations from formula (25) have been allowed to avoid unrealistic values.

(i) Linear extrapolation of the mixing ratio is capable of producing very small or even negative values and therefore a minimum relative humidity of 10% is adopted for air at the boundary layer top.

(ii) When an explicit inversion exists between the model's two lowest levels (i.e., $T_{0.9} < T_{0.7}$) the structure on which the boundary layer model is based is inappropriate. The circumstance seems likely to occur only over high latitude land surfaces in winter, when one would expect it to be associated with very cold ground temperatures and a steep surface inversion. The assumption in the model under this condition is

$$\Theta_T = \Theta_{0.9} - C(1 - b), \qquad C = 10\,°\text{K} \tag{26}$$

For mean wind values within the boundary layer ($\mathbf{V}_B$), the formula is

$$\mathbf{V}_B = \mathbf{V}_{0.9} - (\Delta\mathbf{V}/\Delta\phi)\phi'_9(1 - b) \tag{27}$$

The value of $b$ changes as a result of a number of processes: (i) by large scale explicit convergence or divergence within the boundary layer, these leading to increase or decrease respectively; (ii) in response to the surface heat flux, a positive flux producing growth by the entrainment of air from above, a negative flux being accompanied in general by thinning of the layer; (iii) by environmental subsidence associated with convection; (iv) by mechanical mixing, which however only sets a lower limit to $b$; (v) by smoothing, caused by subgrid-scale eddies.

Symbolically we can write

$$\Delta b = \Delta b_D + \Delta b_H + \Delta b_C + \Delta b_M + \Delta b_S \tag{28}$$

At a particular time step each increment (discussed in more detail below) is calculated separately and added to $b$ in the order shown.

(i) *Large-Scale Dynamics* ($\Delta b_D$)

This component is deduced from the equation for mass conservation in the boundary layer, which in finite difference form is

$$\delta_t\overline{p_* b}^t + (1/a\cos\theta)[\delta_\lambda\overline{U_B b}^\lambda + \delta_\theta(\cos\theta\overline{V_B b}^\theta)] = 0 \tag{29}$$

(ii) *Heat Content* ($\Delta b_H$)

Models of a simple, dry, convectively unstable boundary layer, with a capping inversion in which the vertical gradient of potential temperature is $\gamma$, lead (Betts, 1973; Carson, 1973), in the absence of vertical motion (which here is treated separately) and with the closure

$$H(z_i) = -AH_*, \tag{30}$$

to the following equation for the change of the inversion height $z_i$

$$\frac{dz_i}{dt} = \frac{(1 + 2A)H_*}{\rho c_p z_i \gamma} \tag{31}$$

$H_*$ and $H(z_i)$ are, respectively, the upward heat flux at the ground and the downward heat flux at the inversion, and $\rho$ is the density of air in the boundary layer. Carson has found that the value of $A$ required to give a best fit to the O'Neill site data on a diurnal basis varies with the time of day, from negligible values soon after sunrise (in association with a large value of $\gamma$) to about 0.5 when the surface heat flux is near its maximum value and $\gamma$ is about $6 \times 10^{-3}$ °K m$^{-1}$. Betts gives typical tropical afternoon values of $A = 0.25$ and $\gamma = 3.8 \times 10^{-3}$ °K m$^{-1}$. Here we wish to apply a single formula throughout the day and have adopted what seem to be acceptable mean values for this purpose—namely, $A = 0.2$ and $\gamma = 6 \times 10^{-3}$ °K m$^{-1}$. Converting the equation to apply to $b$ we have

$$\Delta b_H/\Delta t = (1 + 2A)H_* \rho/(\gamma c_p b m^2) \tag{32}$$

where $m = p_* \Delta\sigma_n/g$, the mass of air per unit area in the lowest sigma layer and $\rho z_i \fallingdotseq bm$. $\Delta T_n/\Delta t$, the temperature change at the lowest level due to surface heating, may be written $H_*/mc_p$ and $\rho/m \fallingdotseq 0.5 \times 10^{-3}$ m$^{-1}$ whence we have

$$\Delta b_H/\Delta t = (1 + 2A)(\rho/m)(\Delta T_n/\Delta t)/\gamma b \tag{33}$$

In order to take account, to some extent, of the effects of radiation, $\Delta T_n/\Delta t$ has in practice been replaced by the temperature increment due to surface heating and radiation combined. Since the radiation scheme takes no account of the implied structure of the boundary layer, however, this does not go very far toward a realistic treatment of the role of radiation in the processes we are simulating.

When $\Delta T_n/\Delta t$ is negative the boundary layer is no longer convectively unstable. As previously explained, in this circumstance the boundary layer top is allowed to sink. The same equation is again used with $A = 0.8$. Generally speaking $\Delta T_n/\Delta t$ is much smaller in stable than in unstable conditions, and therefore the rate of sinking observed as a result of "entrainment" in the general circulation model is considerably less than the rate of ascent.

The method that has been used for estimating the entrainment can be regarded as only a first approximation to what is required in a general circulation model. The theories on which it is based deal only with dry conditions, and the constants that occur have been estimated from relatively few observations at a handful of sites, mainly to fit diurnal effects. In the general circulation model much more varied circumstances are encountered. Especially over the oceans, the presence of water vapor and the release of latent heat must influence the entrainment rate significantly. It is clearly desirable to include moisture, but it is not yet clear how it should be done. At this initial stage the best course appeared to be to include the process to the extent to which it has been adequately described in the literature.

(iii) *Convection* ($\Delta b_C$)

Betts (1973) regarded the environmental vertical velocity in an atmosphere in which convection was proceeding as composed of two parts, *viz.*, a large-scale vertical velocity determined by the convergence or divergence of the large-scale winds, and a subsidence which was such as to balance the upward mass flux taking place within the clouds. The convective scheme (see Section IX) calculates explicitly a fraction of the lowest model slab $F_n{}^{\uparrow}$ which is to be involved in convective interchange during a time step $\Delta t$, and this is interpreted as a vertical mass flux out of the boundary layer. If it is additionally assumed that, in a grid area, the environmental subsidence balances the upward flux within clouds, then there is on average over the area a reduction of boundary layer thickness given by

$$\Delta b_C = -F_n{}^{\uparrow} \tag{34}$$

per time step.

(iv) *Mechanical Mixing* ($\Delta b_M$)

The height of the boundary layer to be expected as a result of mechanical stirring by the wind has been demonstrated in a number of theoretical and observational studies to be proportional to $u_*/f$. The constant of proportionality, however, differs among the various authors; the value of 0.14 chosen here is in approximate agreement with some observational estimates (Sheppard, 1969; Hanna, 1969; Carson, analysis of O'Neill data, unpublished); but is substantially lower than the value 0.35 assumed by Deardorff (1972). To make it possible to apply this limit even close to the equator, the expression has been manipulated as follows:

$$\begin{aligned}\text{Minimum height} = 0.14u_*/f &= 0.14(C_M)_G^{1/2}G^2/(fG) \\ &= 0.14(C_M)_G^{1/2}G^2/(|\nabla_p\phi|)\end{aligned} \tag{35}$$

where $G$ is the geostrophic wind speed, $(C_M)_G$ is the geostrophic drag coefficient, and $\nabla_p\phi$ is the gradient of the geopotential in the pressure surface at which the wind is geostrophic. It is then assumed that the lowest level of the model is sufficiently close to this level and hence $G$ is replaced by $|\mathbf{V}_n|$ and the height gradient by $|\nabla_p \phi_n|$, the lowest level values. Also to transform to an equivalent expression for $b$, the minimum height is divided by $\bar{z}_i$, the depth of the boundary layer when $b = 1$, taken as 2000 m. Hence the minimum value of $b$ becomes

$$b_{\min} = 0.14(C_M)_G^{1/2}(\mathbf{V}_{0.9})^2/\{(|\nabla_p \phi_{0.9}|)\bar{z}_i\} \quad (36)$$

Values of $(C_M)_G$ are considered in Section VII on surface exchanges. $\Delta b_M$ is zero unless the increments due to other processes would make $b < b_{\min}$.

(v) *Smoothing* $(\Delta b_S)$

A diffusion term similar to that used for other variables is applied to $b$. The finite difference form is

$$\delta_{2t} p_* b = (K_b / p_* a^2 \cos^2 \theta)(s)[\delta_\lambda \bar{p}_*^\lambda \, \delta_\lambda b + \cos\theta \, \delta_\theta \bar{p}_*^\theta \cos\theta \, \delta_\theta b]^2 \quad (37)$$

where $K_b$ is the diffusion coefficient and $(s)$ stands for the sign of the term within the brackets. In this formulation of the smoothing the value to be taken for the coefficient depends on the units used. Since $b$ is of order 1 and $u$ is of order 10 m sec$^{-1}$, $K_b$ has been taken as 10 $K_u$ to maintain uniformity of effect. This smoothing is, in practice, very light and steep gradients of $b$ occur during integrations.

## VII. Representation of Surface Exchange Processes

### A. Momentum Exchange

An eddy friction proportional to the square of the wind speed is well established and is known to behave reasonably well in a model. That is

$$(\tau_* / \rho_*) = -C_M |\mathbf{V}| \mathbf{V} \quad (38)$$

where $\rho_*$ is the surface density, $\mathbf{V}$ is the wind vector close to the ground, and $C_M$ is a drag coefficient appropriate to the roughness of the ground, the stability, and the height above the ground at which $\mathbf{V}$ is measured. It is usual to write $(|\tau|/\rho_*)^{1/2}$ as $u_*$, the friction velocity. The model does not carry or predict a near surface wind. We appeal therefore to the structure that has been assumed for the boundary layer and to Rossby similarity theory of the planetary boundary layer to enable $\boldsymbol{\tau}_*$ to be expressed in terms of the explicit model variables.

The boundary layer is by definition that region of the atmosphere close to the ground where the wind departs from its frictionless value; this is generally its surface geostrophic or gradient value, though there may be significant deviations from this if the boundary layer is nonbarotropic. Now Rossby similarity theory (see, for example, Monin and Zilitinkevich, 1967) enables $(C_M)_G = (u_*/G)^2$ the geostrophic drag coefficient, $G$ being the geostrophic wind speed, to be expressed in terms of the surface Rossby number $R_0$ ($R_0 = G/fz_0$, where $z_0$ is the surface roughness length) and the stability. It also provides a relation between the turning of the wind through the boundary layer ($\alpha$) and the same variables. Thus (38) can be rewritten

$$\begin{aligned}(\tau_\lambda)_* &= -\rho_*(C_M)_G|\mathbf{V}_T|(u_T \cos\alpha - v_T \sin\alpha)\\(\tau_\theta)_* &= -\rho_*(C_M)_G|\mathbf{V}_T|(v_T \cos\alpha + u_T \sin\alpha)\end{aligned} \tag{39}$$

where, as before, the subscript $T$ indicates a value at the top of the boundary layer.

In practice, no attempt has so far been made in the model to exploit the full generality of the above approach. To do so involves a considerable computational penalty since the geostrophic drag coefficient cannot be expressed explicitly in terms of the model variables [for details of procedures involved see description of the type-II boundary layer parameterization of Clarke (1970)]. We have preferred at this stage to make simple assumptions, distinguishing only between land and sea and between stable and unstable atmospheric conditions. For land and sea, Rossby numbers of $10^6$ and $10^8$, respectively were taken; typically these imply roughness lengths of 10 cm and 0.1 cm. Using curves published by Monin and Zilitinkevich (1967) $(C_M)_G$ and $\alpha$ values, representative of stable and unstable conditions for the assumed surface Rossby numbers, were adopted (Table I).

TABLE I

VALUES OF $(C_M)_G$, $\alpha$

| | Land | | Sea | |
|---|---|---|---|---|
| | $10^3(C_M)_G$ | $\alpha$(rad) | $10^3(C_M)_G$ | $\alpha$(rad) |
| Stable | 0.3 | $\pi/6$ | 0.2 | $\pi/9$ |
| Unstable | 4.0 | $\pi/18$ | 2.0 | 0 |

## B. Sensible and Latent Heat Fluxes

The turbulent flux of virtual heat at the surface may be written

$$H_*^v = \rho_* c_p C_H |\mathbf{V}| \, \Delta\Theta^v \tag{40}$$

where $\mathbf{V}$ is the wind close to the surface and $\Delta\Theta^v$ is the difference in virtual potential temperature across the surface layer. As with the momentum flux it is necessary to replace $\mathbf{V}$ and $C_H$ with values appropriate to the model's explicit variables. It is consistent with the treatment of the surface stress to replace $\mathbf{V}$ by $\mathbf{V}_T$ and $C_H$ by $(C_H)_G$, the geostrophic heat transfer coefficient. Although the transfer coefficients for momentum and heat must, in general, be expected to differ, there is considerable scatter in the observational evidence. The interpretation of observations in terms of Rossby similarity theory of the boundary layer enables estimates of $(C_H)_G$ and $(C_M)_G$ to be compared. Thus, Clarke (1970, in graphs reproduced by Fiedler and Panofsky, 1972) shows values for given Rossby number and stability which differ by a fairly small factor ($<2$). In view of the degree of approximation accepted in discriminating between different surface types and stabilities, different transfer coefficients for heat and momentum did not seem warranted, at least in neutral and stable conditions and the assumption was made that $(C_H)_G = (C_M)_G$.

It is well known, however, that in an unstable regime the effective eddy coefficient for heat is greater than that for momentum, because elements with a positive buoyancy departure are biased in favor of upward motion. Dimensional arguments (Priestley, 1954; Monin and Obhukov, 1954) lead to a "free convection" virtual heat flux proportional to $(\Delta\Theta^v/\Theta^v)^{3/2}$. This effect has been introduced into the model formulation by writing the vertical heat flux as

$$H_*^v = \rho_* c_p (C_H)_G [|\mathbf{V}_T| + A(\Delta\Theta^v/\Theta)^{1/2}] \, \Delta\Theta^v \tag{41}$$

that is, by considering the free convective element in the flux as additional to that due to mechanical turbulence. It is by no means obvious that this is the best way of dealing with the problem. The Rossby similarity approach would indicate, for example, that an increased value of $(C_H/C_M)_G$ is more correct. There is, however, no universally accepted scientific description of the matter and the solution adopted here has virtues in the context of a general circulation integration; in particular, it allows heat exchange to continue in unstable conditions when $\mathbf{V}_T$ is zero, and, remembering that this is not a point value but an average over an area of about $10^{11}$ m$^2$ it is realistic to allow this to happen. The value taken for $A$ is 50 m sec$^{-1}$ if $\Delta\Theta^v$

is positive, and zero otherwise. Thus with $\Delta\Theta^v = 1\ ^\circ\mathrm{K}$, the free convective element in the heat flux is equivalent approximately to an additional $3\ \mathrm{m\ sec}^{-1}$ on the wind, and this rises to about $10\ \mathrm{m\ sec}^{-1}$ when the conditions are very unstable with $\Delta\Theta^v = 10\ ^\circ\mathrm{K}$.

Separating the virtual heat flux into its sensible and latent components we have for the sensible flux and the evaporation

$$\begin{aligned} H_* &= \rho_* c_p (C_H)_G[|\mathbf{V}_T| + A(\Delta\Theta^v/\Theta^v)^{1/2}]\,\Delta\Theta \\ M_* &= \rho_* (C_H)_G[|\mathbf{V}_T| + A(\Delta\Theta^v/\Theta^v)^{1/2}]\,\Delta q \end{aligned} \tag{42}$$

where in calculating potential temperature $p_*$ is used as the reference pressure,

$$\begin{aligned} \Delta\Theta^v &= (\Theta_*^v - \Theta_T{}^v) = \Theta_*(1 + 0.61q_*) - \Theta_T(1 + 0.61q_T) \\ \bar{\Theta}^v &= \tfrac{1}{2}(\Theta_*^v + \Theta_T{}^v) \\ \Delta\Theta &= (\Theta_* - \Theta_T) \\ \Delta q &= (q_* - q_T) \end{aligned} \tag{43}$$

and $q_*$, the mixing ratio of air close to the surface is taken as a constant times the saturation mixing ratio at temperature $T_*$. The constant is 0.8 over the sea and 0.5 over land.

One of the problems that has arisen in dealing with deep boundary layers is that, using formula (41), no upward transfer of sensible heat can occur until a dry adiabatic lapse rate has been established between the ground and the top of the boundary layer. This is reasonable when the air is dry; when $q_* < q_T^{\mathrm{sat}}$, i.e., when a parcel of air ascending from the surface to the top of the layer would not condense, it is appropriate to take $\Theta_* = T_*$. However, it is observed, on perhaps a majority of occasions, that clouds, possibly layered, can form within the boundary layer, and the normal situation is one in which there is a dry adiabatic lapse rate up to the cloud base but a much reduced lapse, probably approximating to the wet adiabatic, above. Failure to take this into account leads to poor simulation of conditions in the lower troposphere, particularly over tropical oceans. To allow for this when condensation is expected, i.e., when $q_* > q_T^{\mathrm{sat}}$, $\Theta_*$ is approximated by

$$\Theta_* = T_* + (L/c_p)(q_* - q_T^{\mathrm{sat}}). \tag{44}$$

The additional term here can be regarded as an estimate of the increase in potential temperature due to the release of latent heat in a parcel moving from the surface to the top of the boundary layer; a small factor involving the pressure at which the release takes place and which might alter the value of the term by up to 5% has been neglected. The implied difference in temperature when the surface heat flux is zero (i.e., $\Theta_T = \Theta_*$) is a good approximation to the difference expected if there is a dry adiabatic up to the condensation level followed by a wet adiabatic up to the top of the layer.

## VIII. Treatment of Land and Ice Surfaces

The temperature change of land and ice surfaces is determined from the equation

$$C_*(\partial T_*/\partial t) = -R_L + (1 - \alpha)R_S - H_* - LM_* + H_I \tag{45}$$

where $R_L$ and $R_S$ are the long- and short-wave radiative contributions. $H_*$ and $LM_*$ are the sensible and latent heat fluxes at the surface, which are discussed in Section VII. $C_*$ is the effective heat capacity of the ground. It depends on the thermal conductivity as well as on the specific heat of the surface materials. Some approximate values of $C_*$, which allow the surface temperature to respond realistically to imposed heat inputs on time scales of at least a day, are new snow, 4; dry soil, 8; wet soil, 25; and ice, 35 J cm$^{-2}$ °K$^{-1}$. No attempt has been made to model a variety of surface types, and therefore $C_*$ has been given the value 12.5 J cm$^{-2}$ °K$^{-1}$ representing a compromise for average conditions, for all land and ice surfaces.

The term $H_I$ represents the conduction of heat from unfrozen water below sea ice. It is formulated, following Holloway and Manabe (1971) as

$$H_I = kI^{-1}(T_W - T_*) \tag{46}$$

where the thermal conductivity of ice $k = 2.1 \times 10^{-2}$ J cm$^{-1}$ °K$^{-1}$ sec$^{-1}$; $T_W$, the temperature of the underlying water is taken as 271.2 °K and the ice depth $I$ is assumed to be 2 m, a compromise between values characteristic of old ice, 3 or more meters thick, and new ice which is less than 1 m.

$\alpha$ is the surface albedo for solar radiation. It is an important element in the radiative balance at the surface. The atmospheric simulation is particularly sensitive to the choice of $\alpha$ in the tropics since there the solar radiation absorption is approximately balanced by the net atmospheric and surface long-wave cooling, and whether there is an energy surplus available for export in vertical circulations or an energy deficit that has to be made good by local descent depends on a relatively small residual. It is therefore desirable to specify the surface albedo as accurately as possible.

As a first step toward a detailed geographical specification of albedo, average values at 10° intervals of latitude were calculated using geographical surveys of vegetation and published values of albedo of different types of surface as a basis. The results are shown in Table II; values for the model were obtained from these by linear interpolation.

TABLE II

ZONALLY AVERAGED ALBEDOS OVER SNOW-FREE LAND

| Latitude | 70° | 60° | 50° | 40° | 30° | 20° | 10° | 0° |
|---|---|---|---|---|---|---|---|---|
| N. Hemisphere | 0.180[a] | 0.149 | 0.170 | 0.207 | 0.212 | 0.225 | 0.188 | 0.146 |
| S. Hemisphere | — | 0.190[a] | 0.190 | 0.190 | 0.224 | 0.201 | 0.173 | |

[a] Land above 1000 m north of 66° N and all land south of 60° S is assumed ice-covered, with appropriate albedo.

It may be noted that values in the tropics are higher than those sometimes proposed (e.g., Posey and Clapp, 1964), mainly because a higher albedo has been assumed for tropical forests. Further discussion and details of the calculations are given in Rowntree (1975).

The effect of snow on the surface albedo is very variable, particularly because the upper surfaces of forests do not remain snow-covered. We have used 0.5, a value derived from Posey and Clapp (1964), who take this effect fully into account. Since snow depth is not carried by the model, the criterion for snow-covered land was taken as $T_* < 269$ °K; a more gradual transition to the snow albedo is probably more realistic and has been incorporated in recent versions of the model. Sea ice and the permanent ice of Greenland and Antarctica were assumed to have an albedo of 0.8, or if $T_*$ rose above 271.2 °K the lower value of 0.5; their surface temperatures were not allowed to exceed 273 °K.

## IX. Convective Interchange

The convective exchange scheme of Corby *et al.* (1972) proceeded on the premise that, over an area as large as that represented by a grid area of the model, the process could be considered as a vertical diffusion in which the diffusion coefficients depended on the degree of vertical instability. Essentially, it involved determining a fractional rate of mixing $F_{k-1/2}$ per time step between two contiguous layers, $k$ the lower, $k-1$ the upper, which had been deemed to be mutually unstable. Instability was considered to exist when a parcel of air at level $k$, warmer than the mean environmental temperature by an amount $\varepsilon^T$ was still buoyant when carried up to level $k - 1$. If the temperature excess of the parcel over that of the environment was $\Delta T_k{}^1$, the fraction of the layers to be interchanged was taken to be proportional to $\Delta T_k{}^1$ and to the length of the time step; that is

$$F_{k-1/2} = A\, \Delta T_k{}^1\, \Delta t/(\Delta\sigma)^2 \tag{47}$$

the $(\Delta\sigma)^2$ in this formula arising from the replacement of derivatives in the statement of the process as a vertical diffusion by finite differences. $A$ was determined empirically by requiring that stabilization of an initially unstable vertical profile should take place in a reasonable time, *viz.*, an hour or two.

The scheme employed in the version of the model described here is a generalization of the original. The latter allowed mixing between contiguous layers only and therefore omitted effective parameterization of deep convection, in which undiluted air from low levels of the atmosphere can penetrate into the upper troposphere. In the modified scheme, the fraction of the lower or base layer taking part in convection is determined essentially as before but allowance is now made for some part of it to reach and mix with air in layers above $k - 1$, providing they are also unstable with respect to this base layer. Compensating downward motion with adiabatic warming is also included.

We consider convection between a base level $k$, and higher levels $k - 1$, $k - 2$, etc. It is assumed that air from the base level taking part in any exchange is warmer and wetter than the environment at level $k$ by amounts $\varepsilon^T$ and $\varepsilon^Q$, respectively, where $\varepsilon^T = 0.002p_*)$°K, where $p_*$ is in mb and $\varepsilon^Q = 0.4q_k(1 - r_k)$, where $q_k$ and $r_k$ are the environment mixing ratio and relative humidity. To simplify expression of the scheme, $T_k$ and $q_k$ used in the following description of the derivation of the fractions of layers taking part in convection are now to be taken as applying to the convecting air and not to the environment at the base level $k$.

For the sake of clarity, consider first of all dry convection. Then, if air is raised adiabatically from level $k$ to level $k - j$, the temperature difference between it and the environment is

$$\sigma_{k-j}^{\kappa}[(T_k/\sigma_k{}^{\kappa}) - (T_{k-j}/\sigma_{k-j}^{\kappa})] \tag{48}$$

while the corresponding expression for air taken from level $k - j$ to level $k$ is

$$-\sigma_k{}^{\kappa}[(T_k/\sigma_k{}^{\kappa}) - (T_{k-j}/\sigma_{k-j}^{\kappa})] \tag{49}$$

Since these are not equal in magnitude, a mixing process strictly dependent on such adiabatic formulas would fail to conserve potential energy. The discrepancy arises because the realization of potential energy in buoyant convection generates some (vertical) kinetic energy. It is convenient to avoid having to allow for this kinetic energy by assuming that it appears as additional heat divided between the layers. This is achieved by replacing expressions like (48) by

$$\sum_{s=1}^{j} [\sigma^{\kappa}_{k-\overline{s-1/2}}\{(T_{k-s}/\sigma^{\kappa}_{k-s}) - (T_{k-\overline{s-1}}/\sigma^{\kappa}_{k-\overline{s-1}})\}]$$

$$= \sum_{s=1}^{j} \{\Delta\sigma\, \sigma^{\kappa}\, \delta_{\sigma}(T/\sigma^{\kappa})\}_{k-\overline{s-1/2}} = \Delta T_k{}^j \quad (50)$$

$\Delta T_k{}^j$ is the estimate of the temperature excess of a parcel of air raised from level $k$ to level $k - j$. For the purpose of the following discussion, level $k - j$ is considered unstable with respect to level $k$ if $\Delta T_k{}^j > 0$. Thus, if $\Delta T_k{}^1 > 0$, then air from level $k$ will be involved in convective interchange. If this condition is satisfied $\Delta T_k{}^1$ is increased if necessary to a minimum value of 0.5 °K in order to avoid recurring small adjustments. Having found a positive $\Delta T_k{}^1$, then the fraction of the mass of the base layer which will mix with the atmosphere above is

$$F_k{}^{\uparrow} = A\, \Delta T_k{}^1\, \Delta t/(\Delta\sigma)^2 \qquad A = 0.5 \times 10^{-6}\ {}^{\circ}\mathrm{K}^{-1}\ \mathrm{sec}^{-1} \quad (51)$$

where $A$ is a constant determining the rate of stabilization. The revised scheme differs from the original in that $F_k{}^{\uparrow}$ is not mixed directly with environmental air in layer $k - 1$. Instead, a test is made for instability between level $k$ and levels $k - 2$, $k - 3$, etc., proceeding until a stable level $(k - \overline{m + 1})$ is reached. $F_k{}^{\uparrow}$ is then partitioned into fractions $F^{\uparrow}_{k(k-j)}$, which are to be mixed with environmental air in layer $k - j$. Thus

$$F_k^{\uparrow} = \sum_{j=1}^{m} F^{\uparrow}_{k(k-j)} \quad (52)$$

The fraction to be mixed with environmental air in layer $k - j$ is given by

$$F^{\uparrow}_{k(k-j)} = \{a/(\Delta T_k^{j+1} + b)\}[F_k^{\uparrow} - F^{\uparrow}_{k(k-1)} - F^{\uparrow}_{k(k-2)} - \cdots - F^{\uparrow}_{k(k-\overline{j-1})}] \quad (53)$$

$$a = 0.45\ {}^{\circ}\mathrm{K} \qquad b = 0.5\ {}^{\circ}\mathrm{K}$$

that is, it is a proportion of the total mass originating from level $k$ which reaches level $k - j$, the proportion being dependent on the degree of instability between the base level and the next higher level. The fraction absorbed in the layer $k - m$ is

$$F^{\uparrow}_{k(k-m)} = F_k^{\uparrow} - F^{\uparrow}_{k(k-1)} - \cdots - F^{\uparrow}_{k(k-\overline{m-1})} \tag{54}$$

To balance the ascent from layer $k$, subsidence is assumed to take place. Denoting the fractional exchange for this by $F^{\downarrow}$ we have by mass conservation requirements

$$\begin{aligned} F^{\downarrow}_{(k-m)(k-\overline{m-1})} &= F^{\uparrow}_{k(k-m)} \\ F^{\downarrow}_{(k-\overline{m-1})(k-\overline{m-2})} &= F^{\uparrow}_{k(k-m)} + F^{\uparrow}_{k(k-\overline{m-1})} \\ \vdots \qquad & \qquad \vdots \\ F^{\downarrow}_{(k-1)k} &= F^{\uparrow}_{k(k-m)} + F^{\uparrow}_{k(k-\overline{m-1})} + \cdots + F^{\uparrow}_{k(k-1)} \\ &= F^{\uparrow}_k \end{aligned} \tag{55}$$

It is to be noted that mixing, which is a consequence of subsidence within the environment outside the convecting regions, occurs only between contiguous layers.

Having determined the fractions of the layers involved, the convecting or subsiding air is mixed with its new environment to give new temperatures. Thus, the temperature change at level $k - j$ due to convection starting from layer $k$ is

$$\begin{aligned} F^{\uparrow}_{k(k-j)} \sum_{s=1}^{j} \{\Delta\sigma\, \sigma^{\kappa}\, \delta_{\sigma}(T/\sigma^{\kappa})\}_{(k-\overline{s-1/2})} \\ - F^{\downarrow}_{(k-\overline{j+1})(k-j)}\{\Delta\sigma\, \sigma^{\kappa}\, \delta_{\sigma}(T/\sigma^{\kappa})\}_{(k-\overline{j+1/2})} \end{aligned} \tag{56}$$

At this mixing stage, of course, $T_k$, occurring in the expanded form of this expression, again refers to the environmental temperature of the base layer, rather than to the enhanced parcel temperature.

Condensation complicates the procedure in practice, but not in principle. Thus the expression for $\Delta T_k{}^j$ is

$$\sum_{s=1}^{j} \{\sigma^{\kappa}\, \Delta\sigma\, \delta_{\sigma}(T/\sigma^{\kappa})\}_{(k-\overline{s-1/2})} + (L/c_p)\{q_k - (q_j^{\text{sat}})'\} \tag{57}$$

where $(q_j^{\text{sat}})'$ is the saturation mixing ratio at temperature

$$T_{(k-j)} + \Delta T_k{}^j \tag{58}$$

for the determination of which an approximate method based on the Clausius–Clapeyron relation is used. The expressions for the change of temperature and humidity at level $k - j$ due to convection originating at level $k$ may be written

$$F^{\uparrow}_{k(k-j)}\,\Delta T_k^{\,j} - F^{\downarrow}_{(k-\overline{j+1})(k-j)}\{\Delta\sigma\,\sigma^{\kappa}\,\delta_{\sigma}(T/\sigma^{\kappa})\}_{(k-\overline{j+1/2})} \tag{59}$$

$$F^{\uparrow}_{k(k-j)}(q_k - q_{k-j}) + F^{\downarrow}_{(k-\overline{j+1})(k-j)}(q_{k-\overline{j+1}} - q_{k-j}) \tag{60}$$

The condensation in the layer $k - j$,

$$(p_*\,\Delta\sigma_{k-j}/g)[F^{\uparrow}_{k(k-j)}\{q_k - (q_j^{\text{sat}})'\}] \tag{61}$$

in units of mass per unit area per time step, is assumed to form liquid precipation. It is allowed to evaporate into the originating layer and successive lower layers, the amount of evaporation being proportional to the subsidence in the layer and its dryness. That is, using the values of variables updated by results of convection so far described, the evaporation from precipitation falling through the layer $k - j$ is

$$E_j = F^{\downarrow}_{(k-\overline{j+1})(k-j)}[q^{\text{sat}}_{k-j} - q_{k-j}], \tag{62}$$

(where $q^{\text{sat}}_{k-j}$ is the saturation mixing ratio of the environmental air) or the total precipitation, whichever is the lesser.

The above convective process is carried out from successive layers starting with the lowest, mixing being assumed complete before convection from a new base level is considered.

Values of the empirical parameters in the scheme, i.e., $A$, $\varepsilon^T$, $\varepsilon^Q$, $a$ and $b$ were fixed by considering the temperature and humidity changes in a series of single column experiments in which initially unstable profiles were allowed to stabilize as a result of convective motions. A subjective assessment indicated that in general the values chosen gave reasonable rates of stabilization and reasonable post-convective vertical profiles.

One of the main reasons for carrying a variable representing the position of the top of the boundary layer in the general circulation model is to allow it to control, in some measure, the extent and depth of convection, insofar as this is the model's method of effecting an exchange of heat and moisture between the boundary layer and the free atmosphere. The process in the real atmosphere is clearly a very complicated one of which there is, at the present time, no comprehensive account that can be adapted to form the basis for treatment within a model. In any case, for the present purposes, a drastically simplified parameterization is needed. In the model, therefore, the boundary layer has been allowed to influence convection in ways that seem qualitatively realistic, but whose formulation cannot on present evidence be substantiated quantitatively. Thus, a low boundary layer top, which would usually imply a marked inversion between the boundary layer and the free atmosphere, would generally be associated with, at most, weak convection, since there is a decreased likelihood of air parcels significantly more buoyant than the environment being present within the boundary layer, and also because convective interchange is likely to be inhibited with only the most vigorous convecting parcels capable of breaking through the capping inversion. It is clearly possible to make an allowance for these two effects in the scheme that has just been described in simple ways. The first can be introduced by reducing $\Delta T_n^1$ for the lowest base layer by an amount that increases as the boundary layer falls, and the second by reducing $F_n^{\uparrow}$ for low boundary layer values. For the former effect there is no obvious reason to go beyond a linear relation, and a reduction which varies from about 2 °K when $b = 0$ to a value of zero when $b = 1$ has been used. For the inhibiting effect of the capping inversion on convection, however, a quadratic relation seems more appropriate. Here we may consider the convecting air to originate within the boundary layer, and the mass convergence within the layer on which it might be expected to depend is a function not only of the depth of the layer but of the mean wind through it, and this, obtained by extrapolation from the sigma level values, is also a function of $b$. The fraction of the lowest layer taking part in convection has therefore been taken as $b^2 F_n^{\uparrow}$, where $F_n^{\uparrow}$ is the fraction calculated ignoring the presence of the boundary layer. The parameterizations chosen appear to give acceptable results, but no consistent attempt has been made to optimize the formulation.

## X. Radiation Scheme

Radiation calculations are computationally expensive. Even those that proceed by using emissivities, thus avoiding the need to deal with different parts of the spectrum individually, take up a substantial fraction of the total computing time for a model. There is, therefore, a strong incentive to simplify the formulation drastically, if this is reasonable for the particular experiment in mind. Indeed there is a need for simple, noninteractive but effective schemes suitable for use when radiation is not the essence of the problem and able to provide a standard of simulation against which to judge the more complete formulations. The scheme described here is deliberately kept simple, being based on long-wave cooling and solar heating rates derived from mean climatological data. Nevertheless, in our experience the simulations using it compare favorably with those involving much more elaborate calculations. Its range of applicability is, of course, limited, but it appears capable of giving good results when the area of interest is the troposphere and the climate being simulated is at most a variation on the presently observed climate. It may be noted, however, that versions of the model incorporating full radiation calculations have run successfully (Mattingly, 1974; Hunt, 1976); the scheme was based on emissivity calculations and was basically similar to the method of Manabe and Strickler (1964), but included variable cloud amounts dependent on the model humidity and vertical velocity.

The treatment of radiation in the original version of the model was based on proposals by C. D. Walshaw (personal communication). The use of empirical emissivities was combined with the cooling to boundaries approximation to give long-wave cooling rates in the form

$$(\Delta T_k/\Delta t) = -a_k(\theta)T_k^{\ 4} + b_k(\theta)(T_*^4 - T_k^{\ 4}) \tag{63}$$

where $T_k$ is the temperature of the $k$th layer and $T_*$ is the temperature of the earth's surface. The parameters $a_k$ and $b_k$ are functions of latitude and pressure only. In practice, the second term on the right was generally small and was neglected in the computations except when considering the lowest level ($k = n = 5$). The effects of cloud on the long-wave radiation were entirely omitted.

The zonal mean climatological cooling rates deduced from the scheme showed substantial differences from those of more elaborate radiation models (e.g., Dopplick, 1970; Rodgers, 1967), and some shortcomings of the original simulations were traced to this deficiency. In particular, the radiative cooling at low levels in the tropics appeared excessive, and that at high latitudes too small. It was decided therefore that, while retaining the basic computational form, the parameters for the scheme should be adjusted to reproduce the zonal mean cooling rates obtained by Rodgers, when applied to zonal mean climatological temperatures. The values of $a_k(\theta)$ required to achieve this are shown in Table III, in terms of the effective emissivities $\varepsilon_k(\theta)$ for

TABLE III

LONG-WAVE EFFECTIVE EMISSIVITIES

| Latitude (N) | 0° | 10° | 20° | 30° | 40° | 50° | 60° | 70° |
|---|---|---|---|---|---|---|---|---|
| 200–0 mb | 0.122 | 0.119 | 0.115 | 0.124 | 0.134 | 0.137 | 0.134 | 0.125 |
| 400–200 mb | 0.230 | 0.220 | 0.213 | 0.202 | 0.193 | 0.184 | 0.179 | 0.171 |
| 600–400 mb | 0.204 | 0.194 | 0.189 | 0.188 | 0.186 | 0.184 | 0.167 | 0.161 |
| 800–600 mb | 0.163 | 0.158 | 0.156 | 0.170 | 0.173 | 0.176 | 0.197 | 0.194 |
| 1000–800 mb | 0.108 | 0.126 | 0.126 | 0.112 | 0.127 | 0.141 | 0.154 | 0.155 |

cooling to space. The relation between $a_k(\theta)$ and $\varepsilon_k(\theta)$ is

$$\varepsilon_k(\theta)s = a_k(\theta)C_k \tag{64}$$

where $s$ is Stefan's constant and $C_k$ is the heat capacity of the layer (205 J cm$^{-2}$ °K$^{-1}$ if the layer depth is 200 mb). $\varepsilon_k$ is determined from Table III according to the pressure at level $k$.

In determining the $\varepsilon$'s certain assumptions were necessary: (i) the cooling rates were constant through the layers for which Rodgers gave values; (ii) the cooling rates for 20–10 mb could be taken as applying also to 10–0 mb for which there were few data; (iii) in calculating values for the lowest layer, allowance was made for the term $b_5(T_*^4 - T_5{}^4)$ by estimating its value from the climatological data taking the original value of $b_5$, $T_* = T_{1000}$ and $T_5 \doteqdot T_{900} \doteqdot (\frac{2}{3}T_{850} + \frac{1}{3}T_{1000})$ where the subscripts indicate the pressure levels at which the temperatures applied.

The values of $\varepsilon$ so found do not vary much with latitude; there is a small fall poleward at 500 mb and 300 mb and a change in the opposite direction at 700 mb and 900 mb. This variation is as expected since the effective radiating level rises as the water content of the atmosphere increases.

An effective emissivity for the ground may also be estimated from Rodgers' values of the long wave flux at the surface and the estimates of the back radiation from the lowest level calculated as above. The sum of the effective emissivities from the atmosphere and from the ground $S(\theta)$ then has the values shown in Table IV.

TABLE IV

SUM OF EFFECTIVE EMISSIVITIES $S(\theta)$

| Latitude (N) | 0° | 10° | 20° | 30° | 40° | 50° | 60° | 70° |
|---|---|---|---|---|---|---|---|---|
| $S(\theta)$ | 0.926 | 0.941 | 0.949 | 0.951 | 0.955 | 0.958 | 0.963 | 0.972 |

It may be shown that $S(\theta)$ should be less than unity.

Integrations for January for the Northern Hemisphere using the above values of $\varepsilon$, the values of $b$ used by Corby *et al.* (1972), and constraining the surface emissivity so that $S(\theta)$ had the value obtained by interpolation from Table IV gave reasonable climate simulations (Rowntree, 1976a, b). The results indeed encouraged the conviction that it would be useful to be able to apply the same scheme to other months and to the Southern Hemisphere.

For this purpose, a relation was sought between the $\varepsilon$'s, and the climatological zonal mean humidity at 850 mb, $q_{850}(\theta)$. The linear relations in Table V (with $q$ in units gm/kg) were found to reduce the residual variances to an acceptable level:

TABLE V

LINEAR RELATIONS FOR EFFECTIVE EMISSIVITIES

| | Proportion of variance accounted for (%) | Residual variance ($10^{-4}$) |
|---|---|---|
| $\varepsilon_1 = 0.132$ | | 1.06 |
| $\varepsilon_2 = 0.170 + 0.0075q_{850}$ | 94 | 0.35 |
| $\varepsilon_3 = 0.171 + 0.0027q_{850}$ | 53 | 0.77 |
| $\varepsilon_4 = 0.198 - 0.0042q_{850}$ | 80 | 0.60 |
| $\varepsilon_5 = 0.147 - 0.0032q_{850}$ | 72 | 1.01 |
| $\varepsilon_* + \sum_1^5 \varepsilon_k = 0.974 - 0.0040q_{850}$ | 78 | 0.53 |

These equations imply a relation for $\varepsilon_*(\theta)$, the surface values, but the fit to the values deduced directly from the data is not acceptable. Investigations showed that this was due to the influence of cloud.
A relation of the form

$$\varepsilon_*^c(\theta) = \varepsilon_*^0(\theta)(1 - c) \tag{65}$$

where $\varepsilon_*^0$ is the appropriate value of $\varepsilon_*$ in the absence of cloud and $c$ is the cloud amount, was therefore considered. It seems likely to give reasonable results if clouds radiate as black bodies and at a temperature sufficiently close to the ground temperature. The relation

$$\varepsilon_*^0(\theta) = 0.4 - 0.08q_{850}^{1/2} \tag{66}$$

with $q$ in units of gm/kg provides an excellent linear fit, accounting for 95% of the variance, with a residual variance of $1.75 \times 10^{-4}$. It may be noted that it gives clear sky emissivities in close agreement with those used in the original experiment of Corby *et al.* (1972). In estimating $c$ from the climatological data, high cloud was ignored, since for it the requirement that the cloud radiating temperature should not be greatly different from the ground temperature is particularly suspect.

The long wave radiation scheme therefore proceeds as follows: (i) the surface "cooling to space" is derived using (65); in integrations so far carried out $c$ represented the climatological mean zonal cloud amount; it seems possible, however, to extend the scheme so that $c$ is made dependent on the model humidities; (ii) the exchange of heat between the surface and the lowest level of the model is calculated and allowed for in the coefficient $\varepsilon_5$; (iii) the effective emissivities deduced from the 850 mb mean humidities given in Table V are amended equally at each level so that

$$\sum_{k=1}^{5} \varepsilon_k(\theta) = S(\theta) - \varepsilon_*^c \tag{67}$$

and the cooling rates are calculated.

For consistency, the same assumptions should be made about clouds in the solar and long-wave treatments. The heating rates for January and July calculated by Rodgers (1967) have therefore been taken with the same conventions for the calculation of layer means as before. The surface flux over land has been derived from Rodger's zonally averaged values of the net radiative flux.

Variations in $p_*$ have been allowed for in the solar heating by using the value for the 200 mb pressure layer in which the level lies, while the downward flux at the surface has been multiplied by a factor $1 + 0.433(1 - p_*/1000)$, deduced from the attenuation between 700 and 1000 mb in Rodgers' results.

For a model incorporating a boundary layer top, which is capable of suppressing convection, it was thought desirable to introduce a diurnal variation of the solar radiation. This was done by assuming that the time of day was the same everywhere and allowing the solar radiation to vary sinusoidally $(0 - \pi)$ between "sunrise" and "sunset" (calculated as for the Greenwich meridian).

## XI. Model's January and July Simulations

Two global general circulation integrations made with the model detailed in the preceding sections will now be described. They are for January and July, respectively; that is, the sea–surface temperatures and the radiation constants were set to the climatic mean values appropriate to the middle of these months, and were then kept constant in time through the period of integration. To avoid lengthy warm-up phases, the integrations were started from day 30 of integrations for the same months which had previously been completed with a different version of the model. These earlier integrations had been started from a data set consisting of isothermal temperatures and zero wind speeds everywhere; the temperature chosen (220 °K) was near the minimum to be expected in the troposphere, so that the atmosphere

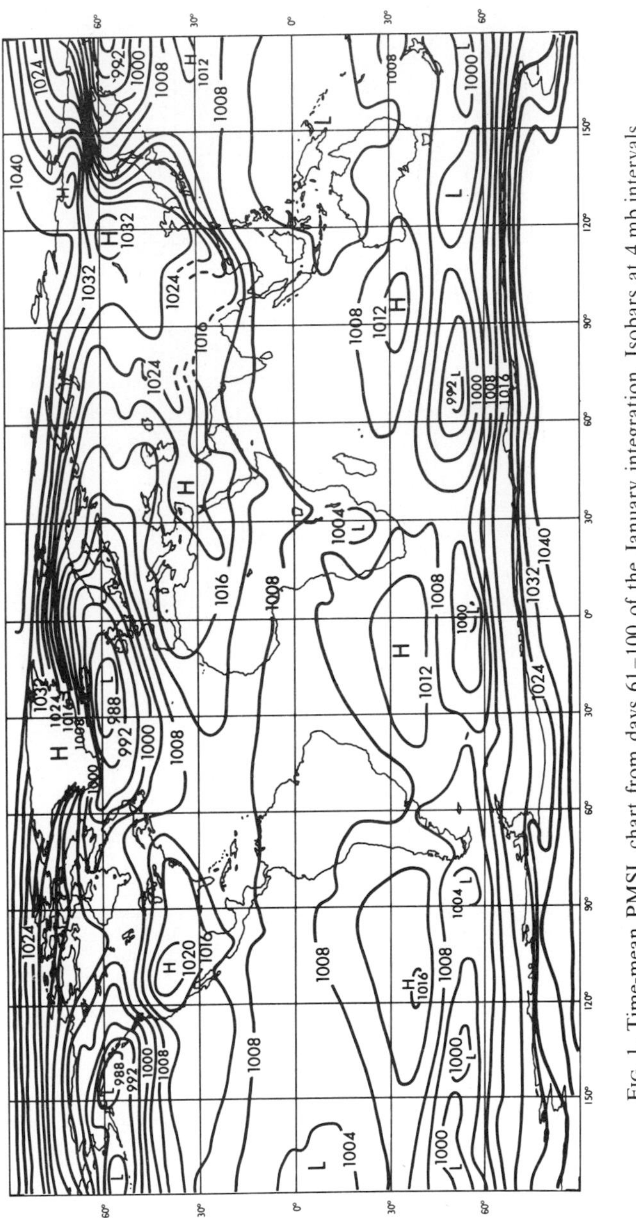

FIG. 1. Time-mean PMSL chart from days 61–100 of the January integration. Isobars at 4 mb intervals.

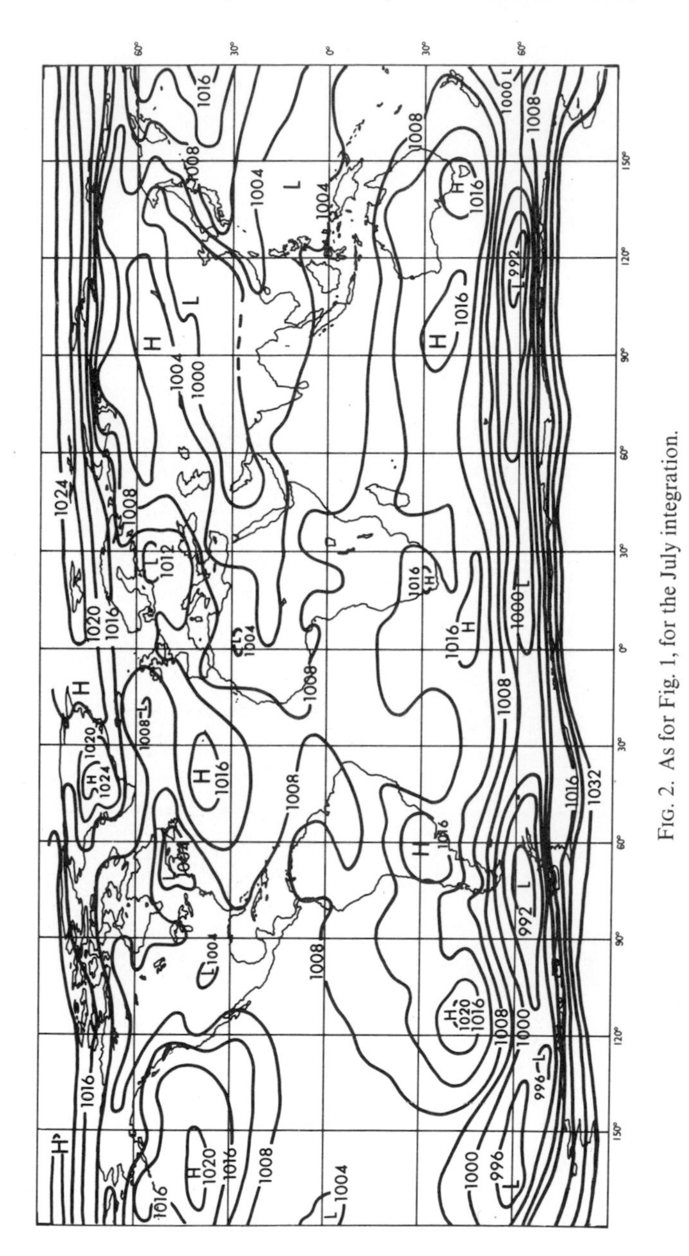

FIG. 2. As for Fig. 1, for the July integration.

would warm up rapidly as a result of heating from the oceans. The integrations now to be described were run out for a further 70 days (i.e., to day 100), and the last 40 days have been taken as representative of the general circulation for the months in question.

We first look at the model's average sea-level pressure for January and July (Figs. 1 and 2), which one would expect to resemble climatological mean maps. There are reasonable simulations of all the main climatological features. The depressions over the Atlantic and Pacific in winter are deeper than the climatology shows, but this is not unexpected since within an integration for 40 days with fixed lower boundary conditions over the oceans there must be less variability of synoptic type than within the periods contributing to a long-term average in which a greater variety of states is possible in part at least due to the fact that the sea–surface temperatures differ significantly from one year to another. The model simulates "a westerly type" which over the oceans is probably the most frequently occurring. At this stage it is not possible to quantify the extent to which the model is deficient in its capacity to reproduce a realistic variety of synoptic types.

It will be noted that, at least qualitatively and to some extent quantitatively, the model obtains the change-over from the dominant winter anticyclone over Asia to the low-pressure system characteristic of the summer monsoon. The difference in latitude of the subtropical anticyclones between winter and summer is again in the correct sense, though it does seem that their intensity is systematically underestimated.

In the Southern Hemisphere, one of the most elusive features to simulate is the depth of the depressions around the Antarctic continent and the westerly flow to the north of the main depression track. The model is not capable of reproducing these features to the extent apparent in the climatology, but the results here are better than those from previous versions of the model. Results from other models shown, for instance, by Manabe and Terpstra (1974) show a similar deficiency.

The wind fields at the lowest level of the model are shown in Figs. 3 and 4 for January and July, respectively. As expected, they are in general agreement with the mean sea-level pressure distribution, with cyclonic and anticyclonic flows around the main low and high pressure centers, respectively. The wind charts are particularly appropriate for showing the simulated flow in the tropics, however. A comparison with the climatology (see, e.g., Atkinson and Sadler, 1970; Newell *et al.*, 1972) shows that here also the model is able to simulate the main features of the flow patterns. Thus, in January, considering the winds over and south of the Asian continent, the flow shows up the northeasterly monsoon over southern Asia, the westerly flow close to the equator, and the return to easterlies on the northern flank of the Southern Hemisphere subtropical anticyclone further

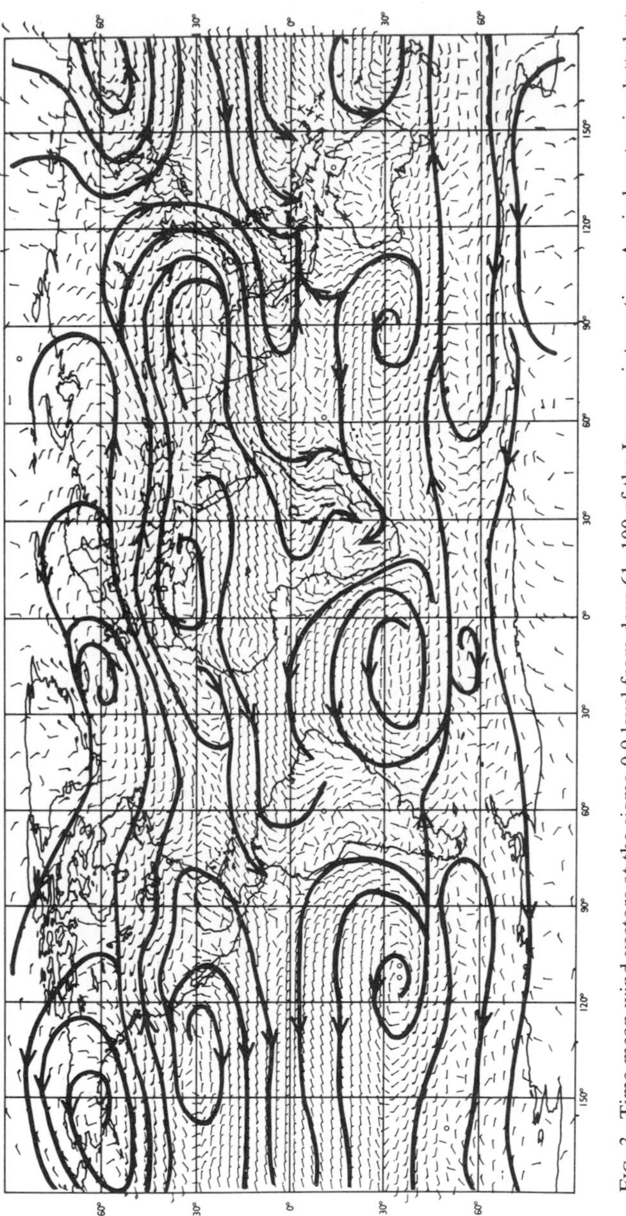

FIG. 3. Time-mean wind vectors at the sigma 0.9 level from days 61–100 of the January integration. A wind vector is plotted at each grid point; speed is indicated thus: ⌊ 5 m $sec^{-1}$, ◣ 25 m $sec^{-1}$. Large arrows indicate the directions of the main flows.

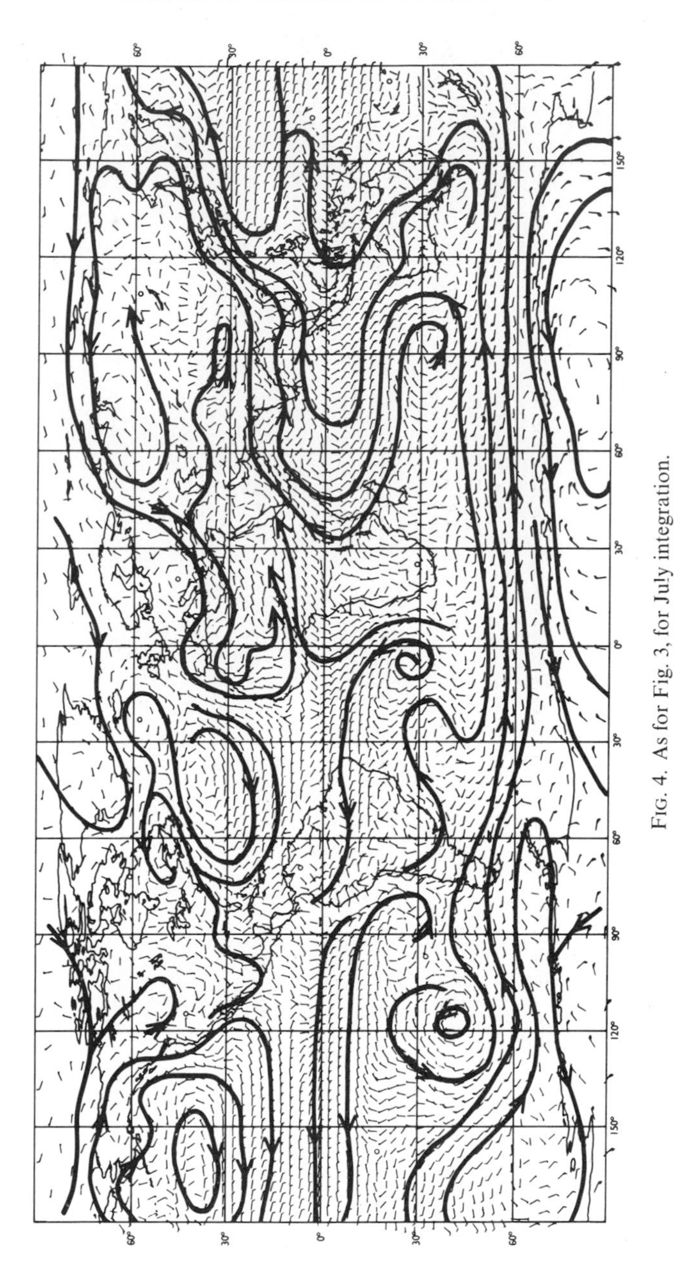


FIG. 4. As for Fig. 3, for July integration.

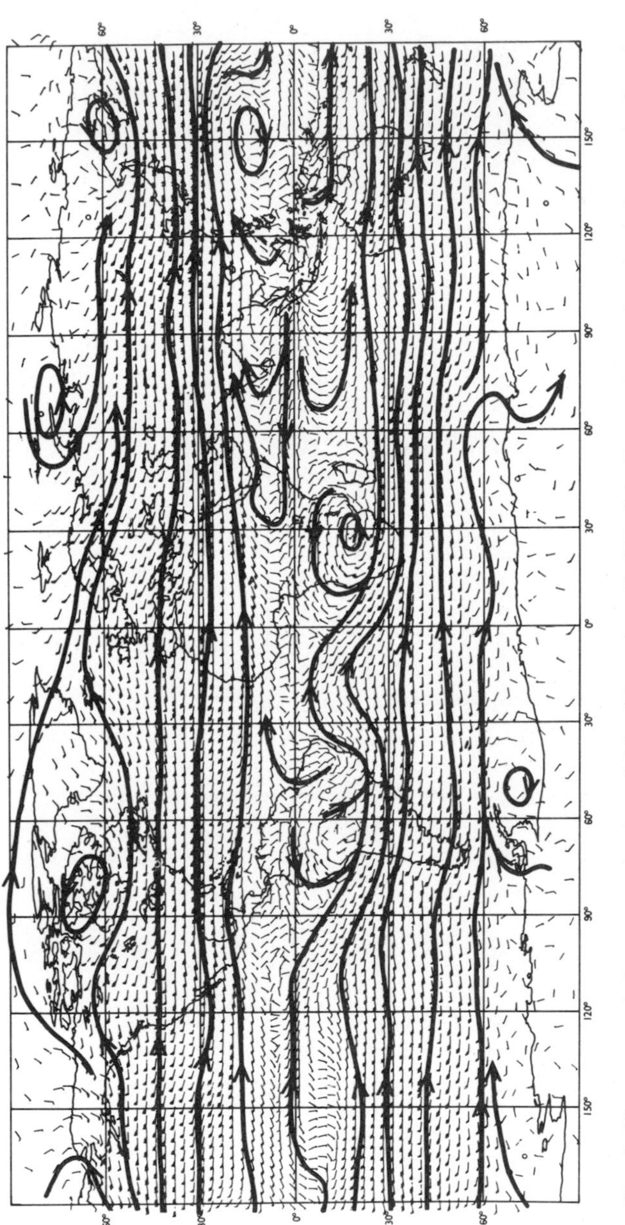


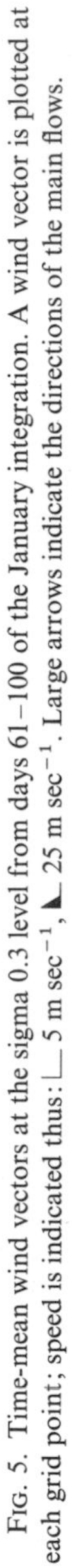

FIG. 5. Time-mean wind vectors at the sigma 0.3 level from days 61–100 of the January integration. A wind vector is plotted at each grid point; speed is indicated thus: ⎿ 5 m sec$^{-1}$, ◣ 25 m sec$^{-1}$. Large arrows indicate the directions of the main flows.

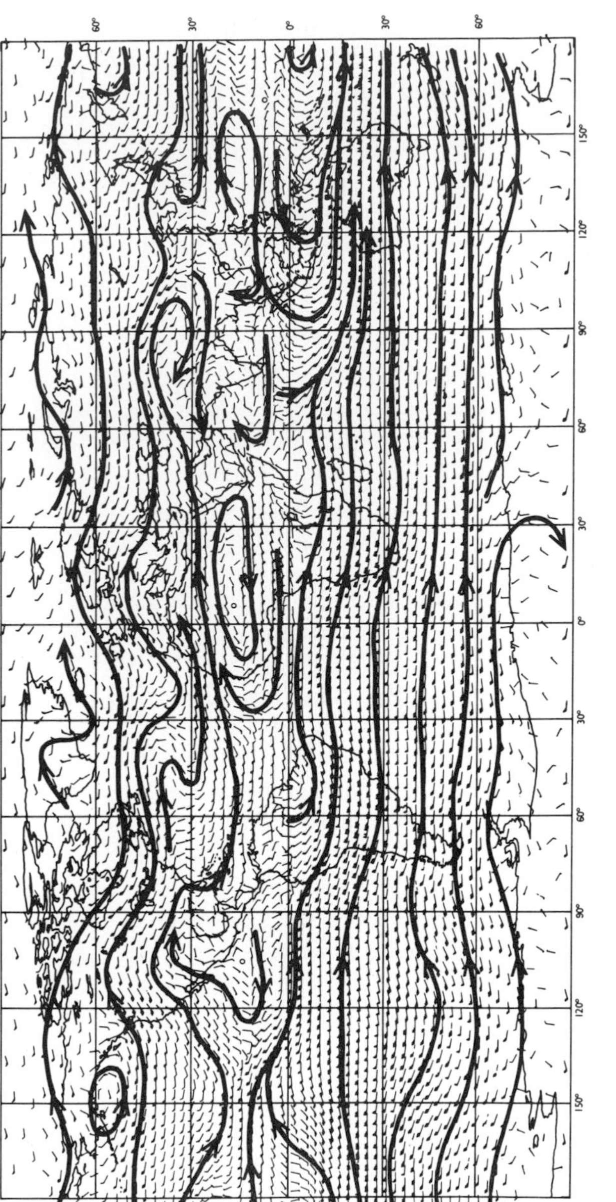


FIG. 6. As for Fig. 5, for the July integration.

south; while, in the July simulation, the representation of the south-westerly monsoon equally shows a number of realistic features, for example, the broad cross equatorial southerly current becoming a rather strong south westerly south of Arabia and then turning into the trough over India.

The winds at about 300 mb (sigma 0.3) are illustrated in Figs. 5 and 6 for January and July, respectively. The Northern Hemisphere winter troughs are close to their climatological positions, with peak mean wind speeds in the jet streams south of them about 45 and 55 m $sec^{-1}$ over eastern America and Asia, respectively. The ridge between these two troughs is weaker than that shown on climatological maps. The model's dominant feature in this month in the Southern Hemisphere is the strong westerly flow around the globe at about 40–50°S. It reaches its highest speeds between Africa and Australia of about 40 m $sec^{-1}$. The model tends to have too weak easterlies in the tropics, and this is apparent to some extent in the January chart; however, we notice that the areas where it obtains easterlies, to the north of upper anticyclonic circulations over South America and South Africa and also south of Asia, are the areas in which easterlies are most likely to be observed.

In July, the westerly flow around the Northern Hemisphere is considerably weaker than in January. The strongest winds are at about 45°N, 40–50°W. The main climatological troughs are evident, but there are also in the simulation a number of short wavelength features that are absent in the climatology. It is interesting to note that the strong upper anticyclone over the Himalayas is realistically reproduced. In the Southern Hemisphere the most significant feature is again the strong westerly jet stream, which is more or less continuous around the earth at about 30–40°S. It reaches its highest speed of about 55 m $sec^{-1}$ just east of Australia. A second jet stream of lesser intensity exists in most longitudes further south at about 60°S. In the July tropical simulation there is a deficiency of easterly winds, for example, across the Atlantic and the eastern Pacific. It is not clear at this stage how far this is due to vertical truncation effects; it is to be observed, however, that the model achieves an adequate representation of the tropical easterly flow at the highest level (sigma 0.1).

The main features of the January rainfall charts are well reproduced in the simulation (Fig. 7). In the Atlantic and Pacific sectors there are well-marked maxima along the main depression tracks. The effects of these penetrate a short way over the continents from the west, but the continental interiors are notably dry. These dry regions continue into the subtropical dry zones, observable over the sea as well as over land. Further south the rainfall increases toward the intertropical convergence zone (ITCZ) which is generally a well-marked simple maximum mostly south of the equator. However it is north of the equator in the Atlantic sector, and also over the

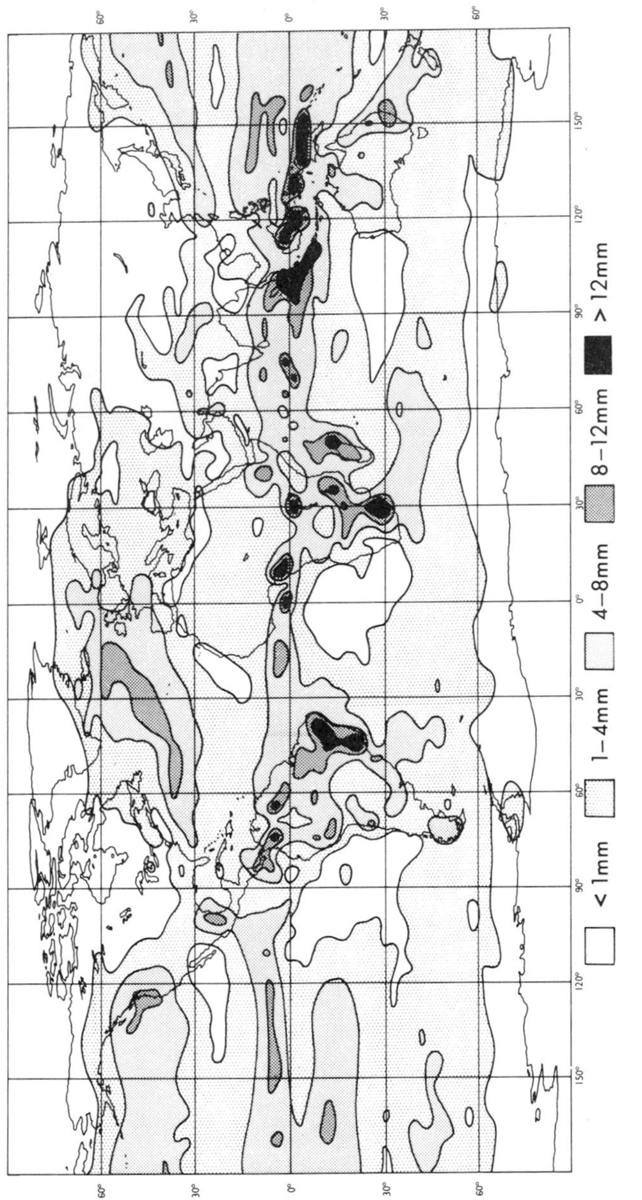


FIG. 7. Global distribution of the mean rate of precipitation (mm day$^{-1}$) for the January integration. From Mason (1976).

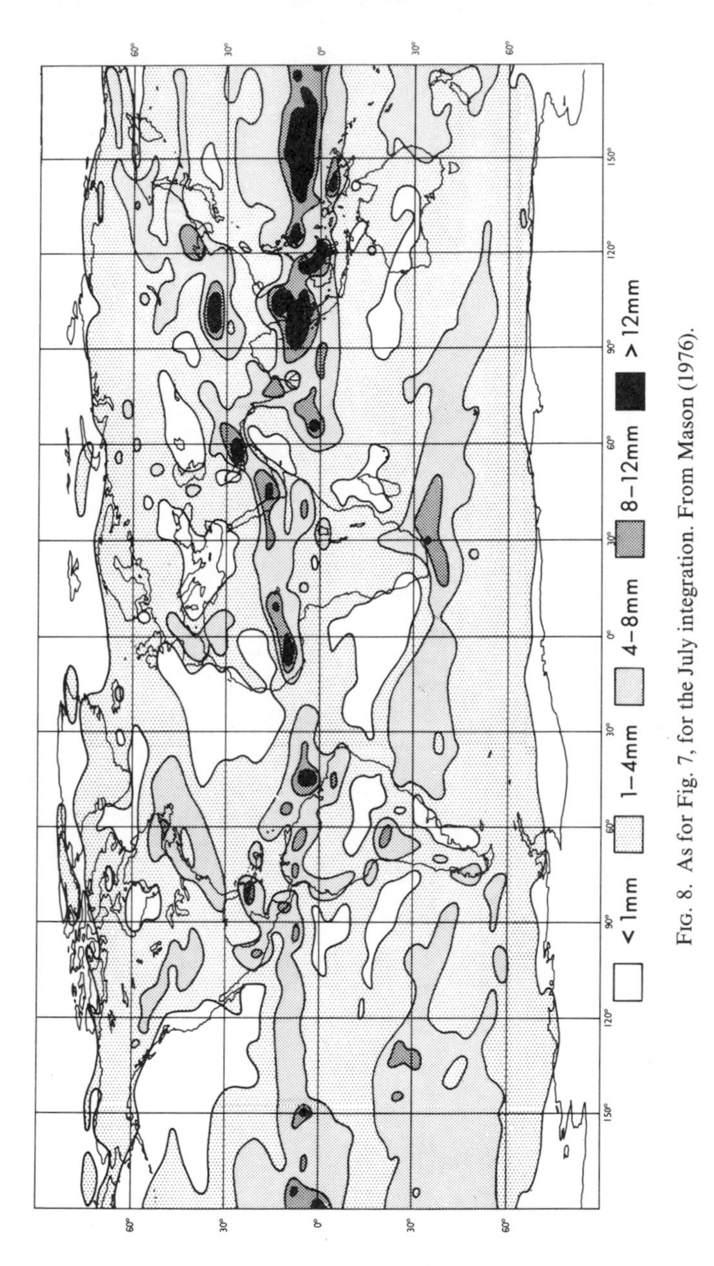


FIG. 8. As for Fig. 7, for the July integration. From Mason (1976).

eastern Pacific, for part of which a subsidiary maximum is found to the south. Over South America heavy rain extends too far north and the maximum south of the equator is displaced eastwards in comparison with observations. Subtropical latitudes in the Southern Hemisphere are notably dry except along the east coast of the continents. Rainbelts in the latter areas merge with the more diffuse maximum further south associated with the main Southern Hemisphere depressions.

The July rainfall simulation (Fig. 8) shows many realistic features. In particular, the rainfall belt associated with the ITCZ is evident at all longitudes, and, mostly, its position corresponds well with observation. Over the Atlantic and Pacific Oceans, it is generally at about 8°N. Over continental land masses it is at higher latitudes. The regions where it is least well simulated are over Africa, where it is too far north, and further east, where rain centers, clearly caused by the Red Sea and the Persian Gulf, are produced in what ought to be desert regions. There is a good maximum over India, associated with the Indian monsoon, but immediately to the east there is an area where rainfall is notably deficient. A second maximum at about 35°N, 100°E has no close parallel on climatological charts. Over the Atlantic and Pacific there are well-marked rainbelts along the primary depression tracks. The extensive dry areas of the eastern parts of the Atlantic and Pacific are well simulated though they extend too far north, particularly in the Pacific.

In agreement with observation, the tropical rainfall maximum extends into the Southern Hemisphere in the Indian Ocean and the Western Pacific. Minima in Southern Hemisphere subtropical latitudes are well-marked, while further south the maximum of the depression belt is broad with highest rainfall associated with time-mean cyclonic centers.

It has been possible to reproduce here only a few of the model results to give some indication of their characteristics. Further results are available in internal notes of the U.K. Meteorological Office, particularly Gilchrist (1974, 1975a, b, c, 1976).

## List of Symbols

|  |  |
|---|---|
| $a$ | Radius of the earth |
| $\theta$ | Latitude |
| $\lambda$ | Longitude |
| $p$ | Pressure |
| * | Subscript indicating a surface value |
| $k$ | Subscript indicating a layer (or level) of the model with $1 \leqslant k \leqslant n$, where $n$ applies to the lowest layer |
| $(k + \frac{1}{2})$ | Subscript indicating the boundary between layers $k$ and $k + 1$ |
| $\mathbf{l}$ | Unit vertical vector |
| $\sigma$ | $p/p_*$ the vertical coordinate |
| $\dot{\sigma}$ | $d\sigma/dt$ |

|  |  |
|---|---|
| $\omega$ | $dp/dt$ |
| $\mathbf{V} \equiv (u, v)$ | Lateral vector wind |
| $\mathbf{V}_3 \equiv (u, v, \dot{\sigma})$ | Three-dimensional motion vector |
| $u_*$ | Friction velocity |
| $U, V$ | $p_* u$, $p_* v$ |
| $T$ | Temperature |
| $\Theta$ | Potential temperature |
| $\rho$ | Density |
| $\phi$ | Geopotential |
| $q$ | Specific humidity |
| $f$ | Coriolis parameter |
| $f'$ | $f + (u \tan \theta/a)$ |
| $\nabla$ | Lateral gradient operator in a sigma surface |
| $\nabla \cdot$ | Lateral divergence operator in a sigma surface |
| $\nabla_3 \cdot$ | Three-dimensional divergence operator |
| $\boldsymbol{\tau} \equiv (\tau_\lambda, \tau_\theta)$ | Upward flux of momentum (zonal and meridional) due to vertical eddy mixing |
| $H$ | Upward flux of sensible heat due to vertical eddy mixing |
| $M$ | Upward flux of moisture due to vertical eddy mixing |
| $\dot{Q}_R$ | Rate of heating per unit mass due to net radiation |
| $L$ | Latent heat of condensation |
| $\dot{P}$ | Rate of precipitation per unit mass made up of $\dot{P}_L$ (large scale) and $\dot{P}_c$ (convective) |
| $R$ | Gas constant |
| $c_p$ | Specific heat at constant pressure |
| $g$ | Acceleration due to gravity |
| $D_u, D_v, D_T, D_q$ | Rate of dissipation of $U, V, p_* T, p_* q$ by horizontal subgrid-scale eddy motions |
| $R_L$ | Net upward flux of long wave radiation at the ground |
| $R_s$ | Net downward flux of solar radiation at the ground |
| $\delta_x A$ | $\{A[x + (\Delta x/2)] - A[x - (\Delta x/2)]\}/\Delta x$ |
| $\bar{A}^x$ | $\{A[x + (\Delta x/2)] + A[x - (\Delta x/2)]\}/2$ |
| $\widetilde{AB}^x$ | $\{A[x + (\Delta x/2)] \times B[x - (\Delta x/2)] + A[x - (\Delta x/2)] \times B[x + (\Delta x/2)]\}/2$ where $A$, $B$ are arbitrary variables specified at grid points |

## References


Arakawa, A. (1966). *J. Computational Phys.* **1**, 119–143.

Asselin, R. (1972). *Mon. Weather Rev.* **100**, 487–490.

Atkinson, G. D., and Sadler, J. C. (1970). "Mean Cloudiness and Gradient-level Wind Charts over the Tropics," Tech. Rep. No. 215. US Air Weather Serv. Washington, D.C.

Ball, F. K. (1960). *Q. J. R. Meteorol. Soc.* **86**, 483–494.

Betts, A. K. (1973). *Q. J. R. Meteorol. Soc.* **99**, 178–196.

Bryan, K. (1966). *Mon. Weather Rev.* **94**, 39–40.

Carson, D. J. (1973). *Q. J. R. Meteorol. Soc.* **99**, 450–467.

Charnock, H. and Ellison, T. H. (1967). "The Boundary Layer in Relation to Large-scale Motions of the Atmosphere and Ocean," Report on the GARP Study Conference, 1967, Stockholm. ICSU/IUGG and WMO.

Clarke, R. H. (1970). *Aust. Meteorol. Mag.* **18**, 51–73.

Corby, G. A., Gilchrist, A., and Newson, R. L. (1972). *Q. J. R. Meteorol. Soc.* **98**, 809–832.

Deardorff, J. W. (1972). *Mon. Weather Rev.* **100**, 93–106.

Dopplick, T. C. (1970). "Global Radiative Heating of the Earth's Atmosphere," Planet. Circ. Proj., Rep. No. 24. Dept. Meteorol., Massachusetts Institute of Technology, Cambridge.

Fiedler, F., and Panofsky, H. A. (1972). *Q. J. R. Meteorol. Soc.* **98**, 213–220.
Francis, P. E. (1975). *Q. J. R. Meteorol. Soc.* **101**, 567–582.
Gilchrist, A. (1974). "A General Circulation Model of the Atmosphere Incorporating an Explicit Boundary Layer Top," Met. O. 20 Tech. Note II/29. UK Meteorological Office, Bracknell, England.
Gilchrist, A. (1975a). "General Circulation Models with Examples Taken from the Meteorological Office Model," Met. O. 20 Tech. Note II/54. UK Meteorological Office, Bracknell, England.
Gilchrist, A. (1975b). "Tropical Results from a July Integration of a General Circulation Model Incorporating an Explicit Boundary Layer Top," Met. O. 20 Tech. Note II/46. UK Meteorological Office, Bracknell, England.
Gilchrist, A. (1975c). "The Meteorological Office General Circulation Model," Seminar on Scientific Foundation of Medium Range Weather Forecasts, Part II, pp. 594–661. European Centre for Medium-Range Weather Forecasting, Bracknell, England.
Gilchrist, A. (1976). "Tropical Results from a January Integration of a General Circulation Model Incorporating an Explicit Boundary Layer," Met. O. 20 Tech. Note II/67. UK Meteorological Office, Bracknell, England.
Gilchrist, A., Corby, G. A., and Newson, R. L. (1973). *Q. J. R. Meteorol. Soc.* **99**, 2–34.
Grimmer, M., and Shaw, D. B. (1967). *Q. J. R. Meteorol. Soc.* **93**, 337–349.
Hanna, S. R. (1969). *Atmos. Environ.* **3**, 519–536.
Holloway, J. L., Jr., and Manabe, S. (1971). *Mon. Weather Rev.* **99**, 335–370.
Hunt, G. E. (1976). "A January Climatology, Zonal Heat Balance and Cloudiness Simulated by the Meteorological Office 5-level General Circulation Model," Met. O. 20 Tech. Note II/63. U.K. Meteorological Office, Bracknell, England.
Kurihara, Y. (1965). *Mon. Weather Rev.* **93**, 399–415.
Leith, C. E. (1969). "Two-dimensional Eddy Viscosity Coefficients," Tokyo, Tech. Rep. No. 67, pp. 41–44. Japan Meteorol. Agency.
Lilly, D. K. (1965). *Mon. Weather Rev.* **93**, 11–26.
Lilly, D. K. (1967). *Proc. IBM Sci. Comput. Symp. Environ. Sci., 1967* pp. 195–210.
Manabe, S., and Strickler, R. F. (1964). *J. Atmos. Sci.* **21**, 361–385.
Manabe, S., and Terpstra, T. B. (1974). *J. Atmos. Sci.* **31**, 3–42.
Mason, B. J. (1976). *Endeavour* **25**, 51–57.
Mattingly, S. R. (1974). *Proc. Anglo-Fr. Symp. Meteorol. Effects Stratos. Aircraft*, 24–26 Sept. 1974, Oxford Vol. 2, Pap. XXVIII, pp. 1–22.
Monin, A. S., and Obukhov, A. M. (1954). "Basic Regularity in Turbulent Mixing in the Surface Layer of the Atmosphere," No. 24, pp. 163–187. USSR Acad. Sci. Geophys. Inst., Moscow.
Monin, A. S., and Zilitinkevich, S. S. (1967). "Planetary Boundary Layer and Large-scale Atmospheric Dynamics," Report on the GARP Study Conference, 1967, Stockholm, ICSU/IUGG and WMO.
Newell, R. E., Kidson, J. W., Vincent, D. G., and Boer, G. J. (1972). "The General Circulation of the Tropical Atmosphere and Interactions with Extratropical Latitudes," Vol. I. MIT Press, Boston, Massachusetts.
Newson, R. L. (1974a). *Proc. CIAP Conf., 3rd, 1974* sponsored by U.S. Dep. Transportation, Rep. No. DOT-TSC-OST-74-15, pp. 461–474.
Newson, R. L. (1974b). *Proc. Anglo-Fr. Symp. Meteorol. Effects Stratos. Aircraft*, 24–26 Sept. *1974* Oxford, Vol. 2, Pap. XXVII, pp. 1–19.
Phillips, N. A. (1957). *J. Meteorol.* **14**, 184–185.
Posey, J. W., and Clapp, P. F. (1964). *Geofis. Int.* **4**, 33–48.
Priestley, C. H. B. (1954). *Aust. J. Phys.* **7**, 176–201.
Richtmyer, R. D., and Morton, K. W. (1967). "Difference Methods for Initial-Value Problems." Wiley (Interscience), New York.

Rodgers, C. D. (1967). "The Radiative Heat Budget of the Troposphere and Lower Stratosphere," Planet. Circ. Proj., Rep. No. A2. Dept. Meteorol., Massachusetts Institute of Technology, Cambridge.

Rowntree, P. R. (1975). "The Representation of Radiation and Surface Heat Exchange in a General Circulation Model," Met. O. 20 Tech. Note II/58. UK Meteorological Office, Bracknell, England.

Rowntree, P. R. (1976a). *Q. J. R. Meteorol. Soc.* **102**, 607–625.

Rowntree, P. R. (1976b). *Q. J. R. Meteorol. Soc.* **102**, 583–605.

Shapiro, R. (1971). *J. Atmos. Sci.* **28**, 523–531.

Sheppard, P. A. (1969). "The Atmospheric Boundary Layer in Relation to Large-scale Dynamics." Global Circulation of the Atmosphere, R. Meteorol. Soc., Bracknell, England.

Shuman, F. G. (1970). *J. Appl. Meteorol.* **9**, 564–570.

Tennekes, H. (1973). *J. Atmos. Sci.* **30**, 558–567.

# A Description of the NCAR Global Circulation Models

Warren M. Washington and David L. Williamson

NATIONAL CENTER FOR ATMOSPHERIC RESEARCH*
BOULDER, COLORADO

I. Origin and Development of the NCAR Global Circulation Models . . . . . . . 111
II. Continuous Equations . . . . . . . . . . . . . . . . . . . . . . . 115
A. Prognostic Equations . . . . . . . . . . . . . . . . . . . . . . . 115
B. Diagnostic Equations . . . . . . . . . . . . . . . . . . . . . . . 117
C. Boundary Conditions . . . . . . . . . . . . . . . . . . . . . . . 119
D. Physical Processes . . . . . . . . . . . . . . . . . . . . . . . 122
III. Numerical Approximations . . . . . . . . . . . . . . . . . . . . . 131
A. Grid Structure . . . . . . . . . . . . . . . . . . . . . . . 131
B. Discrete Operators . . . . . . . . . . . . . . . . . . . . . . . 136
C. Discrete Equations . . . . . . . . . . . . . . . . . . . . . . . 137
D. Filtering . . . . . . . . . . . . . . . . . . . . . . . 162
IV. Application of NCAR Models . . . . . . . . . . . . . . . . . . . . 164
References . . . . . . . . . . . . . . . . . . . . . . . 169

## I. Origin and Development of the NCAR Global Circulation Models

THIS ARTICLE DETAILS THE various National Center for Atmospheric Research (NCAR) global circulation models (GCM), showing their development for climate simulation and short-range weather forecasting. When modeling began in 1964 at NCAR, experience elsewhere with global or hemispheric primitive equation models was limited (see Kasahara in this volume). For example, in 1964, work on such models in the United States was progressing at the Geophysical Fluid Dynamics Laboratory (GFDL) (Smagorinsky *et al.*, 1965; Manabe *et al.*, 1965), Lawrence Radiation Laboratory (LRL) (Leith, 1965), and the University of California, Los Angeles (UCLA) (Mintz, 1964; Arakawa, 1966).

The GFDL and UCLA models used variants of the sigma vertical coordinate system devised by Phillips (1957) for incorporating orography. The LRL model used pressure as the vertical coordinate with no attempt to include orography. It was apparent that there were difficulties with the

* The National Center for Atmospheric Research is sponsored by the National Science Foundation.

sigma system, so a different approach was tried by NCAR using Richardson's (1922) z-system formulation.

The first NCAR GCM was made possible in 1965 by the acquisition of a Control Data Corporation 3600 computer which allowed solution of a low-resolution GCM. This early GCM had two 6-km thick layers in the vertical and a horizontal resolution of 5° in latitude and longitude with longitudinal grid skipping poleward of 60°. The physical processes included were quite simple compared to recent versions. For example, the mean climatological distribution of January surface temperature was specified over the entire globe, and the earth's surface was assumed to be saturated. Orography was not included nor was an explicit calculation of moisture. The latent heat release was assumed proportional to the upward vertical velocity, and cloudiness for the radiation calculation was specified from zonal climatology. The horizontal eddy viscosity was constant. An economical infrared radiation computer program, based on radiation-chart approximations of fluxes, was developed by Sasamori (1968a, b). Details of this model, as well as the basic framework for later versions, are described in Kasahara and Washington (1967).

During these early years, a great deal of attention was given to numerical schemes (Kasahara, 1965; Houghton *et al.*, 1966; Williamson, 1966). In fact, we have continually examined numerical schemes to improve the model with particular emphasis on the problems associated with spherical geometry (Williamson, 1968, 1970; Williamson and Browning, 1973).

In 1968–1969, an explicit prediction of moisture was added because the earlier assumption of saturation had led to continuous atmospheric warming. With inclusion of a full hydrological cycle, a balance between evaporation and precipitation is maintained. A nonconstant horizontal eddy viscosity was added as a function of flow deformation, greatly reducing the over-damping of the baroclinic disturbances that occurred in the earlier model with constant eddy viscosity. Washington and Kasahara (1970) compared the results of the two-layer model with these additions with observed data and also computed momentum, energy, and moisture budgets in terms of eddy and mean circulations.

Development of a more general code began in 1968 when the earlier version proved inflexible in programming additional features. The major improvements incorporated were more efficient programming, diagnostically determined cloudiness, surface temperature calculation over non-ocean regions based upon a local surface energy balance, and orography. We refer to this as the z-system model. Since this version uses geometric height as the vertical coordinate and the earth's surface is not a coordinate surface, a special procedure is used to incorporate the earth's orography into the model. Orography was first included in a two-layer version of the model (Kasahara and Washington, 1969) and later an improved treatment was developed for a six-layer version (Kasahara and Washington, 1971). Computational details of the z-system model appear in Oliger *et al.* (1970). This model has six layers with a thickness of 3 km and a horizontal resolution of 5° in latitude and longitude with longitudinal grid skipping poleward of 60°. Various eddy viscosity formulations, convective parameterizations, surface hydrology, seasonal changes, and boundary-layer parameterizations have been tested (Table II, Section IV). Kasahara and Washington (1971) discussed the results of a January simulation with the six-layer version including momentum, energy, and moisture budgets.

To investigate the lower stratosphere, the z-system model was expanded by adding six layers at the top, thus extending the model height from 18 to 36 km, but retaining the same vertical grid increment as the tropospheric version. The method of calculating solar and infrared radiation is described in Sasamori *et al.* (1972). Ozone heating was included in the radiation calculation in the lower stratosphere. Because the ozone processes are not known well enough to formulate a prediction equation, a latitude–height climatological distribution of ozone was specified for the radiation calculation. The description of this model is contained in Kasahara *et al.* (1973) and Kasahara and Sasamori (1974), along with computed results for a January simulation with and without large-scale mountain effects.

The 5° horizontal resolution tropospheric model is too coarse for simulation of many features of regional interest. To alleviate this, the horizontal resolution was reduced from 5° to 2.5°. This higher resolution model includes several more elaborate parameterization schemes such as convective parameterization (Kuo, 1965; Krishnamurti and Moxim, 1971) and predictive equations for snow cover and soil moisture (Washington, 1974).

For short-range weather prediction and parameterization studies on very fine grids over less-than-global domains, a limited-area version of the z-system model has been developed. The inflow lateral boundary conditions for this model are obtained from a forecast with a low-resolution global model. The outflow boundary conditions are provided by Lagrangian extrapolation (Williamson and Browning, 1974).

Development of an improved, more flexible NCAR GCM began in 1972. Reducing grid points in the longitudinal direction at high latitudes in the z-system model results in low accuracy. Williamson and Browning (1973) showed that substantial improvement could be made by restoring skipped grid points and eliminating computationally unstable modes by Fourier filtering. The new model also uses either second- or fourth-order horizontal finite-difference approximations. To run efficiently on the next-generation parallel processor or pipeline computers, the new model adopts transformed height (Kasahara, 1974) as the vertical coordinate. The program is designed so that with very minor modifications the user can choose a different vertical increment or number of levels, vary the height of the model top, or specify a different vertical transformation. Thus far, the options in the program have been tested only for geometric height with a configuration corresponding to the standard z-system model. Various physical processes in this model are essentially the same as in the z-system model; however, the solar and infrared radiation programs are generalized to allow cloudiness at all levels rather than at just two levels as in the z-system model. This new version is referred to as the s-system model.

A schematic of the physical processes in the models is given in Fig. 1. The two upper boxes marked stratosphere and troposphere represent separate portions of the atmosphere. As mentioned earlier, most model versions were primarily for tropospheric forecasting and simulation. The straight lines from the Sun represent solar radiation that is partly absorbed by ozone, water vapor, and carbon dioxide in the stratosphere and troposphere. Solar radiation is also reflected by clouds and absorbed at the earth's surface. The wavy lines indicate infrared radiation that is of terrestrial origin. The figure also shows precipitation in the form of rain or snow. The two lower boxes represent interactions between the atmosphere and earth's surface. The earth's surface is divided into two separate classes, one for the continent and ice-covered polar regions and the second for oceans. Wind stress or momentum exchange and sensible and latent heat transport are included.

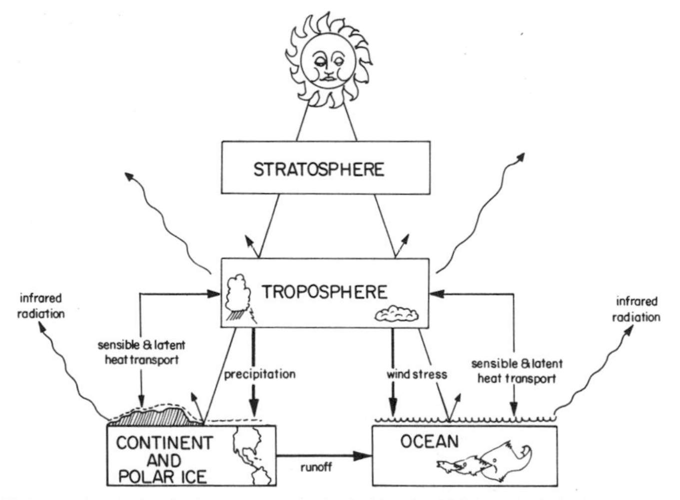


FIG. 1. Schematic of physical processes included in the NCAR global circulation models.

## II. Continuous Equations

For convenience, we divide the governing equations of the models into two categories, prognostic and diagnostic. The prognostic equations forecast the future state of the model and thus the variables forecasted—the prognostic variables—define the state of the model at any time. Additional diagnostic relations relate auxiliary variables to the prognostic variables at any one time.

### A. Prognostic Equations

Following Kasahara (1974), we write the horizontal equations of motion in terms of a transformed $z$-coordinate system. We denote the transformed variable by $s$ and require that the transformation be independent of time and that the top of the model $s_T$ be a constant height $z$. The advective form of the horizontal equations of motion in spherical coordinates with $\lambda$, $\phi$, $s$, and $t$ as the independent variables in the longitudinal, latitudinal, vertical, and temporal directions, respectively, is

$$\frac{\partial u}{\partial t} = -\mathbf{V} \cdot \nabla u - \dot{s}\frac{\partial u}{\partial s} + \left(f + \frac{u}{a}\tan\phi\right)v - \frac{RT}{a\cos\phi}\frac{\partial}{\partial\lambda}\ln p - \frac{g}{a\cos\phi}\frac{\partial z}{\partial\lambda} + \frac{1}{\rho}F_\lambda \tag{2.1}$$

$$\frac{\partial v}{\partial t} = -\mathbf{V} \cdot \nabla v - \dot{s}\frac{\partial v}{\partial s} - \left(f + \frac{u}{a}\tan\phi\right)u - \frac{RT}{a}\frac{\partial}{\partial\phi}\ln p - \frac{g}{a}\frac{\partial z}{\partial\phi} + \frac{1}{\rho}F_\phi \tag{2.2}$$

where $\mathbf{V}$ is the horizontal velocity vector with zonal and meridional components $u$ and $v$, respectively, $\dot{s}$ is the generalized vertical velocity related to vertical velocity $w$, and $z$ the height of the $s$ surface above mean sea level. The horizontal gradient operator $\nabla$ along the $s$ surface acting on some scalar $\psi$ in spherical coordinates is

$$\nabla\psi = \frac{\mathbf{i}}{a\cos\phi}\frac{\partial\psi}{\partial\lambda} + \frac{\mathbf{j}}{a}\frac{\partial\psi}{\partial\phi} \tag{2.3}$$

where $\mathbf{i}$ and $\mathbf{j}$ are the unit vectors in the longitudinal and latitudinal directions, respectively. The state variables are pressure $p$, temperature $T$, and density $\rho$.

The acceleration due to gravity is denoted by $g$, the earth's radius by $a$, and the gas constant for dry air by $R$. The Coriolis parameter $f$ is given by $2\Omega \sin \phi$, where $\Omega$ is the earth's angular velocity. The explicit forms of the horizontal frictional terms $F_\lambda$, $F_\phi$ are given later.

These prognostic equations for horizontal motion combined with the continuity equation can be rewritten in flux form using a modified density

$$\rho^* = \rho(\partial z/\partial s) \tag{2.4}$$

as

$$\frac{\partial}{\partial t}(\rho^* u) = -\nabla \cdot (\rho^* u \mathbf{V}) - \frac{\partial}{\partial s}(\rho^* u \dot{s}) + \left(f + \frac{u}{a}\tan \phi\right)\rho^* v - \frac{1}{a \cos \phi}\frac{\partial z}{\partial s}\frac{\partial p}{\partial \lambda} - \frac{g\rho^*}{a \cos \phi}\frac{\partial z}{\partial \lambda} + \frac{\partial z}{\partial s}F_\lambda \tag{2.5}$$

$$\frac{\partial}{\partial t}(\rho^* v) = -\nabla \cdot (\rho^* v \mathbf{V}) - \frac{\partial}{\partial s}(\rho^* v \dot{s}) - \left(f + \frac{u}{a}\tan \phi\right)\rho^* u - \frac{1}{a}\frac{\partial z}{\partial s}\frac{\partial p}{\partial \phi} - \frac{g\rho^*}{a}\frac{\partial z}{\partial \phi} + \frac{\partial z}{\partial s}F_\phi \tag{2.6}$$

where $\nabla \cdot$ is the horizontal divergence operator along the $s$ surface in spherical coordinates given by

$$\nabla \cdot \mathbf{V} = \frac{1}{a \cos \phi}\left[\frac{\partial u}{\partial \lambda} + \frac{\partial(v \cos \phi)}{\partial \phi}\right]. \tag{2.7}$$

The pressure tendency equation is

$$\frac{\partial p}{\partial t} = \left(\frac{\partial p}{\partial t}\right)_{\mathrm{T}} + g\rho^* \dot{s} - g \int_s^{s_{\mathrm{T}}} \nabla \cdot (\rho^* \mathbf{V})\, ds \tag{2.8}$$

where $s_{\mathrm{T}}$ (constant) denotes the top of the model atmosphere at a constant height $z_{\mathrm{T}}$. The pressure change at the top $(\partial p/\partial t)_{\mathrm{T}}$, a diagnostic quantity determined so that the boundary condition $w = 0$, or $\dot{s} = 0$, is satisfied at the top, is given later.

The fourth prognostic equation of the models is for the water vapor field. The variable used in the NCAR models is specific humidity $q$. As with horizontal motion, the moisture equation can be written in both advective and flux forms

$$\frac{\partial q}{\partial t} = -\mathbf{V} \cdot \nabla q - \dot{s}\frac{\partial q}{\partial s} + \frac{1}{\rho}M + E + \frac{1}{\rho}CT \tag{2.9}$$

$$\frac{\partial}{\partial t}(\rho^* q) = -\nabla \cdot (\rho^* q\mathbf{V}) - \frac{\partial}{\partial s}(\rho^* q\dot{s}) + \frac{\partial z}{\partial s}M + \rho^* E + \frac{\partial z}{\partial s}CT. \tag{2.10}$$

$M$ is the rate of condensation of water vapor per unit volume, $E$ is the rate of change of water vapor per unit mass due to vertical and horizontal diffusion of water vapor, and $CT$ is the rate of moisture transport by cumulus clouds. The explicit forms of $M$, $E$, and $CT$ in (2.9) and (2.10) are discussed later.

### B. Diagnostic Equations

Pressure $p$, specific humidity $q$, and horizontal velocity $u$ and $v$, or alternatively $\rho^* u$ and $\rho^* v$, are sufficient to specify the model state at any time. Additional variables appearing in the prognostic equations can be determined from these prognostic variables using the diagnostic relationships given below.

The state variables pressure and density are related by the hydrostatic equation

$$\partial p/\partial s = -g\rho^* \tag{2.11}$$

and temperature, pressure, and density are related by the ideal gas law

$$p = \rho RT. \tag{2.12}$$

The generalized vertical velocity $\dot{s}$ is obtained from the Richardson equation

$$\dot{s}\frac{\partial z}{\partial s} = -\int_{s_B}^{s} \left\{\nabla \cdot \left(\mathbf{V}\frac{\partial z}{\partial s}\right) + \frac{1}{\gamma p}\left[\left(\frac{\partial p}{\partial t}\right)_{\mathrm{T}} + J\right]\frac{\partial z}{\partial s} - \frac{Q}{c_{\mathrm{P}}T}\frac{\partial z}{\partial s}\right\} ds \tag{2.13}$$

where $s_{\mathrm{B}}$ denotes the bottom of the model atmosphere, $\gamma$ is the ratio of specific heat for dry air at constant pressure and constant volume $(c_p/c_v)$, and term $J$ is given by

$$J = \mathbf{V} \cdot \nabla p - g\int_{s}^{s_{\mathrm{T}}} \nabla \cdot (\rho^* \mathbf{V})\, ds. \tag{2.14}$$

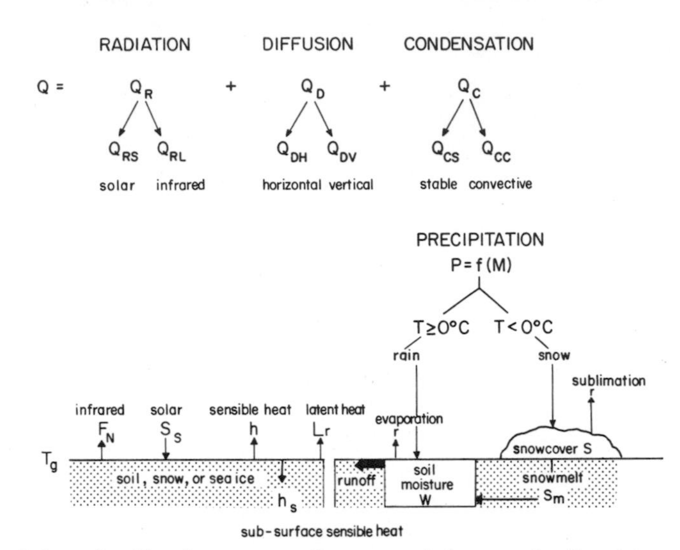


FIG. 2. Schematic of heating terms, surface energy balance, and soil moisture, snow cover processes.

As indicated in Fig. 2, the rate of heating per unit mass $Q$ appearing in the Richardson equation (2.13) is divided into three parts

$$Q = Q_{\mathrm{R}} + Q_{\mathrm{D}} + Q_{\mathrm{C}} \tag{2.15}$$

where $Q_{\mathrm{R}}$ is the heating or cooling due to radiation, $Q_{\mathrm{D}}$ the heating or cooling due to subgrid-scale eddy diffusion, and $Q_{\mathrm{C}}$ the heating due to the release of latent heat by condensation of water vapor. These individual components are described in Sections II, D and III, C, 3. The pressure change at the top of the model $(\partial p/\partial t)_{\mathrm{T}}$ appearing in both the pressure tendency equation (2.8) and Richardson equation (2.13) is obtained by applying the top boundary condition $\dot{s} = 0$ to the Richardson equation and solving for $(\partial p/\partial t)_{\mathrm{T}}$

$$\left(\frac{\partial p}{\partial t}\right)_{\mathrm{T}} = -\int_{s_{\mathrm{B}}}^{s_{\mathrm{T}}} \left[\frac{J}{\gamma p}\frac{\partial z}{\partial s} - \frac{Q}{c_{\mathrm{p}} T}\frac{\partial z}{\partial s} + \nabla\cdot\left(\mathbf{V}\frac{\partial z}{\partial s}\right)\right] ds \cdot \left[\int_{s_{\mathrm{B}}}^{s_{\mathrm{T}}} \frac{1}{\gamma p}\frac{\partial z}{\partial s}\, ds\right]^{-1}. \tag{2.16}$$

The governing equations (2.1)–(2.16) are written in terms of the transformed height coordinate $s$. The equations in terms of height $z$ can be obtained by simply setting $s$ equal to $z$ and noting that $\partial z/\partial s$ is then unity. The generalized vertical velocity $s$ becomes $w$ and the second term in the pressure gradient forces in (2.1), (2.2), (2.5), and (2.6) becomes zero, because in this case the independent variable $z$ is not a function of longitude or latitude. We will use both the $s$ and $z$ coordinates in the following description of the models.

## C. Boundary Conditions

The top boundary condition is taken to be a rigid lid requiring no mass flux in or out of the top of the model atmosphere, that is, $\rho w = 0$ which reduces to $\dot{s} = w = 0$ at $z_T$ since $\rho$ is finite there. We specify free-slip boundary conditions for horizontal momentum, preventing the upper boundary from being a momentum source or sink. In addition, we specify that the vertical sensible and latent heat fluxes vanish there, preventing the loss or gain of sensible heat or moisture through the top of the model atmosphere. These upper boundary conditions can be summarized as

$$\dot{s} = w = \tau_\lambda = \tau_\phi = h = r = 0 \qquad \text{at} \quad z = z_T \tag{2.17}$$

where $\tau_\lambda$ and $\tau_\phi$ are the longitudinal and latitudinal components of Reynolds stress, $h$ is the vertical sensible heat flux, and $r$ is the vertical latent heat flux. $\tau_\lambda$, $\tau_\phi$, $h$, and $r$ are defined in Section II, D, 1 in terms of vertical gradients of momentum, temperature, and moisture.

At the lower boundary, specification of conditions is more involved because many important physical processes take place at the interface between the earth's surface and the atmosphere. In the s system, the dynamical conditions result in

$$\dot{s} = 0 \qquad \text{at} \quad s_B \,(= \text{constant}). \tag{2.18}$$

since we require no mass transport through the earth's surface expressed by a constant $s$ surface in this model.

In the z system where the lower boundary is no longer a coordinate surface, the lower boundary condition becomes

$$w_a = \mathbf{V}_a \cdot \nabla z_B \tag{2.19}$$

Subscript a denotes values at anemometer level in the atmospheric boundary layer. This a level is discussed in more detail when the vertical grid is introduced in Section III, A. $z_B$ is the height of the mountains above mean sea level, $\mathbf{V}_a$ the horizontal wind vector at the a level and $w_a$ the mountain-induced vertical motion.

Above the earth's surface, we define a surface boundary (Prandtl) layer in which surface horizontal stress, sensible heat flux, and moisture flux are assumed constant. The thickness of this layer is not strictly defined; however, typical values are in the range of 20 to 200 m. We represent the flux values at the anemometer level by

$$\left.\begin{aligned} \tau_\lambda &= C_D \rho_a u_a V_a \\ \tau_\phi &= C_D \rho_a v_a V_a \\ h &= -c_p C_D \rho_a (T_a - T_g) V_a \\ r_{ap} &= -C_{DW} \rho_a (q_a - q_g) V_a \end{aligned}\right\} \quad \text{at} \quad z = z_B \tag{2.20}$$

where subscript a refers to variables in the constant flux layer, $C_D$ is the drag coefficient ($C_D = 0.003$), $C_{DW}$ a different coefficient for evaporation ($C_{DW} = 0.7C_D$), $\rho_a$ is density, $r_{ap}$ potential evaporation rate equal to the evaporation rate obtained if the surface were saturated and

$$V_a = (u_a{}^2 + v_a{}^2)^{1/2}. \tag{2.21}$$

$T_g$ is the surface temperature, and $q_g$ the specific humidity for saturated surface with temperature $T_g$. We discuss in Section II, D, 3 how the actual evaporation rate is related to the potential evaporation rate over non-saturated regions. To handle the problem of boundary-layer convection in windless or light wind situations, we define a minimum value of 5 m sec$^{-1}$ for $V_a$. Deardorff (1972) and Benoit (1976) devised a more promising approach to the atmospheric boundary layer by replacing Eqs. (2.20) with improved parameterizations which take into account changes in static stability, the turning of wind with height, and a variable boundary layer height. This approach has been tested in one version of the model, but because it has not yet been included in any standard versions, we do not describe it here.

Surface temperature $T_g$ in the model is either specified or computed depending on location. Over the open ocean, we specify surface temperature from observed climatological distributions (Washington and Thiel, 1970). Use of such climatological means is adequate for seasonal or annual atmospheric simulation or short- and medium-range forecasting because the ocean has a large heat capacity compared to the atmosphere, and surface temperature changes are relatively small on these time scales. Over land and snow-ice surfaces, temperature is calculated from a surface energy balance. If we assume no net energy flux at the earth's surface, the equation which determines the surface temperature is of the following general form:

$$F_N - S_s + h + Lr + h_s + L_f S_m = 0 \tag{2.22}$$

where

$F_N = F\uparrow - F\downarrow$ = net long-wave radiation at the ground
$F\uparrow = \sigma T_g^4$ = upward blackbody long-wave radiation from surface
$F\downarrow$ = downward long-wave radiation from atmosphere
$\sigma$ = Stefan–Boltzmann constant
$S_s = (1 - A_b)F_{RS}$ = absorbed solar flux at surface
$A_b$ = albedo of ground surface
$F_{RS}$ = solar flux arriving at surface
$h$ = sensible heat flux to atmosphere
$r$ = flux of water vapor to atmosphere from evaporation or sublimation
$h_s$ = conduction of heat to subsurface
$S_m$ = melting of snow and ice
$T_g$ = surface temperature
$L_f$ = latent heat of fusion
$L$ = latent heat of evaporation $L_e$ or latent heat of sublimation $L_f + L_e$.

These terms are shown schematically in Fig. 2. The signs of quantities in (2.22) are defined as plus for heat flux out from the surface and minus into the surface.

The calculation of net long-wave and short-wave radiation at the ground is discussed in Section III, C, 3. The surface albedoes in the model differ depending on the version. In early versions they are fixed with time with the values taken mostly from Posey and Clapp (1964). When snow cover is computed in the model, the albedo varies (Section II, D, 3). We do not compute explicitly the heat conducted in and out of the subsurface. From Sasamori's (1970) numerical study of atmosphere–soil interface and adjacent layers, it was found that $h_s$ is approximately proportional to $h$; therefore, we assume

$$h_s = bh \tag{2.23}$$

where $b$ is usually taken as 1/3. For model versions in which soil moisture is not explicitly computed, we assume a Bowen ratio $B$ of unity, so that the evaporative flux used in (2.22) is uncoupled from the hydrological cycle and is given by

$$r = h/B. \tag{2.24}$$

The method of solution of (2.22) is to separate the term $T_g - T_a$ from $h$ in (2.20) and to solve for $T_g$. Because $T_g$ is also included implicitly in several other terms of (2.22), strictly speaking we should iterate the solution of (2.22). We presently only solve one iteration to save computer time. Tests comparing several iterations versus one show that one is sufficiently accurate. More details are given in Oliger *et al.* (1970).

## D. Physical Processes

### 1. *Subgrid-Scale Diffusion*

The longitudinal and latitudinal components of the frictional force $F_\lambda$ and $F_\phi$ are written as a sum of vertical and horizontal terms

$$F_\lambda = F_{\lambda \mathrm{V}} + F_{\lambda \mathrm{H}} \tag{2.25}$$

$$F_\phi = F_{\phi \mathrm{V}} + F_{\phi \mathrm{H}}. \tag{2.26}$$

Similarly, the time rate of change of water vapor per unit mass due to diffusion is divided into vertical and horizontal components

$$E = E_\mathrm{V} + E_\mathrm{H} \tag{2.27}$$

as is the heating or cooling rate due to eddy diffusion

$$Q_\mathrm{D} = Q_\mathrm{DV} + Q_\mathrm{DH}. \tag{2.28}$$

*a. Vertical.* The vertical component of friction is written in terms of Reynolds stress as

$$F_{\lambda \mathrm{V}} = \partial \tau_\lambda / \partial z \tag{2.29}$$

$$F_{\phi \mathrm{V}} = \partial \tau_\phi / \partial z \tag{2.30}$$

where the Reynolds stress may be expressed by

$$\tau_\lambda = \rho K_\mathrm{MV}(\partial u/\partial z) \tag{2.31}$$

$$\tau_\phi = \rho K_\mathrm{MV}(\partial v/\partial z). \tag{2.32}$$

$K_{MV}$, the vertical kinematic eddy viscosity, is defined later. The vertical diffusion of water vapor is given by

$$\rho E_V = -\partial r/\partial z. \tag{2.33}$$

The vertical flux of water vapor $r$ is expressed as

$$r = -\rho K_{WV}(\partial q/\partial z) \tag{2.34}$$

where $K_{WV}$ is the vertical diffusivity of water vapor.

The vertical diffusion of sensible heat has the same form as moisture and momentum

$$Q_{DV} = -(1/\rho)(\partial h/\partial z). \tag{2.35}$$

The vertical flux of sensible heat $h$ is

$$h = -\rho c_p K_{TV}(\partial \theta/\partial z - \gamma_{CG}) \tag{2.36}$$

where $K_{TV}$ denotes the vertical kinematic thermal diffusivity. Potential temperature $\theta$ is given by

$$\theta = T(\tilde{p}_0/p)^\kappa \tag{2.37}$$

where $\tilde{p}_0$ is a reference pressure

$$\tilde{p}_0 = 1013.25 \text{ mb} \tag{2.38}$$

and

$$\kappa = R/c_p. \tag{2.39}$$

The vertical derivative of potential temperature used in (2.36) can be obtained by differentiating (2.37)

$$\frac{\partial \theta}{\partial z} = \left(\frac{\tilde{p}_0}{p}\right)^\kappa \left(\frac{\partial T}{\partial z} + \frac{g}{c_p}\right). \tag{2.40}$$

In (2.36) $\gamma_{CG}$ is a countergradient after Deardorff (1966). The value normally used and specified in Section III, C, 3 has been determined experimentally by Washington and Kasahara (1970).

To allow for the effect of free convection in the atmosphere, vertical kinematic thermal diffusivity $K_{TV}$ is a function of stability. For the unstable case, it has the form

$$K_{TV} = A_1 + A_2\{1 - \exp[A_3(\partial\theta/\partial z - \gamma_{CG})]\} \quad \text{for} \quad \partial\theta/\partial z \le \gamma_{CG} \tag{2.41}$$

while for the stable case

$$K_{TV} = [A_1/(1 + A_4 R_i)] + A_5 \quad \text{for} \quad \partial\theta/\partial z > \gamma_{CG}. \tag{2.42}$$

The actual value of the constants $A_i$ are also given in Section III, C, 3 along with the discrete approximations. The Richardson number $R_i$ in (2.42) is defined by

$$R_i = \frac{(g/T)[(\partial\theta/\partial z) - \gamma_{CG}]}{(\partial u/\partial z)^2 + (\partial v/\partial z)^2 + A_6}. \tag{2.43}$$

$A_6$ (also in Section III, C, 3) provides a minimum value for the vertical shear to ensure that $R_i$ remains finite. The vertical diffusivities of water vapor and momentum are assumed to be equal to the vertical kinematic thermal diffusivity

$$K_{MV} = K_{WV} = K_{TV}. \tag{2.44}$$

*b. Horizontal.* We have used several different forms for the horizontal component of eddy diffusion. One, still used in the z-system model, follows Smagorinsky (1963) and was first used in the NCAR model by Washington and Kasahara (1970). The NCAR form is somewhat different from that of GFDL (Kurihara, 1965; Holloway and Manabe, 1971) but has been found to be suitable for simulation since it is globally dissipative. The differences between the NCAR and GFDL forms are largest near the poles.

$$F_{\lambda H} = \frac{1}{a\cos\phi}\frac{\partial}{\partial\lambda}(\rho K_{MH} D_T) + \frac{1}{a}\frac{\partial}{\partial\phi}(\rho K_{MH} D_S) \tag{2.45}$$

$$F_{\phi H} = \frac{1}{a\cos\phi}\frac{\partial}{\partial\lambda}(\rho K_{MH} D_S) - \frac{1}{a}\frac{\partial}{\partial\phi}(\rho K_{MH} D_T) \tag{2.46}$$

where

$$D_T = \frac{1}{a\cos\phi}\left[\frac{\partial u}{\partial\lambda} - \frac{\partial}{\partial\phi}(v\cos\phi)\right] \tag{2.47}$$

$$D_{\mathrm{S}} = \frac{1}{a \cos \phi} \left[ \frac{\partial v}{\partial \lambda} + \frac{\partial}{\partial \phi} (u \cos \phi) \right] \tag{2.48}$$

and $K_{\mathrm{MH}}$ is the horizontal kinematic eddy viscosity.

$$K_{\mathrm{MH}} = 2k_0{}^2 l^2 D \tag{2.49}$$

where

$$D = (D_{\mathrm{T}}{}^2 + D_{\mathrm{S}}{}^2)^{1/2}. \tag{2.50}$$

The proportionality constant $k_0$ is taken to be as small as possible so as not to damp excessively the large-scale motions. However, to prevent nonlinear instability, $k_0$ must be large enough for terms $F_{\lambda \mathrm{H}}$ and $F_{\lambda \phi}$ to remove a sufficient amount of energy from wavelengths shorter than four grid intervals. In general, $l$ is equal to the distance between grid points. We give its form in Section III, C, 3 when discussing the finite difference approximations.

The horizontal diffusion of moisture $E_{\mathrm{H}}$ is given by

$$\rho E_{\mathrm{H}} = \nabla \cdot (\rho K_{\mathrm{WH}} \nabla q). \tag{2.51}$$

The horizontal diffusivity of water vapor $K_{\mathrm{WH}}$ is taken to be equal to that of momentum $K_{\mathrm{MH}}$. The horizontal diffusion of sensible heat $Q_{\mathrm{DH}}$ has the same form as (2.51)

$$\rho Q_{\mathrm{DH}} = c_{\mathrm{P}} \nabla \cdot (\rho K_{\mathrm{TH}} \nabla \theta). \tag{2.52}$$

The horizontal diffusivity of sensible heat $K_{\mathrm{TH}}$ is assumed to be equal to that of water vapor and momentum

$$K_{\mathrm{TH}} = K_{\mathrm{WH}} = K_{\mathrm{MH}}. \tag{2.53}$$

The horizontal diffusion terms are included in the models to parameterize the effect of subgrid scale motions on the scales of motion that can be resolved by the computational grids. In practice, these diffusion terms also provide a mechanism to eliminate nonlinear instability caused by aliasing and, as mentioned earlier, the coefficient $k_0$ is determined to be the smallest value that allows the model to run stably. For coarser resolution models this value is often too large in that it damps the baroclinic waves too much as hypothesized in Manabe *et al.* (1970) and Wellck *et al.* (1971).

Because the s-system model is designed to use fourth-order horizontal differences, which are more susceptible to aliasing problems than second order, we have generalized to spherical coordinates the higher degree diffusion terms proposed by Kreiss and Oliger (1972) to stabilize centered fourth-order difference approximations. This fourth-degree diffusion is

$$F_{\lambda \mathrm{H}} = -\nabla^2(\rho K_{\mathrm{MH}} \nabla^2 u) \tag{2.54}$$

$$F_{\phi \mathrm{H}} = -\nabla^2(\rho K_{\mathrm{MH}} \nabla^2 v) \tag{2.55}$$

$$\rho E_{\mathrm{H}} = -\nabla^2(\rho K_{\mathrm{WH}} \nabla^2 q) \tag{2.56}$$

$$\rho Q_{\mathrm{DH}} = -c_{\mathrm{P}}\nabla^2(\rho K_{\mathrm{TH}} \nabla^2 \theta) \tag{2.57}$$

where the nonlinear coefficients $K_{\mathrm{MH}}$, $K_{\mathrm{WH}}$, and $K_{\mathrm{TH}}$ are defined as in (2.49) and (2.53).

2. *Radiation*

The heating rate due to radiational sources $Q_{\mathrm{R}}$ appearing in (2.15) is divided into two parts

$$Q_{\mathrm{R}} = Q_{\mathrm{RL}} + Q_{\mathrm{RS}} \tag{2.58}$$

where $Q_{\mathrm{RL}}$ is the rate of heating or cooling due to the long-wave radiation and $Q_{\mathrm{RS}}$ is the rate of heating due to atmospheric absorption of solar radiation.

*a. Long Wave.* The long-wave heating or cooling computation follows the method developed by Sasamori (1968a, b) based on a radiation chart approach using numerical approximations to the transfer equations. To reduce computing time, the absorption functions are expressed as analytic functions. The rate of heating or cooling due to the divergence of the net infrared radiation flux is obtained from

$$Q_{\mathrm{RL}} = -(1/\rho)(\partial/\partial z)(F^{\uparrow} - F^{\downarrow}) \tag{2.59}$$

where $F^{\uparrow}$ and $F^{\downarrow}$ are the upward and downward fluxes of infrared radiation, respectively. The formulas for these fluxes appear in Section III, C, 3 after the vertical grid is introduced.

*b. Short Wave.* The rate of heating due to atmospheric absorption of solar radiation is

$$Q_{\mathrm{RS}} = (1/\rho)(\partial F_{\mathrm{S}}/\partial z) \tag{2.60}$$

where $F_{\mathrm{S}}$ represents the solar radiation flux. The details of the computation of the flux including the effect of cloudiness are set forth in Section III, C, 3.

3. *Hydrological Cycle*

The major heat source driving the tropical circulation and much of the general circulation of the atmosphere is the release of latent heat. This heat is added directly to the atmosphere, whenever rain or snow falls to the earth's surface. The heat release due to condensation can be divided into two parts

$$Q_C = Q_{CS} + Q_{CC} \tag{2.61}$$

where $Q_{CS}$, referred to as stable latent heat release, is caused by widespread ascent of moist air usually associated with warm frontal processes resulting in precipitation under stable conditions. $Q_{CC}$ is precipitation associated with cumulus convection activity and usually dominates in the tropics and higher latitude continental areas in the summer.

*a. Stable Latent Heat Release.* Specific humidity $q$ can be written as

$$q = \frac{\varepsilon e}{p - (1 - \varepsilon)e}, \tag{2.62}$$

where $e$ is the partial pressure of water vapor and $\varepsilon$ the ratio of molecular weight of water vapor to that of dry air ($\varepsilon = 0.622$). For saturated air over water or ice, the saturation vapor pressure is designated by $e_{st}$. The corresponding saturation specific humidity is

$$q_{st} = \frac{\varepsilon e_s}{p - (1 - \varepsilon)e_{st}}. \tag{2.63}$$

Whenever the value of specific humidity $q$ exceeds the saturation value $q_{st}$, the excess water vapor over saturation is available for condensation as liquid water. Latent heat of condensation is released to warm the surrounding air when the water vapor changes to liquid water. This calculation is described in more detail in Section III, C, 3.

*b. Convective Latent Heat Release.* This method of cumulus parameterization was developed by Kuo (1965), later modified by Krishnamurti and Moxim (1971), and adapted to the NCAR global circulation model by Kanamitsu (1971).

The heating rate due to cumulus convection in the Kuo scheme is

$$Q_{CC} = \frac{(ar)c_p(T_{ma} - T)}{\Delta\tau} \tag{2.64}$$

where $Q_{CC}$ is the heating rate per unit mass, $ar$ is a parameter related to the area covered by cumulus convection, $\Delta\tau$ the lifetime of convection (assumed to be 30 minutes), $c_p$ the specific heat at constant pressure, $T_{ma}$ the temperature along the moist adiabat which passes through the lifting condensation level, and $T$ the model temperature. The quantity $ar$ is defined as

$$ar = I/S, \tag{2.65}$$

where $I$ is the net convergence of moisture in an air column extending from the cloud base $z_{CB}$ to the cloud top $z_{CT}$, i.e.,

$$I = -\int_{z_{CB}}^{z_{CT}} \nabla \cdot \rho q \mathbf{V} \, dz - \rho_{CT} q_{CT} w_{CT} + \rho_{CB} q_{CB} w_{CB}. \tag{2.66}$$

Subscripts CT and CB in (2.66) refer to the top and bottom of the cloud, respectively. In the present formulation in the NCAR model, we assume that the first two terms on the right of (2.66) are negligible compared to the third. The denominator $S$ of (2.65) is the amount of energy needed to saturate the entire grid volume from the cloud bottom to top. For our model, $S$ takes the form

$$S = \frac{1}{\Delta\tau} \int_{z_{CB}}^{z_{CT}} \left[ \frac{c_p}{L} (\rho_{ma} T_{ma} - \rho T) + (\rho_{ma} q_{ma} - \rho q) \right] dz \tag{2.67}$$

where $L$ is the latent heat of condensation and the subscript ma refers to temperature, density, and specific humidity along the moist adiabat.

The rate of moisture transport by cumulus clouds is

$$CT = (ar/\Delta\tau)(\rho_{ma} q_{ma} - \rho q). \tag{2.68}$$

We assume that all moisture for the cumulus heating and moisture transport comes from the bottom layer of the model. Experiments by Washington and Daggupaty (1975) indicate that the original scheme proposed by Kuo underestimates precipitation in the tropics. We have tried to remedy this shortcoming by making $ar$ equal to 1 if the lower layers become supersaturated. Another change having an effect on the simulation is to increase the vertical eddy diffusivity $K_{TV}$ to values of the order of $10^7$ cm$^2$ sec$^{-1}$ whenever the lower layers become supersaturated. The increased $K_{TV}$ has the beneficial effect of drying out the lower layers by increasing subgrid-scale transports of moisture, momentum, and sensible heat in regions of active condensation.

*c. Convective Adjustment.* In most versions of the models, moist convective adjustment is used to simulate subgrid-scale cumulus convection instead of the scheme described above. Dry convective adjustment is included in all versions. If the temperature lapse rate $-\partial T/\partial z$ becomes greater than the dry adiabatic lapse rate $\gamma_d$ in the unsaturated atmosphere, or greater than the moist adiabatic lapse $\gamma_w$ in the saturated atmosphere, the motion becomes gravitationally unstable. These critical lapse rates are given by

$$\gamma_d = g/c_p \tag{2.69}$$

and

$$\gamma_w = \frac{(g/c_p)[1 + L(q_s/R\bar{T})]}{(1 + 0.622L^2 q_s)/c_p R\bar{T}^2} \tag{2.70}$$

where $\bar{T}$ is the density-weighted vertical average of $T$. In nature, this instability takes place on scales much smaller than the grid sizes of the models; however, the models must make the adjustment on grid scales. Our method of convective adjustment is somewhat different in detail from that used by Smagorinsky *et al.* (1965), Manabe *et al.* (1965), and Mintz (1964), but it is based on the same general principles.

We require that (1) the internal energy remain constant during the adjustment process

$$\int \rho T \, dz = \text{constant} \tag{2.71}$$

where density is taken to be the same before and after the adjustment, (2) the new lapse rate equal the appropriate critical lapse rate

$$\partial T/\partial z = -\gamma_d \qquad \text{if} \quad w \leqslant 0$$

or (2.72)

$$\partial T/\partial z = -\gamma_w \qquad \text{if} \quad w > 0,$$

and (3) the mean temperature remain invariant in the adjustment process

$$\bar{T} = \int \rho T \, dz \bigg/ \int \rho \, dz. \tag{2.73}$$

Thus, in the scheme described here any released latent heat goes into the heating rate of the Richardson equation only and is redistributed in the vertical by the convective adjustment scheme.

*d. Soil Moisture, Snow Cover, and Albedo.* The early versions of the NCAR models assumed that land surfaces were always saturated and that the snow–ice line was geographically fixed. These limitations were first removed from GCMs by Manabe (1969a, b) and Bryan (1969), later by Holloway and Manabe (1971) for the GFDL model, and in 1971 for some versions of the NCAR models (Washington, 1974). The basis for the soil moisture formulation in the NCAR model is from Budyko (1956). Figure 2 shows schematically the processes involved.

The soil moisture equation is

$$\partial W/\partial t = P - r + S_m \tag{2.74}$$

where $W$ is the total soil moisture in centimeters stored in a surface layer of soil, $P$ the rain precipitation rate, $r$ the evaporation rate, and $S_m$ the snowmelt rate also used in the surface energy balance (2.22). From field studies of evaporation rates, it has been found that the evaporation rate dependence on soil moisture can be parameterized in the following manner:

$$\text{if} \quad W \geq W_C, \qquad \text{then} \quad r = r_{ap} \tag{2.75}$$

and

$$\text{if} \quad W < W_C, \qquad \text{then} \quad r = r_{ap}(W/W_C) \tag{2.76}$$

where $W_C$ is a critical value of soil moisture and $r_{ap}$ refers to the potential evaporation rate from a saturated surface defined in (2.20). Essentially, the above equations state that if the soil moisture is greater than $W_C$, the evaporation rate is a maximum $r_{ap}$, whereas if the soil moisture is less than $W_C$, the evaporation rate is reduced linearly as a function of $W$. Manabe (1969a) chose 15 cm of water for the field capacity of the soil moisture layer; we use the same value. Any excess over this amount is termed runoff and is assumed to drain off to the oceans. The critical value $W_C$ is assumed to be 75% of the field capacity. As seen, this type of parameterization does not include any detailed soil hydrology, differing types of soils, vegetation, etc.

The snow cover equation is

$$\partial S/\partial t = P - r - S_m \tag{2.77}$$

where $S$ is the snow cover amount measured in terms of liquid water equivalent (centimeters), $P$ the snow precipitation rate computed from the rate of condensation of water vapor $M$ in (2.9) and (2.10), $r$ the rate of sublimation given following (2.22), and $S_m$ the snowmelt rate appearing also in (2.74). It is possible for the snow cover to melt and increase soil moisture. Also, if rain is falling onto the snow, we assume that the snow cannot hold the water and it seeps through the snow into the ground to increase the soil moisture. We do not allow evaporation from the ground when it is snow-covered;

however, we do allow sublimation of the snow itself. The judgment whether rain or snow is falling is based on the temperature in the first model layer above the ground; if the temperature there is above freezing, rain is assumed to fall and if below freezing, snow. Figure 2 indicates the interactions between the different terms in the surface hydrology.

Before leaving this section on snow cover, we mention a modification of the surface albedo depending on the snow depth. This feedback mechanism is important to the surface heat balance equation and is parameterized simply. Over snow or ice surfaces, surface albedo is given by

$$A_b = 0.2 + 0.4S \tag{2.78}$$

with the restriction that $A_b$ cannot exceed 0.6, the value obtained with 1 cm liquid equivalent which is approximately equal to 10 cm snow depth.

*e. Cloudiness.* Cloudiness is computed from a simple dependence on relative humidity. We compute only two layers of cloudiness—low and high. In the z-system model, these are taken at 3 and 9 km, while in the s-system model they are taken at corresponding levels. The radiation has been formulated to allow more layers of clouds in the future. A linear dependence between cloudiness and relative humidity is assumed where the coefficients are adjusted to fit the observed data of London (1957). The empirical formulas are

$$C_l = 2.4\ \text{R.H.} - 1.6, \qquad 0.2 \leqslant C_l \leqslant 0.8 \tag{2.79}$$

and

$$C_h = 0.24\ \text{R.H.} - 0.16, \qquad 0.0 \leqslant C_h \leqslant 0.4 \tag{2.80}$$

where $C_l$ and $C_h$ are the low and high cloudiness, respectively, and R.H. is relative humidity at that level. We have the additional restrictions that vertical velocity $w$ be $> -2\ \text{cm sec}^{-1}$ and R.H. be $> 0.75$ for clouds to exist. Results from this method are shown in Kasahara and Washington (1971).

## III. Numerical Approximations

### A. Grid Structure

#### 1. *z-System Model*

The governing equations in Section II are written in terms of horizontal spherical coordinates ($\lambda$, $\phi$) and a transformed height coordinate ($s$). As described at the end of that section, the z-system equations are obtained from this set by setting $s = z$ and therefore $\partial z/\partial s = 1$. The discrete grid for the z-system model is uniform in this $\lambda$, $\phi$, $z$ coordinate system. Let $I$ and $J$ be the number of grid points in the longitude and latitude directions, respectively. We require $I$ to be even and $J$ to be odd. The longitudinal and latitudinal grid intervals are then given by

$$\Delta\lambda = 2\pi/I \tag{3.1}$$

$$\Delta\phi = \pi/(J - 1). \tag{3.2}$$

The longitudes and latitudes of the grid points are given by

$$\lambda_i = -\pi + (i - 1)\,\Delta\lambda; \qquad i = 1, 2, \ldots, I \tag{3.3}$$

$$\phi_j = -\pi/2 + (j - 1)\,\Delta\phi; \qquad j = 1, 2, \ldots, J. \tag{3.4}$$

The variables are placed on the mesh such that they are staggered in time. All variables are defined at the same grid points, but these points change with time (Fig. 3). Let $\Delta t$ denote the discrete time increment so that $t = n\,\Delta t$. In Fig. 3, the circles denote points at which the variables are defined for even $n$, and the crosses denotes points for odd $n$. Since the z-system model uses second-order centered space and time differences, it is possible to use the staggered grid, rather than computing the variables at all points each time step. This staggered grid reduces the computer storage and time by a factor of two without introducing any significant additional computational error.

The horizontal grid of the z-system model is modified near the pole to maintain a nearly constant geographical distance between mesh points. This is accomplished by decreasing the number of points around the latitude

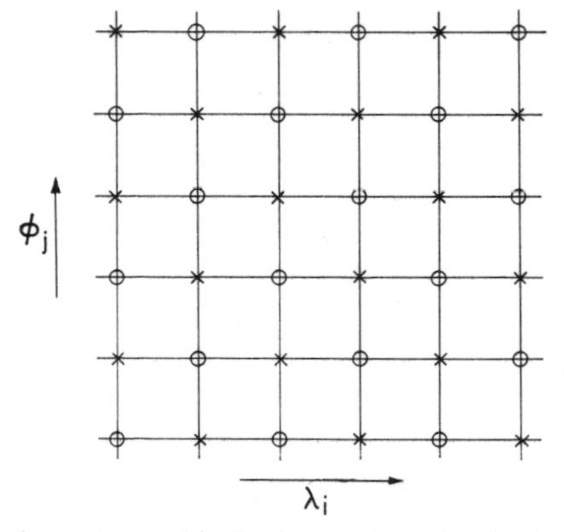


FIG. 3. Time staggered z-system grid. ○ denotes time level $n\Delta t$; × denotes time level $(n + 1)\Delta t$ or $(n - 1)\Delta t$.

circles poleward of 60° so that the distance between points is close to the distance at 60°. When neighbors $\Delta\phi$ above or below a point are needed for the latitudinal differences, they are obtained by linear interpolation between the two nearest points on the appropriate latitude line at the appropriate level in time.

The poles themselves are singular points in this coordinate system. The prognostic equations are not used at the poles since these points are singular in the spherical coordinate system. Inspection of the difference equations at points $\Delta\phi$ from the poles reveals that except in the diffusion terms the only quantity needed there is pressure. This is so because in all other terms the variables are multiplied by the factor $\cos\phi$ which is zero at the poles. We use the average of the pressure on the latitude circle next to the pole for the required pressure value at the pole. The diffusion terms need $\rho u$, $\rho v$, and $\rho q$ at the poles. These quantities are taken to be zero there for that calculation.

The vertical domain of the z-system model is divided into $K$ layers with

$$\Delta z = z_T/K \tag{3.5}$$

where $z_T$ denotes the height of the model top. Variables are staggered on the vertical mesh. Figure 4 shows the placement of variables in the vertical.

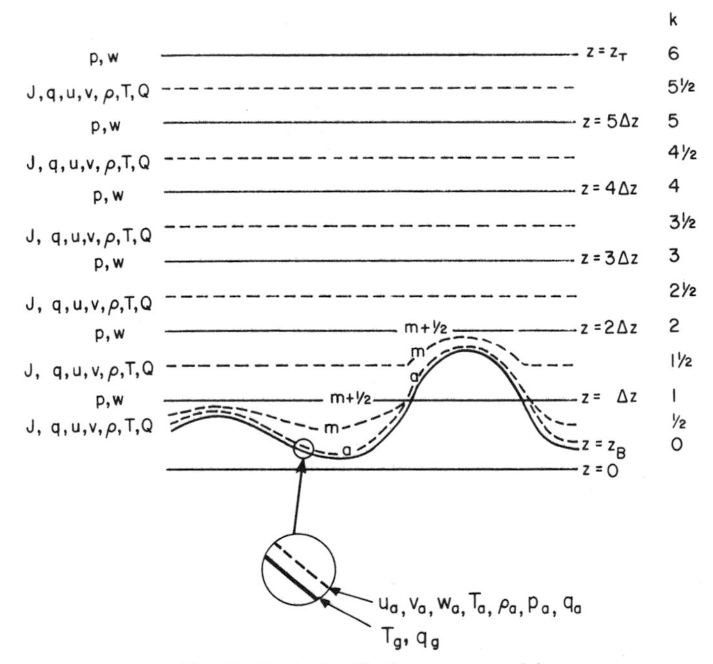


FIG. 4. Vertical grid of z-system model.

Pressure $p$ and vertical velocity $w$ are placed at integer $k$ levels while $u$, $v$, $\rho$, $T$, $q$, and $Q$ are at half-integer $k$ levels. At the lower boundary, an extra level denoted by subscript a, is placed at anemometer height (approximately 10 m). The earth's surface is denoted $z_B$ and variables at this level are distinguished by subscript g. The numerical values of $z_B$ are the surface height above mean sea level. It is extremely important that the g and a levels be conceptually distinct in the formulation of the surface boundary conditions. However, the difference in height between these two levels is so small compared to $\Delta z$ that it does not have computational significance. Therefore, in the difference equations we consider the a level to be at height $z_B$.

The height $z_B$ above mean sea level may exceed the grid interval $\Delta z$ (Fig. 4). When this occurs, variables are not computed at points where $z_k < z_B$. We assume $z_B \leqslant 2\Delta z$ in the specific details of the following description but this condition can be easily relaxed.

To incorporate the earth's orography into the governing equations, it is convenient to distinguish an additional level $m$ above the a level (Fig. 4). The half-integer mesh level immediately above the earth's surface is the $m$ level. The first horizontal mesh level immediately above the earth's surface is denoted as the $m + \frac{1}{2}$ level. The height of the $m + \frac{1}{2}$ level is $(m + \frac{1}{2})\,\Delta z$, where $m$ can be $\frac{1}{2}$ or $1\frac{1}{2}$ depending on whether $z_B < \Delta z$ or $\Delta z \leqslant z_B < 2\Delta z$. The height of the $m$ level is $z = (\Delta z + Z_B)/2$ for $z_B < \Delta z$ and $z = (2\Delta z + z_B)/2$ for $\Delta z \leqslant z_B < 2\Delta z$. Note that the $m$ level is not horizontal. This distinction is made because the momentum equations take a slightly different form at the first level above the ground in order to incorporate the lower boundary condition (Section III, C, 1).

Subscripts $i$, $j$, and $k$ denote a grid point. In addition, superscript $n$ denotes the time, $t = n\,\Delta t$, where $\Delta t$ is the discrete time increment. Thus, for any variable we write

$$\psi^n_{ijk} = \psi(\lambda_i, \phi_j, z_k, n\,\Delta t). \tag{3.6}$$

2. *s-System Model*

The discrete grid for the s-system model is uniform in the $\lambda$, $\phi$, $s$-coordinate system. The horizontal grid is shifted with respect to the grid for the z-system model in latitudinal direction by $\Delta\phi/2$ so that there are no grid points at the poles or on the equator. Let $I$ and $J$, both even, be the number of grid points in the longitude and latitude directions, respectively. The horizontal grid intervals are then

$$\Delta\lambda = 2\pi/I \tag{3.7}$$

$$\Delta\phi = \pi/J \tag{3.8}$$

and the longitudes and latitudes of the grid points are

$$\lambda_i = -\pi + (i - 1)\,\Delta\lambda; \qquad i = 1, 2, \ldots, I \tag{3.9}$$

$$\phi_j = -\pi/2 + (j - \tfrac{1}{2})\,\Delta\phi; \qquad j = 1, 2, \ldots, J. \tag{3.10}$$

In the s-system model, the variables are not staggered on the horizontal mesh; all variables are defined at each grid point each time step. The staggered mesh is not used in this version because fourth-order approximations are used for the horizontal derivatives and thus the variables are needed at time $n$ at $\lambda_{i\pm 1}$, $\lambda_{i\pm 2}$, $\phi_{j\pm 1}$, and $\phi_{j\pm 2}$. In the staggered mesh, the values are not known at all these points at the same time.

The horizontal mesh is not modified near the poles in the s-system model as it is in the z-system model. Instead, the high wave number spectral components of the prognostic variables are filtered in the longitudinal direction to remove the short wavelength linearly unstable modes for the chosen time step. This filtering is discussed in Section III, D.

For the transformed height coordinate, it is required that $s = 0$ at the bottom boundary ($z_B$) and $s = 1$ at the top horizontal boundary ($z_T$). Let $K$ denote the number of layers in the vertical, then

$$\Delta s = 1/K. \tag{3.11}$$

The finite difference mesh (Fig. 5) is staggered in the vertical as in the z-system model. Note that $z$ and $\partial z/\partial s$ are known at all levels. As before, we use

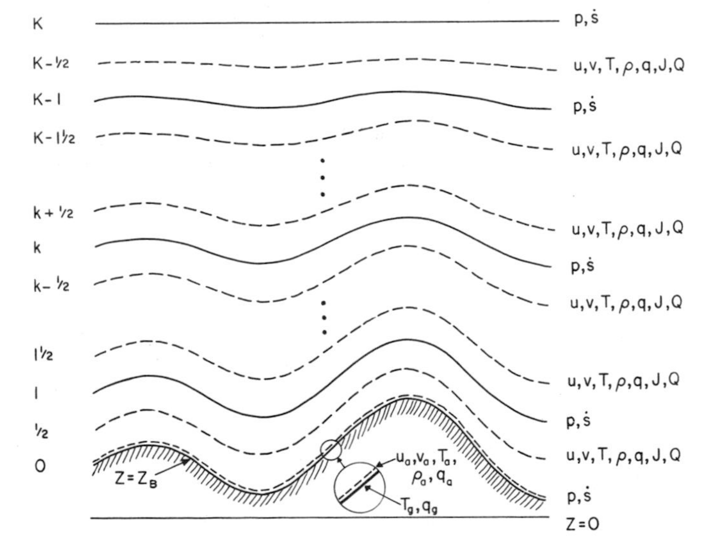


FIG. 5. Vertical grid of s-system model.

subscript a to denote variables at anemometer height and g at the ground $s_B$, but we have no need to distinguish an *m* level.

Subscripts *i*, *j*, and *k* and superscript *n* again denote a grid point. For any variable, we write

$$\psi_{ijk}^n = \psi(\lambda_i, \phi_j, s_k, n\,\Delta t). \tag{3.12}$$

## B. Discrete Operators

In listing the discrete operators used in the models, we adopt the usual convention that any missing subscripts are meant to be *i*, *j*, *k*, or *n* so that, for example, $\psi_{j+1}$ denotes $\psi_{i,\,j+1,\,k}^n$, etc.

The discrete operators involving time are

$$\delta_{2t}\psi = (\psi^{n+1} - \psi^{n-1})/2\Delta t \tag{3.13}$$

$$\overline{\psi}^{2t} = \tfrac{1}{2}(\psi^{n+1} + \psi^{n-1}). \tag{3.14}$$

Those involving longitude and latitude are

$$\delta_{2\lambda}\psi = (\psi_{i+1} - \psi_{i-1})/2\Delta\lambda \tag{3.15}$$

$$\delta_{4\lambda}\psi = (\psi_{i+2} - \psi_{i-2})/4\Delta\lambda \tag{3.16}$$

$$\overline{\psi}^{2\lambda} = \tfrac{1}{2}(\psi_{i+1} + \psi_{i-1}) \tag{3.17}$$

$$\overline{\psi}^{4\lambda} = \tfrac{1}{2}(\psi_{i+2} + \psi_{i-2}) \tag{3.18}$$

$$\delta_{2\phi}\psi = (\psi_{j+1} - \psi_{j-1})/2\Delta\phi \tag{3.19}$$

$$\delta_{4\phi}\psi = (\psi_{j+2} - \psi_{j-2})/4\Delta\phi \tag{3.20}$$

$$\overline{\psi}^{2\phi} = \tfrac{1}{2}(\psi_{j+1} + \psi_{j-1}) \tag{3.21}$$

$$\overline{\psi}^{4\phi} = \tfrac{1}{2}(\psi_{j+2} + \psi_{j-2}). \tag{3.22}$$

The vertical operators are

$$\delta_s\psi = (\psi_{k+1/2} - \psi_{k-1/2})/\Delta s \tag{3.23}$$

$$\overline{\psi}^s = \tfrac{1}{2}(\psi_{k+1/2} + \psi_{k-1/2}). \tag{3.24}$$

In the z system

$$\delta_z\psi = (\psi_{k+1/2} - \psi_{k-1/2})/\Delta z \tag{3.25}$$

$$\overline{\psi}^z = \tfrac{1}{2}(\psi_{k+1/2} - \psi_{k-1/2}). \tag{3.26}$$

We also use a $\delta_z$ in the s system for some physical approximations. In this case, we define

$$\delta_z \psi = (\psi_{k+1/2} - \psi_{k-1/2})/(z_{k+1/2} - z_{k-1/2}) \tag{3.27}$$

which is just the same as (3.25) with $\Delta z = z_{k+1/2} - z_{k-1/2}$.

In the following sections, when the mesh operators are applied at arbitrary grid points, some indices are out of the range of their definition. The rules below are used to evaluate an arbitrary quantity $\psi$ with such an index.

If $i < 1$ or $i > I$, $\psi$ is evaluated using longitudinal periodicity

$$\begin{aligned} \psi_i &= \psi_{I+i} \quad \text{for} \quad i < 1 \\ \psi_i &= \psi_{i-I} \quad \text{for} \quad i > I \end{aligned} \tag{3.28}$$

If $j < 1$ or $j > J$, $\psi$ is evaluated using points across the pole following Williamson and Browning (1973). This method is used in the s-system model only. The z-system model finite differences do not need values across the pole.

$$\begin{aligned} \psi_{ij} &= \pm \psi_{i+I/2,\, 1-j} \quad \text{for} \quad j < 1 \\ \psi_{ij} &= \pm \psi_{i+I/2,\, 2J+1-j} \quad \text{for} \quad j > J \end{aligned} \tag{3.29}$$

The positive sign is used for scalar quantities such as $p$ and the negative sign for the vector components $u$ and $v$ and $\cos \phi$. If $k < 1$ or $k > K$, the quantity $\psi$ can be arbitrary in the dynamical equations since it is always multiplied by zero in these cases. Other cases arise in the physical processes and the necessary boundary conditions will be given in the appropriate sections as needed.

## C. Discrete Equations

### 1. *z-System Model*

We consider first the discrete momentum equations for the z-system model. The prognostic equations are modified at the first level above the surface ($m$ level) to include explicitly the kinematic lower boundary condition (2.19). The method essentially considers the integral of the equations over the first layer from $z_B$ to $z_{m+1/2}$. This has been discussed in detail in Kasahara and Washington (1971); however, we briefly discuss it here so that it can be compared with the s-system model.

Consider first the integrated form of the three-dimensional mass divergence defined as

$$I = \int_{z_B}^{z_{m+1/2}} \left[ \nabla \cdot (\rho \mathbf{V}) + \frac{\partial(\rho w)}{\partial z} \right] dz \tag{3.30}$$

where for now the limits of integration $z_B$ and $z_{m+1/2}$ are considered functions of $\lambda$ and $\phi$. If we reverse the order of differentiation and integration in (3.30), noting that the integration limits can be functions of $\lambda$ and $\phi$, we obtain

$$\begin{aligned} I = \nabla \cdot \int_{z_B}^{z_{m+1/2}} \rho \mathbf{V}\, dz - (\rho \mathbf{V})_{m+1/2} \cdot \nabla z_{m+1/2} \\ + (\rho \mathbf{V})_a \cdot \nabla z_B + (\rho w)_{m+1/2} - (\rho w)_a. \end{aligned} \tag{3.31}$$

Referring to Fig. 4, we define the variable thickness of the first layer due to orography to be $\sigma$ so that $\sigma = z_{m+1/2} - z_B$. Note that both $\sigma$ and $z_B$ are functions of $\lambda$ and $\phi$ but $z_{m+1/2}$ is not. Therefore, $\nabla z_{m+1/2} = 0$ in (3.31). From the boundary condition (2.19), $I$ reduces to just the first and fourth terms on the right-hand side of (3.31), which can be approximated by

$$I = \nabla \cdot (\sigma \rho \mathbf{V})_m + (\rho w)_{m+1/2}. \tag{3.32}$$

If $I$ is substituted into the vertical integral of the horizontal momentum equations (2.5) and (2.6), the pressure tendency equation (2.8), the moisture equation (2.10), and the Richardson equation (2.16), we obtain the equations from which the discrete z-system equations at the $m$ level given below are derived. It should be pointed out that the z-system modifications for orography are the same in principle as the more general s system, except that the transformation is confined to just the lowest layer in the model.

Using the operators (3.13)–(3.26), momentum equations are written at the first level above the surface, the $m$ level, as

$$\begin{aligned} \delta_{2t}(\rho u)_m = -\frac{1}{\sigma}\Bigg\{ & \frac{1}{a \cos\phi} [\delta_{2\lambda}(\sigma \rho u^2)_m + \delta_{2\phi}(\sigma \rho u v \cos\phi)_m] \\ & + \tfrac{1}{2}[\overline{(\rho u)}_m^{2\lambda 2\phi} + \overline{(\rho u)}_{m+1}^{2\lambda 2\phi}] w_{m+1/2} \Bigg\} - \frac{1}{a\cos\phi} \tfrac{1}{2}[\delta_{2\lambda} p_{m+1/2} + \delta_{2\lambda}\hat{p}] \\ & + \left( 2\Omega \sin\phi + \frac{\tan\phi}{a} \overline{u}^{2\lambda 2\phi} \right) \overline{\rho v}^{2t} + F_\lambda^{n-1} \end{aligned} \tag{3.33}$$

and

$$\delta_{2t}(\rho v)_m = -\frac{1}{\sigma}\left\{\frac{1}{a\cos\phi}\left[\delta_{2\lambda}(\sigma\rho uv)_m + \delta_{2\phi}(\sigma\rho v^2\cos\phi)_m\right]\right.$$

$$\left. + \tfrac{1}{2}\left[\overline{(\rho v)}_m^{2\lambda 2\phi} + \overline{(\rho v)}_{m+1}^{2\lambda 2\phi}\right]w_{m+1/2}\right\} - \frac{1}{a}\tfrac{1}{2}\left[\delta_{2\phi}p_{m+1/2} + \delta_{2\phi}\hat{p}\right]$$

$$-\left(2\Omega\sin\phi + \frac{\tan\phi}{a}\,\overline{u}^{2\lambda 2\phi}\right)\overline{\rho u}^{2t} + F_\phi^{n-1}. \qquad (3.34)$$

Quantity $\hat{p}$ is introduced in (3.33) and (3.34) to evaluate the horizontal pressure gradient at the a level. This is necessary because, in general, the a level is not horizontal.

Consider the evaluation of the pressure gradients at point 0 in Fig. 6. Point 1 and 2 refer to the location of two immediately neighboring points in either the east–west or north–south directions. $\hat{p}$ denotes an interpolated pressure at point 1 at height $z_B(0)$ or possibly an extrapolated pressure at point 2 at height $z_B(0)$. The horizontal pressure gradient at height $z_B(0)$ at point 0 can easily be calculated once these $\hat{p}$ values are obtained. To evaluate $\hat{p}$ at points 1 and 2, it is convenient to define pressure $p_a^*$, which is the sea level pressure at $z = 0$, extrapolating from above assuming the same temperature in the mountain as immediately above, $T_m$. Quantity $p_a^*$ is actually stored in the model instead of the surface pressure $p_a$ since this minimizes conversions between the two quantities. The hydrostatic equation (2.11) written in the z system and the ideal gas law (2.12) can be combined to give

$$T = -(g/R)(\partial z/\partial \ln p). \qquad (3.35)$$

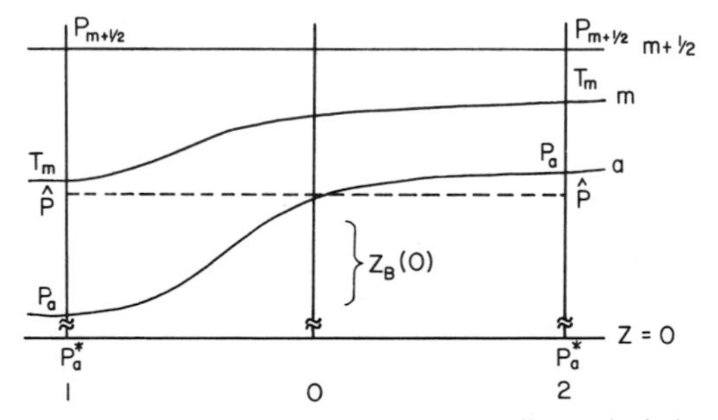


FIG. 6. Variable placement for horizontal pressure gradient calculation at the $m$ level.

Assuming $T$ to be constant (equal to $T_m$) and integrating the above equation from $z = 0$ to $z = z_B$ one obtains

$$p_a^* = p_a \exp(gz_B/RT_m) \tag{3.36}$$

Due to the assumption that the temperature in the mountain is equal to $T_m$, there is a linear relationship between ln $p$ and $z$. At point 1 or 2, we have

$$\frac{\ln \hat{p} - \ln p_a^*}{(z_B)_0} = \frac{\ln p_{m+1/2} - \ln p_a^*}{(m + 1/2)\,\Delta z}. \tag{3.37}$$

Rearranging terms and exponentiating, we then have

$$\hat{p} = p_a^*(p_{m+1/2}/p_a^*)^{(z_B)_0/(m+1/2)\,\Delta z}. \tag{3.38}$$

At all the levels above the $m$ level, the finite-difference forms of the momentum equations are

$$\begin{aligned}\delta_{2t}(\rho u) = & -\frac{1}{a\cos\phi}[\delta_{2\lambda}(\rho u^2) + \delta_{2\phi}(\rho uv\cos\phi)] \\ & - \delta_z(\overline{\rho u}^{2\lambda 2\phi z} w) + \left(2\Omega\sin\phi + \frac{\tan\phi}{a}\,\overline{u}^{2\lambda 2\phi}\right)\overline{\rho v}^{2t} \\ & - \frac{1}{a\cos\phi}\,\delta_{2\lambda}(\overline{p}^z) + F_\lambda^{n-1}\end{aligned} \tag{3.39}$$

and

$$\begin{aligned}\delta_{2t}(\rho v) = & -\frac{1}{a\cos\phi}[\delta_{2\lambda}(\rho uv) + \delta_{2\phi}(\rho v^2\cos\phi)] \\ & - \delta_z(\overline{\rho v}^{2\lambda 2\phi z} w) - \left(2\Omega\sin\phi + \frac{\tan\phi}{a}\,\overline{u}^{2\lambda 2\phi}\right)\overline{\rho u}^{2t} \\ & - \frac{1}{a}\,\delta_{2\phi}(\overline{p}^z) + F_\phi^{n-1}.\end{aligned} \tag{3.40}$$

Examination of (3.33) and (3.34) or (3.39) and (3.40) shows that these equations are implicit in $(\rho u)^{n+1}$ and $(\rho v)^{n+1}$ because of the time averaging of the Coriolis terms. These systems of equations have the general form

$$\begin{aligned}\frac{1}{2\Delta t}[(\rho u)^{n+1} - (\rho u)^{n-1}] &= E_1 + E_3[(\rho v)^{n+1} + (\rho v)^{n-1}] \\ \frac{1}{2\Delta t}[(\rho v)^{n+1} - (\rho v)^{n-1}] &= E_2 - E_3[(\rho u)^{n+1} + (\rho u)^{n-1}]\end{aligned} \tag{3.41}$$

Rearranging gives

$$\begin{aligned}-E_3(\rho v)^{n+1} + \frac{1}{2\Delta t}(\rho u)^{n+1} &= \frac{1}{2\Delta t}(\rho u)^{n-1} + E_1 + E_3(\rho v)^{n-1} \\ \frac{1}{2\Delta t}(\rho v)^{n+1} + E_3(\rho u)^{n+1} &= \frac{1}{2\Delta t}(\rho v)^{n-1} + E_2 - E_3(\rho u)^{n-1}\end{aligned} \tag{3.42}$$

which can be solved for $(\rho u)^{n+1}$ and $(\rho v)^{n+1}$ by Cramer's rule.

The discrete form of the pressure tendency equation at the a level or surface is

$$\begin{aligned}\delta_{2t}p_{\mathrm{a}} = \left(\frac{\partial p}{\partial t}\right)_{\mathrm{T}} - \frac{g}{a\cos\phi}\Bigg\{&\delta_{2\lambda}[\sigma(\rho u)_m] + \delta_{2\phi}[\sigma(\rho v\cos\phi)_m] \\ &+ \Delta z \sum_{l=m+1}^{K-1/2}[\delta_{2\lambda}(\rho u)_l + \delta_{2\phi}(\rho v\cos\phi)_l]\Bigg\}.\end{aligned} \tag{3.43}$$

The finite difference form of $(\partial p/\partial t)_{\mathrm{T}}$ is given with the diagnostic equations. At levels $k \geqslant m + \frac{1}{2}$ above the surface, the pressure tendency equation is written

$$\begin{aligned}\delta_{2t}p_k = \left(\frac{\partial p}{\partial t}\right)_{\mathrm{T}} &+ g\overline{\rho}_k^{2\lambda 2\phi z} w_k \\ &- \frac{g\,\Delta z}{a\cos\phi}\sum_{l=k+1/2}^{K-1/2}[\delta_{2\lambda}(\rho u)_l + \delta_{2\phi}(\rho v\cos\phi)_l].\end{aligned} \tag{3.44}$$

The remaining prognostic equation is the moisture equation. At the first level above the surface—the $m$ level—Eq. (2.10) takes the form

$$\begin{aligned}\delta_{2t}(\rho q)_m = -\frac{1}{\sigma}\Bigg\{&\frac{1}{a\cos\phi}[\delta_{2\lambda}(\sigma\rho q u)_m + \delta_{2\phi}(\sigma\rho q v\cos\phi)_m] \\ &+ \overline{(\rho q)}_{m+1/2}^{2\lambda 2\phi z} w_{m+1/2}\Bigg\} + M_m + \rho E_m^{n-1} + (CT)_m.\end{aligned} \tag{3.45}$$

For the upper levels—those above $m$—the finite-difference form of the moisture equation is

$$\delta_{2t}(\rho q)_k = -\frac{1}{a \cos \phi}[\delta_{2\lambda}(\rho q u)_k + \delta_{2\phi}(\rho q v \cos \phi)_k] - \delta_z(\overline{\rho q}^{2\lambda 2\phi z} w)_k + M_k + \rho E_k^{n-1} + (CT)_k. \tag{3.46}$$

The discrete forms of the condensation term $M$ and diffusion term $E$ are described later in Section III, C, 3, while the cumulus transport CT is described earlier in Section II, D.

The remaining variables in the model are calculated from diagnostic relations. To calculate $w$ from the Richardson equation, we first calculate term $J$. The finite difference form for $J$ at the a level is

$$J_{\mathrm{a}} = \frac{1}{a \cos \phi}[\overline{u}_a^{2\lambda} \delta_{2\lambda} \hat{p} + \overline{v_a \cos \phi}^{2\phi}\, \delta_{2\phi} \hat{p}] - g(\rho w)_a - \frac{g}{a \cos \phi}\Bigg\{\delta_{2\lambda}[\sigma(\rho u)_m] + \delta_{2\phi}[\sigma(\rho v \cos \phi)_m] + \Delta z \sum_{l=m+1}^{K-1/2} [\delta_{2\lambda}(\rho u)_l + \delta_{2\phi}(\rho v \cos \phi)_l]\Bigg\} \tag{3.47}$$

for level $K > k \geqslant m + \frac{1}{2}$, the finite-difference form is

$$J_k = \frac{1}{a \cos \phi}[\overline{u}_k^{2\lambda z}\, \delta_{2\lambda} p_k + \overline{(v \cos \phi)}_k^{2\phi z}\, \delta_{2\phi} p_k] - \frac{g\,\Delta z}{a \cos \phi} \sum_{l=k+1/2}^{K-1/2} [\delta_{2\lambda}(\rho u)_l + \delta_{2\phi}(\rho v \cos \phi)_l]. \tag{3.48}$$

At the top $K$, the $J$ term reduces to

$$J_K = \frac{1}{a \cos \phi}[\overline{u}_{K-1/2}^{2\lambda}\, \delta_{2\lambda} p_K + \overline{v_{K-1/2} \cos \phi}^{2\phi}\, \delta_{2\phi} p_K]. \tag{3.49}$$

The Richardson equation for $w$ at each level $m + \frac{1}{2} \leqslant k < K$ is written

$$w_k = -\sum_{l=m}^{k-1/2} \frac{1}{a \cos \phi}[\delta_{2\lambda}(\sigma_l u_l) + \delta_{2\phi}(\sigma_l v_l \cos \phi)] + \sum_{l=m}^{k-1/2} \left\{\left[\left(\frac{\partial p}{\partial t}\right)_{\mathrm{T}} + J_l^z\right] \Big/ (\gamma \overline{p}_l^{2\lambda 2\phi z}) - \frac{1}{c_{\mathrm{p}}} \frac{Q^{n-1}}{T_l}\right\} \sigma_l \tag{3.50}$$

where it is understood that $m - \frac{1}{2} = a$. $\sigma_l$ is defined as

$$\sigma_l = \begin{cases} \Delta z - z_B & \text{if } l = m \text{ and } z_B < \Delta z \\ 2\Delta z - z_B & \text{if } l = m \text{ and } \Delta z \leqslant z_B < 2\Delta z \\ \Delta z & \text{otherwise.} \end{cases} \tag{3.51}$$

If we extend the sums in (3.50) to $k = K$, then $w_K$ must vanish due to the upper boundary condition. The resulting equation can be solved for the unknown $(\partial p/\partial t)_T$ which is needed in the Richardson equation (3.50) and the pressure tendency equation (3.44).

State variables $T$ and $\rho$ are obtained diagnostically from $p$. At the a level, density is determined using the equation of state (2.12)

$$\rho_a = p_a/RT_a \tag{3.52}$$

where $T_a$ is determined from the lower boundary condition. At upper levels, density is computed from the hydrostatic equation (2.11). At the $m$ level,

$$\rho_m = -(p_{m+1/2} - p_a)/g\sigma \tag{3.53}$$

where $\sigma$ is defined by (3.51). At half-integer levels above the $m$ level,

$$\rho_k = -(1/g)\, \delta_z p_k. \tag{3.54}$$

The anemometer and surface level computations of temperature are described in Section II, C. At levels above the surface, temperature is computed by combining the hydrostatic and state equations (2.11) and (2.12). At the $m$ level,

$$T_m = g\sigma/R \ln (p_a/p_{m+1/2}) \tag{3.55}$$

where again $\sigma$ is defined by (3.51). At half-integer levels above the $m$ level,

$$T_k = g\, \Delta z/R \ln (p_{k-1/2}/p_{k+1/2}). \tag{3.56}$$

### 2. *s-System Model*

The discrete equations for the horizontal wind components in the s-system model are obtained from the advective form of Eqs. (2.1) and (2.2). The horizontal derivatives are approximated by fourth-order centered differences, yielding

$$
\begin{aligned}
\delta_{2t}u = &-\frac{u}{a\cos\phi}\left(\frac{4}{3}\delta_{2\lambda}-\frac{1}{3}\delta_{4\lambda}\right)u-\frac{v}{a}\left(\frac{4}{3}\delta_{2\phi}-\frac{1}{3}\delta_{4\phi}\right)u \\
&-\delta_s(\dot{s}\overline{u}^s)+u\delta_s(\dot{s})+\left(f+\frac{u}{a}\tan\phi\right)\overline{v}^{2t} \\
&-\frac{R}{a\cos\phi}\left(\frac{4}{3}\overline{T}^{2\lambda}\,\delta_{2\lambda}\ln p-\frac{1}{3}\overline{T}^{4\lambda}\,\delta_{4\lambda}\ln p\right) \\
&-\frac{g}{a\cos\phi}\left(\frac{4}{3}\delta_{2\lambda}-\frac{1}{3}\delta_{4\lambda}\right)z+\frac{1}{\rho}F_\lambda^{n-1}
\end{aligned}
\tag{3.57}
$$

$$
\begin{aligned}
\delta_{2t}v = &-\frac{u}{a\cos\phi}\left(\frac{4}{3}\delta_{2\lambda}-\frac{1}{3}\delta_{4\lambda}\right)v-\frac{v}{a}\left(\frac{4}{3}\delta_{2\phi}-\frac{1}{3}\delta_{4\phi}\right)v \\
&-\delta_s(\dot{s}\overline{v}^s)+v\delta_s(\dot{s})-\left(f+\frac{u}{a}\tan\phi\right)\overline{u}^{2t} \\
&-\frac{R}{a}\left(\frac{4}{3}\overline{T}^{2\phi}\,\delta_{2\phi}\ln p-\frac{1}{3}\overline{T}^{4\phi}\,\delta_{4\phi}\ln p\right) \\
&-\frac{g}{a}\left(\frac{4}{3}\delta_{2\phi}-\frac{1}{3}\delta_{4\phi}\right)z+\frac{1}{\rho}F_\phi^{n-1}.
\end{aligned}
\tag{3.58}
$$

Note that (3.57) and (3.58) use a pressure value at the same level as $u$ and $v$. These auxiliary values are defined later, along with $T$ and $\rho$ at that level. As with the z-system model, the Coriolis term is averaged in time so that the equations are implicit in $u^{n+1}$ and $v^{n+1}$ at point $(i, j)$. The method of solution for $u$ and $v$ is similar to that of the $z$-system model (3.41) and (3.42).

The pressure tendency equation takes the discrete form

$$
\begin{aligned}
\delta_{2t}p_k = &\left(\frac{\partial p}{\partial t}\right)_{\mathrm{T}}+g\overline{\rho_k^*}^s\dot{s}_k \\
&-g\,\Delta s\sum_{l=k+1/2}^{K-1/2}\frac{1}{a\cos\phi}\left[\left(\frac{4}{3}\delta_{2\lambda}-\frac{1}{3}\delta_{4\lambda}\right)\rho_l^*u_l\right. \\
&\left.+\left(\frac{4}{3}\delta_{2\phi}-\frac{1}{3}\delta_{4\phi}\right)\rho_l^*v_l\cos\phi\right].
\end{aligned}
\tag{3.59}
$$

The prognostic equation for specific humidity is

$$\delta_{2t}q = -\frac{u}{a\cos\phi}\left(\frac{4}{3}\delta_{2\lambda} - \frac{1}{3}\delta_{4\lambda}\right)q - \frac{v}{a}\left(\frac{4}{3}\delta_{2\phi} - \frac{1}{3}\delta_{4\phi}\right)q$$

$$- \delta_s(\dot{s}\overline{q}^s) + q\,\delta_s\dot{s} + \frac{1}{\rho}M + \frac{1}{\rho}(\rho E)^{n-1}. \tag{3.60}$$

The discrete condensation term $M$ and diffusion term $E$ are described later.

The $J$ term needed in the Richardson equation for $\dot{s}$ and in the equation for the pressure change at the top of the model is

$$J_k = \frac{u_k}{a\cos\phi}\left(\frac{4}{3}\delta_{2\lambda} - \frac{1}{3}\delta_{4\lambda}\right)p_k + \frac{v_k}{a}\left(\frac{4}{3}\delta_{2\phi} - \frac{1}{3}\delta_{4\phi}\right)p_k$$

$$- g\,\Delta s \sum_{l=k}^{K-1/2} \alpha_{kl}\frac{1}{a\cos\phi}\left[\left(\frac{4}{3}\delta_{2\lambda} - \frac{1}{3}\delta_{4\lambda}\right)(\rho_l^* u_l)\right.$$

$$\left. + \left(\frac{4}{3}\delta_{2\phi} - \frac{1}{3}\delta_{4\phi}\right)(\rho_l^* v_l\cos\phi)\right] \tag{3.61}$$

where

$$\alpha_{kl} = \begin{cases} 1 & \text{for} \quad l > k \\ \frac{1}{2} & \text{for} \quad l = k. \end{cases} \tag{3.62}$$

Recall that $J$ is defined at the half-integer levels so that $k$ and $l$ are half integers in (3.61). The $\frac{1}{2}$ factor enters in the vertical sum in $J$ because the first term represents only half a layer. Using the $J$ term, the Richardson equation is written

$$\dot{s}_k = \left\{ -\sum_{l=1/2}^{k-1/2} \frac{1}{a\cos\phi}\left[\left(\frac{4}{3}\delta_{2\lambda} - \frac{1}{3}\delta_{4\lambda}\right)u_l + \left(\frac{4}{3}\delta_{2\phi} - \frac{1}{3}\delta_{4\phi}\right)(v_l\cos\phi)\right]\left(\frac{\partial z}{\partial s}\right)_l \right.$$

$$- \sum_{l=1/2}^{k-1/2}\left[\frac{u_l}{a\cos\phi}\left(\frac{4}{3}\delta_{2\lambda} - \frac{1}{3}\delta_{4\lambda}\right) + \frac{v_l}{a}\left(\frac{4}{3}\delta_{2\phi} - \frac{1}{3}\delta_{4\phi}\right)\right]\left(\frac{\partial z}{\partial s}\right)_l$$

$$\left. - \frac{1}{\gamma}\sum_{l=1/2}^{k-1/2}\left[\left(\frac{\partial p}{\partial t}\right)_T + J_l\right]\left(\frac{\partial z}{\partial s}\right)_l \Big/ \overline{p}_l^s + \frac{1}{c_p}\sum_{l=1/2}^{k-1/2}\frac{Q_l^{n-1}}{T_l^{n-1}}\left(\frac{\partial z}{\partial s}\right)_l \right\} \Delta s\left(\frac{\partial z}{\partial s}\right)_k^{-1}. \tag{3.63}$$

The pressure change at the top is obtained from the discrete Richardson equation (3.63) by setting $\dot{s}_K = 0$. The result is

$$\left(\frac{\partial p}{\partial t}\right)_{\mathrm{T}} = \left\{\frac{1}{\gamma}\sum_{l=1/2}^{K-1/2}\frac{J_l}{\bar{p}_l^s}\left(\frac{\partial z}{\partial s}\right)_l - \frac{1}{c_{\mathrm{p}}}\sum_{l=1/2}^{K-1/2}\frac{Q_l^{n-1}}{T_l^{n-1}}\left(\frac{\partial z}{\partial s}\right)_l\right.$$
$$+\sum_{l=1/2}^{K-1/2}\frac{1}{a\cos\phi}\left[\left(\frac{4}{3}\delta_{2\lambda} - \frac{1}{3}\delta_{4\lambda}\right)u_l + \left(\frac{4}{3}\delta_{2\phi} - \frac{1}{3}\delta_{4\phi}\right)(v_l\cos\phi)\right]$$
$$\times\left(\frac{\partial z}{\partial s}\right)_l + \sum_{l=1/2}^{K-1/2}\left[\frac{u_l}{a\cos\phi}\left(\frac{4}{3}\delta_{2\lambda} - \frac{1}{3}\delta_{4\lambda}\right) + \frac{v_l}{a}\left(\frac{4}{3}\delta_{2\phi} - \frac{1}{3}\delta_{4\phi}\right)\right]$$
$$\left.\times\left(\frac{\partial z}{\partial s}\right)_l\right\} \times \left[-\frac{1}{\gamma}\sum_{l=1/2}^{K-1/2}\frac{1}{\bar{p}_l^s}\left(\frac{\partial z}{\partial s}\right)_l\right]^{-1} \tag{3.64}$$

The state variables at midlayers (half-integer levels) are determined from adjacent pressures (at integer levels) by assuming the layer is isothermal. Pressure as a function of height is then

$$p = p_k \exp\left[-g(z - z_k)/RT_k\right] \tag{3.65}$$

in the layer $p_{k-1/2} \leqslant p \leqslant p_{k+1/2}$, where $k$ is a half integer. $T_k$ is the constant temperature of the layer, $z_k$ is the height at midlayer, and $p_k$ is the pressure at midlayer. The two unknowns $p_k$ and $T_k$ can be found by solving the two equations obtained by applying (3.65) to $p_{k+1/2}$ at $z_{k+1/2}$ and $p_{k-1/2}$ at $z_{k-1/2}$, which are known from the pressure tendency equation.

$$T_k = \frac{g(z_{k+1/2} - z_{k-1/2})}{R\ln(p_{k-1/2}/p_{k+1/2})} \tag{3.66}$$

$$p_k = p_{k-1/2}{}^{(z_{k+1/2}-z_k)/(z_{k+1/2}-z_{k-1/2})}\,p_{k+1/2}{}^{(z_k-z_{k-1/2})/(z_{k+1/2}-z_{k-1/2})} \tag{3.67}$$

where, again, $k$ is a half-integer. The density at the half-integer levels is obtained from the ideal gas law

$$\rho_k = p_k/RT_k \tag{3.68}$$

and

$$\rho_k^* = \rho_k(\partial z/\partial s)_k. \tag{3.69}$$

3. *Physical Processes*

*a. Subgrid Scale Diffusion.* The discrete form of the vertical component of momentum diffusion is

$$F_{\lambda \mathrm{v}}^{n-1} = \delta_z \tau_\lambda^{n-1} \tag{3.70}$$

$$F_{\phi \mathrm{v}}^{n-1} = \delta_z \tau_\phi^{n-1}. \tag{3.71}$$

This form is used in both the s- and z-system models using (3.25) or (3.27) for definition of the vertical difference operator. Reynolds stress is written

$$\tau_\lambda^{n-1} = \rho^{n-1} K_{\mathrm{MV}}^{n-1}\, \delta_z u^{n-1} \tag{3.72}$$

$$\tau_\phi^{n-1} = \rho^{n-1} K_{\mathrm{MV}}^{n-1}\, \delta_z v^{n-1} \tag{3.73}$$

where the vertical kinematic eddy viscosity $K_{\mathrm{MV}}^{n-1}$ is taken to be equal to the vertical kinematic thermal diffusivity $K_{\mathrm{TV}}^{n-1}$ given later. At the surface, we take

$$[\delta_z(u)]_0 = (u_{1/2} - u_{\mathrm{a}})/(z_{1/2} - z_{\mathrm{B}}) \tag{3.74}$$

for the s-system model and

$$[\delta_z(u)]_0 = (u_{\mathrm{m}} - u_{\mathrm{a}})/(\sigma/2) \tag{3.75}$$

for the z-system model. At the top of both versions,

$$[\delta_z(u)]_K = 0 \tag{3.76}$$

from the upper boundary condition (2.17), (2.31), and (2.32). The vertical differences at the top and bottom boundaries for $v$ are the same as those for $u$ with $u$ replaced by $v$.

The vertical diffusion of water vapor for both versions of the model is given

$$(\rho E_{\mathrm{v}})^{n-1} = \delta_z[\bar{\rho}^z K_{\mathrm{wv}}\, \delta_z(q)]^{n-1}. \tag{3.77}$$

The vertical differences at the surface and top of the model are the same as (3.74)–(3.76) with $u$ replaced by $q$. We assume that $K_{\mathrm{WV}} = K_{\mathrm{MV}} = K_{\mathrm{TV}}$ which is given in (3.82).

The vertical diffusion of sensible heat is given by

$$Q_{\mathrm{DV}}^{n-1} = -(1/\rho^{n-1})(\partial h^{n-1}/\partial z) \tag{3.78}$$

where the vertical flux of sensible heat is expressed as

$$h_k^{n-1} = -c_{\mathrm{p}} K_{\mathrm{TV}}^{n-1} (\bar{\rho}^z)^{n-1}\, [(\partial\theta/\partial z)_k^{n-1} - \gamma_{\mathrm{CG}}]. \tag{3.79}$$

The discrete form of the vertical derivative of potential temperature is given by

$$\left(\frac{\partial\theta}{\partial z}\right)_k^{n-1} = \left(\frac{\tilde{p}_0}{p_k}\right)^{R/c_p} \left\{[\delta_z(T)]_k^{n-1} + \frac{g}{c_p}\right\}. \tag{3.80}$$

The vertical differences at the surface are given by (3.74)–(3.75) with $u$ replaced by $T$. At the top, the boundary condition $h = 0$ in (2.17) applies. The standard reference $\tilde{p}_0$ is given by (2.38). In the z-system model, the term $(\tilde{p}_0/p_k)^{R/c_p}$ is taken to be independent of longitude and latitude to simplify the computation by using an average value $\bar{p}_k$ instead of $p_k$. The values obtained from the U.S. Standard Atmosphere are given in Oliger *et al.* (1970). This simplification is not made in the s system because of the much greater variation in pressure along the s surface over mountains.

The countergradient in (3.79) is defined to be

$$\gamma_{CG} = \begin{cases} 5 \times 10^{-5} \quad {}^\circ\mathrm{K/cm} & \text{at} \quad z_B \\ 1 \times 10^{-5} \quad {}^\circ\mathrm{K/cm} & \text{elsewhere.} \end{cases} \tag{3.81}$$

The discrete form of $K_{TV}$ is

$$\begin{aligned} (K_{TV})_k^{n-1} &= d_k \left\{A_1 + A_2 \left[1.0 - \exp\left\{A_3\left[\left(\frac{\partial\theta}{\partial z}\right)_k - \gamma_{CG}\right]\right\}\right]\right\}^{n-1} \\ &\qquad \text{if} \quad \left(\frac{\partial\theta}{\partial z}\right)_k^{n-1} \leqslant \gamma_{CG} \\ (K_{TV})_k^{n-1} &= d_k\{A_1/[1.0 + A_4(R_i)_k] + A_5\}^{n-1} \\ &\qquad \text{if} \quad \left(\frac{\partial\theta}{\partial z}\right)_k^{n-1} > \gamma_{CG} \end{aligned} \tag{3.82}$$

where

$$\begin{aligned} d_k &= (z_{k+1/2} - z_{k-1/2})/\max_{ij}(z_{k+1/2} - z_{k-1/2}) \\ d_0 &= (z_{1/2} - z_B)/\max_{ij}(z_{1/2} - z_B) \\ A_1 &= 10^5 \quad \mathrm{cm^2\ sec^{-1}} \\ A_2 &= 10^6 \quad \mathrm{cm^2\ sec^{-1}} \\ A_3 &= 1.2 \times 10^5 \quad \mathrm{cm\ {}^\circ K^{-1}} \\ A_4 &= 40.00 \\ A_5 &= \begin{cases} 2.0 \times 10^4 \quad \mathrm{cm^2\ sec^{-1}} & \text{for} \quad k = 0 \\ 1.0 \times 10^3 \quad \mathrm{cm^2\ sec^{-1}} & \text{for} \quad k > 0 \end{cases} \\ A_6 &= 10^{-12}. \end{aligned} \tag{3.83}$$

Experience has shown it necessary to reduce the value of $K_{TV}$ when the thickness of the layer is reduced over the mountains. The factor $d$ has been added to (3.82) for this purpose. In (3.83), $\max_{ij}(z_{k+1/2} - z_{k-1/2})$ is the maximum value of $(z_{k+1/2} - z_{k-1/2})$ over all horizontal grid points at that level.

The discrete computation of the Richardson number used in (3.82) is

$$(R_i)_k = \frac{g/\overline{T}_k^z[(\partial\theta/\partial z)_k - \gamma_{CG}]}{(\delta_z u)_k^2 + (\delta_z v)_k^2 + A_6} \tag{3.84}$$

where the differences at the surface are given by (3.74) and (3.75) or their equivalent with $u$ replaced by $v$ or $T$ as appropriate, and at the top by (3.76) and (2.17). The average at the surface $\overline{T}_0$ is taken to be $T_a$.

We next give the horizontal diffusion in the s-system model where all variables are available at all grid points at time $n - 1$. We define an auxiliary set of grid points denoted $(i', j')$ halfway between those of the normal grid (Fig. 7). These auxiliary points are related to the normal grid by

$$(i', j') = (i \pm \tfrac{1}{2}, j \pm \tfrac{1}{2}). \tag{3.85}$$

$D_S$ and $D_T$ are defined at the auxiliary points by using the following discrete operators

$$\delta_\lambda \psi = (\psi_{i'+1/2} - \psi_{i'-1/2})/\Delta\lambda \tag{3.86}$$

$$\delta_\phi \psi = (\psi_{j'+1/2} - \psi_{j'-1/2})/\Delta\phi \tag{3.87}$$

$$\overline{\psi}^\lambda = \tfrac{1}{2}(\psi_{i'+1/2} + \psi_{i'-1/2}) \tag{3.88}$$

$$\overline{\psi}^\phi = \tfrac{1}{2}(\psi_{j'+1/2} + \psi_{j'-1/2}) \tag{3.89}$$

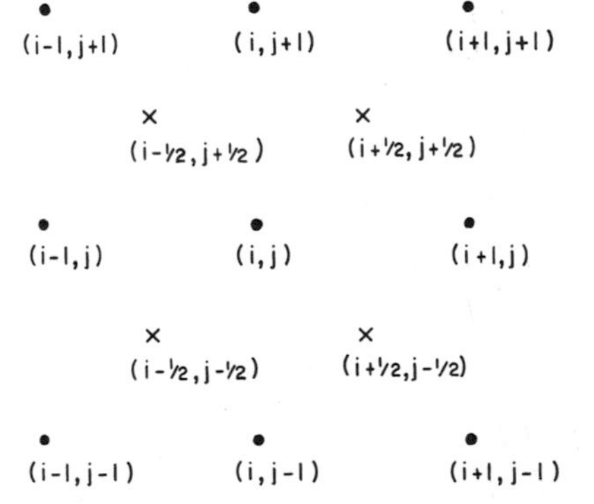


FIG. 7. Auxiliary grid points for horizontal diffusion calculation.

where, as before, only the incremented subscripts are written. Any missing subscripts are meant to be $i'$ or $j'$. Then at point $(i', j')$

$$D_{\mathrm{T}} = \frac{1}{a}\left[\overline{\left(\frac{\delta_{\lambda} u}{\cos\phi}\right)}^{\phi} - \frac{\overline{\delta_{\phi}(v|\cos\phi|)}^{\lambda}}{\overline{|\cos\phi|}^{\phi}}\right] \tag{3.90}$$

$$D_{\mathrm{S}} = \frac{1}{a}\left[\overline{\left(\frac{\delta_{\lambda} v}{\cos\phi}\right)}^{\phi} + \frac{\overline{\delta_{\phi}(u|\cos\phi|)}^{\lambda}}{\overline{|\cos\phi|}^{\phi}}\right] \tag{3.91}$$

$$(\rho K_{\mathrm{MH}})_{i'j'} = 2(k_0 l)^2 \overline{\rho}^{\lambda\phi} D_{i'j'} \tag{3.92}$$

where

$$D_{i'j'} = (D_{\mathrm{T}}{}^2 + D_{\mathrm{S}}{}^2)^{1/2} \tag{3.93}$$

The horizontal length scale $l$ is taken to be $a\Delta\phi$. Note that $D_{\mathrm{T}}$, $D_{\mathrm{S}}$, and $\rho K_{\mathrm{MV}}$ are defined at these auxiliary points $(i', j')$ but are derived from data at the original grid points $(i, j)$. The model is stable for short-range forecasts with $k_0$ as small as 0.1.

The horizontal friction terms are then written at the regular grid point $(i, j)$ as

$$F_{\lambda\mathrm{H}}^{n-1} = \left\{\frac{1}{a\cos\phi}\overline{\delta_{\lambda}[(\rho K_{\mathrm{MH}})D_{\mathrm{T}}]}^{\phi} + \frac{1}{a}\overline{\delta_{\phi}[(\rho K_{\mathrm{MH}})D_{\mathrm{S}}]}^{\lambda}\right\}^{n-1} \tag{3.94}$$

$$F_{\phi\mathrm{H}}^{n-1} = \left\{\frac{1}{a\cos\phi}\overline{\delta_{\lambda}[(\rho K_{\mathrm{MH}})D_{\mathrm{S}}]}^{\phi} - \frac{1}{a}\overline{\delta_{\phi}[(\rho K_{\mathrm{MH}})D_{\mathrm{T}}]}^{\lambda}\right\}^{n-1}. \tag{3.95}$$

The horizontal friction in the z-system model is very similar to that in the s system. The difference, necessitated by the latitude shift between the two grids, is the definition of $D_{\mathrm{T}}$ and $D_{\mathrm{S}}$. In the z-system model, these are

$$D_{\mathrm{T}} = \frac{1}{a\,\overline{\cos\phi}^{\phi}}\left[\overline{\delta_{\lambda} u}^{\phi} - \overline{\delta_{\phi}(v\cos\phi)}^{\lambda}\right] \tag{3.96}$$

$$D_{\mathrm{S}} = \frac{1}{a\,\overline{\cos\phi}^{\phi}}\left[\overline{\delta_{\lambda} v}^{\phi} + \overline{\delta_{\phi}(u\cos\phi)}^{\lambda}\right]. \tag{3.97}$$

Another difference is necessitated by the staggered grid in the z-system model. The winds used in (3.90) and (3.91) of the s-system model are all at time $n - 1$.

In the z-system model, these terms are not available at all grid points at $n-1$, so the time levels are mixed. If we are forcasting from time $n-1$ to $n+1$ at point $(i, j)$, then points of the form $(i \pm 1, j \pm 1)$ are at time $n-1$, and points of the form $(i \pm 1, j)$ and $(i, j \pm 1)$ are at time $n$. The values at these times are used in (3.96) and (3.97). The horizontal length scale $l$ in (3.92) is taken to be $a\Delta\lambda \cos\phi$ in the z-system model and the parameter $k_0$ is generally taken to be around 0.3 to 0.4. Unlike the s-system model, the z-system model is not stable for smaller values since, as discussed in Section III, D, this diffusion term is also used to control the time splitting. The s-system model has a separate time filter to control the time splitting.

The discrete form of the horizontal diffusion of moisture for the s-system model is

$$E_{\mathrm{H}} = \frac{1}{\rho}\frac{1}{a^2 \cos\phi}\left\{\frac{1}{\cos\phi}\overline{\delta_\lambda[(\rho K_{\mathrm{WH}})\,\overline{\delta_\lambda \overline{q}^{\phi}}]}^{\phi} + \overline{\delta_\phi[(\rho K_{\mathrm{WH}})\,\overline{\cos\phi}^{\phi}\,\overline{\delta_\phi \overline{q}^{\lambda}}]}^{\lambda}\right\}^{n-1} \tag{3.98}$$

where we take $(\rho K_{\mathrm{WH}}) = (\rho K_{\mathrm{MH}})$ as defined earlier. The same form holds for the z-system model except, as with the momentum, the time levels are mixed according to which are available with the staggered grid.

The horizontal diffusion of sensible heat for the s-system model has a similar form, namely

$$Q_{\mathrm{DH}} = \frac{1}{\rho}\frac{c_{\mathrm{p}}}{a^2 \cos\phi}\left\{\frac{1}{\cos\phi}\overline{\delta_\lambda[(\rho K_{\mathrm{TH}})\,\overline{\delta_\lambda \overline{\theta}^{\phi}}]}^{\phi} + \overline{\delta_\phi[(\rho K_{\mathrm{TH}})\,\overline{\cos\phi}^{\phi}\,\overline{\delta_\phi \overline{\theta}^{\lambda}}]}^{\lambda}\right\}^{n-1} \tag{3.99}$$

where potential temperature is computed from

$$\theta_k = (\tilde{p}_0/p_k)^{R/c_{\mathrm{p}}} T_k \tag{3.100}$$

and $\tilde{p}_0$ is the standard reference pressure (2.38). We also take $(\rho K_{\mathrm{TH}}) = (\rho K_{\mathrm{MH}})$. As with moisture, (3.99) also holds for the z system but with the time levels mixed.

For the fourth-degree diffusion used in the s-system model (2.54–2.57) we define $D_{\mathrm{T}}$ and $D_{\mathrm{S}}$ as before (3.90, 3.91) on the auxiliary grid. We also define a discrete $\nabla^2$ operator defined on the regular mesh

$$l^2 \nabla^2 \psi = \frac{l_\lambda^2}{a^2 \cos^2 \phi} \delta_\lambda \delta_\lambda \psi + \frac{l_\phi^2}{a^2 \cos \phi} \delta_\phi (\overline{\cos \phi}^\phi \delta_\phi \psi) \tag{3.101}$$

where

$$l_\lambda = a\,\Delta\lambda \cos \phi \qquad l_\phi = a\,\Delta\phi \tag{3.102}$$

The diffusion terms then take the form

$$F_{\lambda \mathrm{H}}^{n-1} = -[(l^2 \nabla^2)\overline{\rho K_{\mathrm{MH}}}^{\lambda\phi}(l^2 \nabla^2)]u^{n-1} \tag{3.103}$$

$$F_{\lambda\phi}^{n-1} = -[(l^2 \nabla^2)\overline{\rho K_{\mathrm{MH}}}^{\lambda\phi}(l^2 \nabla^2)]v^{n-1} \tag{3.104}$$

$$Q_{\mathrm{DH}}^{n-1} = -\frac{c_{\mathrm{p}}}{\rho^{n-1}}[(l^2 \nabla^2)\overline{\rho K_{\mathrm{TH}}}^{\lambda\phi}(l^2 \nabla^2)]\theta^{n-1} \tag{3.105}$$

$$E_{\mathrm{H}}^{n-1} = -\frac{1}{\rho^{n-1}}[(l^2 \nabla^2)\overline{\rho K_{\mathrm{WH}}}^{\lambda\phi}(l^2 \nabla^2)]q^{n-1} \tag{3.106}$$

where

$$K_{\mathrm{MH}} = 2k_0^2 D \tag{3.107}$$

$$D = (D_{\mathrm{T}}^2 + D_{\mathrm{S}}^2)^{1/2} \tag{3.108}$$

$$K_{\mathrm{TH}} = K_{\mathrm{WH}} = K_{\mathrm{MH}} \tag{3.109}$$

With second- or fourth-order finite differences, the s-system model is generally stable for $k_0$ as small as 0.1.

*b. Radiation.* The long-wave heating/cooling computation follows the method developed by Sasamori (1968a, b) based on a radiation chart approach using numerical approximations to the transfer equations. Sasamori's method is modified to include additional radiation by clouds. We describe the details of the radiation calculation as it appears in the s-system model. The principles are the same for the z-system model, but the details are slightly different (Oliger *et al.*, 1970). For radiation calculations, we add an additional isothermal layer extending from the model top to $p = 0$. Temperature and mixing ratio in this layer are assumed to be equal to $T_{K-1/2}$ and $q_{K-1/2}$, respectively. Figure 8 illustrates the variable placement in the

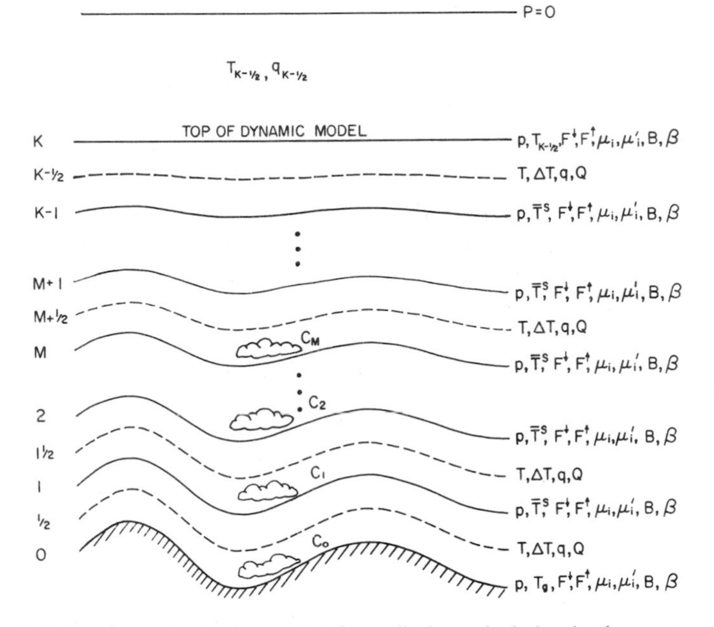


FIG. 8. Variable placement in the vertical for radiation calculation in the s-system model.

vertical for the radiation calculation and the additional isothermal layer at the top for the s-system model.

The models allow up to three absorbing gases—water vapor, carbon dioxide, and ozone, denoted by subscripts 1, 2, and 3, respectively. Index $N$ specifies the number of absorbing gases used in the calculation. The fluxes depend on the path lengths of each absorbing gas.

The pressure-corrected path length of water vapor $\mu_1$ (gm cm$^{-2}$) between a specific level with pressure $p$ and the top of the atmosphere is given by

$$\mu_1(p) = \frac{1}{g\tilde{p}_0} \int_0^p pq(p)\, dp \tag{3.110}$$

where $\tilde{p}_0 = 1.013 \times 10^3$ mb. In finite-difference form, (3.110) becomes

$$\mu_1(k) = \frac{1}{2g\tilde{p}_0} \sum_{l=k}^{K-1} (p_l + p_{l+1}) q_{l+1/2} (p_l - p_{l+1}) + \frac{1}{2g\tilde{p}_0} q_{K-1/2} p_K{}^2; \quad 1 \le k \le K - 1 \tag{3.111}$$

and

$$\mu_1(K) = \frac{1}{2g\tilde{p}_0} q_{K-1/2} p_K{}^2.$$

We assume a constant mixing ratio ($3.2 \times 10^{-4}$ by volume) for carbon dioxide. The path length from the top of the atmosphere to level $k$, expressed in centimeters of a column length of $CO_2$ at standard temperature (273.2 °K) and pressure (1013.25 mb) is then given by

$$\mu_2(k) = 1.25 \times 10^{-10}\, p_k{}^2. \tag{3.112}$$

We assume that the vertical distribution of ozone is a function of the total ozone amount in the vertical air column. The pressure-corrected ozone amount (centimeters at standard temperature and pressure) from the top of the atmosphere to level $k$, expressed in centimeters of a column length of ozone at standard temperature and pressure and denoted by $\mu_3(k)$, is tabulated in the model as a function of latitude, vertical coordinate $s$ or $z$, and month. Values at intermediate times are obtained by cubic spline interpolation. The values used for a particular application of the model are obtained from the formula

$$\mu_3(k) = \bar{\mu}_3(k)\mu_3(\text{total}) \tag{3.113}$$

where $\bar{\mu}_3$, the pressure-corrected normalized ozone path length from the top of the atmosphere to level $k$, is tabulated in the model for representative total ozone amounts as a function of height. Values for other heights and total ozone amounts are obtained by interpolation within the table. The latitudinal and seasonal variation of $\mu_3$ (total) is also tabulated in the model as a function of latitude and month. These and other such values were supplied by T. Sasamori (personal communication).

In general, the path length between two levels $k$ and $l$ is given by

$$\mu_i'(k, l) = |\mu_i(k) - \mu_i(l)|. \tag{3.114}$$

The absorption functions for each atmospheric absorption gas are summarized in Table I. As before, subscripts $i = 1, 2, 3$ denote water vapor, carbon dioxide, and ozone, respectively.

TABLE I

ABSORPTION FUNCTIONS[a]

| $i$ | Range of $\mu_i'$ | $A_i$ |
|---|---|---|
| 1 | $\mu_1' < 0.01$ | $0.846(\mu_1' + 3.59 \times 10^{-5})^{0.243} - 6.90 \times 10^{-2}$ |
| | $\mu_1' \geqslant 0.01$ | $0.240 \log_{10} (\mu_1' + 0.010) + 0.622$ |
| 2 | $\mu_2' \leqslant 1$ | $0.0676(\mu_2' + 0.01022)^{0.421} - 0.00982$ |
| | $\mu_2' > 1$ | $0.0546 \log_{10} \mu_2' + 0.0581$ |
| 3 | $\mu_3' \leqslant 0.01$ | $0.209(\mu_3' + 7.0 \times 10^{-5})^{0.436} - 0.00321$ |
| | $\mu_3' > 0.01$ | $0.0212 \log_{10} \mu_3' + 0.0748$ |
| $i$ | Range of $\mu_i$ | $\alpha_i$ |
| 1 | $\mu_1 < 0.001$ | $3.0\ \mu_1^{0.414}$ |
| | $\mu_1 \geqslant 0.001$ | $0.218 \log_{10} (\mu_1 + 0.001) + 0.758$ |
| 2 | $\mu_2 \leqslant 0.5$ | $0.0825\ \mu_2^{0.456}$ |
| | $\mu_2 > 0.5$ | $0.0461 \log_{10} \mu_2 + 0.0746$ |
| 3 | All $\mu_3$ | $0.0122 \log_{10} (\mu_3 + 6.5 \times 10^{-4}) + 0.0385$ |
| $i$ | Range of $\mu_1'$ | $\tau_i$ |
| 1 | All $\mu_1'$ | 1 |
| 2 | All $\mu_1'$ | $1.33 - 0.832(\mu_1' + 0.0286)^{0.26}$ |
| 3 | All $\mu_1'$ | $\exp(-0.1\mu_1')$ |

[a] $i = 1, 2, 3$ denotes water vapor, carbon dioxide, and ozone, respectively.

The fluxes are computed in three steps. First, the clear sky fluxes are computed. Next, the fluxes for an atmosphere with clouds at one level forming a completely overcast sky are computed. For these fluxes, the clouds provide an additional radiation source. Finally, we compute the fluxes for the general case of fractional cloudiness at every layer.

In the clear sky case, we denote the total absorptivity between two levels $z_1$ and $z_2$ by

$$B(z_1, z_2) = \sum_{i=1}^{N} A_i[\mu_i(z_1, z_2)]\tau_i[\mu_1'(z_1, z_2)]. \tag{3.115}$$

Here $A_i$ is the fractional absorptivity of the individual absorption gas and $\tau_i$ is the transmissivity of water vapor in the infrared frequency range where the absorption bands of carbon dioxide and ozone overlap the water vapor absorption band. The functional forms of $A_i$ and $\tau_i$ are listed in Table I.

We denote the isothermal emissivity by

$$\beta(z) = \sum_{i=1}^{N} \alpha_i[\mu_i(z)]. \tag{3.116}$$

The functional forms of $\alpha_i$ are also listed in Table I. At discrete levels, these equations become

$$\begin{aligned} B(k, l) &= \sum_{i=1}^{N} A_i[\mu_i'(k, l)]\tau_i[\mu_1'(k, l)]; \qquad k \neq l \\ B(k, k) &= 0 \end{aligned} \tag{3.117}$$

and

$$\beta(k) = \sum_{i=1}^{N} \alpha_i[\mu_i(k)]. \tag{3.118}$$

The equations for the clear sky fluxes are

$$F_s^{\uparrow}(z) = \sigma[T(z_B)]^4 + 4\sigma \int_{T(z_B)}^{T(z)} B(z, z')T'^3\, dT' \tag{3.119}$$

$$F_s^{\downarrow}(z) = \sigma[T(z_T)]^4\beta(z) - 4\sigma \int_{T(z)}^{T(z_T)} B(z, z')T'^3\, dT' \tag{3.120}$$

where $T(z_T)$ is the temperature at the top of the atmosphere, $T(z_B)$ is the temperature at the bottom, and $\sigma$ is Stefan's constant equal to $5.67 \times 10^{-5}$ erg cm$^{-2}$ sec$^{-1}$ deg$^{-4}$. $T'$ is the temperature at height $z'$. In finite-difference form, these equations become

$$F_s^{\uparrow}(0) = \sigma(T_g)^4$$

$$F_s^{\uparrow}(k) = \sigma(T_g)^4 + 4\sigma \sum_{l=1/2}^{k-1/2} \tfrac{1}{2}[B(k, l - \tfrac{1}{2}) + B(k, l + \tfrac{1}{2})]T_l^3\, \Delta T_l; \tag{3.121}$$

$$1 \leqslant k \leqslant K$$

$$F_s^{\downarrow}(k) = \sigma(T_{K-1/2})^4\beta(k) - 4\sigma \sum_{l=k+1/2}^{K-1/2} \tfrac{1}{2}[B(k, l - \tfrac{1}{2}) + B(k, l + \tfrac{1}{2})]T_l^3\, \Delta T_l; \qquad 1 \leqslant k \leqslant K - 1 \tag{3.122}$$

$$F_s^{\downarrow}(K) = \sigma(T_{K-1/2})^4\beta(K)$$

where

$$\begin{aligned} \Delta T_{1/2} &= \tfrac{1}{2}(T_{1/2} + T_{3/2}) - T_g \\ \Delta T_k &= \tfrac{1}{2}(T_{k+1} - T_{k-1}); \qquad 1\tfrac{1}{2} \leqslant k \leqslant K - \tfrac{1}{2} \\ \Delta T_{K-1/2} &= \tfrac{1}{2}(T_{K-1/2} - T_{K-3/2}). \end{aligned} \tag{3.123}$$

The $\frac{1}{2}$ factor occurs in the third formula of (3.123) because we assume that the atmosphere is isothermal above level $K$.

In a completely overcast sky, we assume that one level $m$ is completely overcast with clouds that are perfect blackbodies for the infrared radiation. Note that this $m$ level is not the same as the first level above the surface in the z-system model. We denote the fluxes at height $z$ for the case with clouds at level $m$ by $F_m^{\uparrow}(z)$ and $F_m^{\downarrow}(z)$. These fluxes are given by

$$F_m^{\uparrow}(z) = F_s^{\uparrow}(z) \tag{3.124}$$

$$F_m^{\downarrow}(z) = \sigma[T(z_m)]^4 - 4\sigma \int_{T(z)}^{T(z_m)} B(z, z')T'^3\, dT' \tag{3.125}$$

for $z \leqslant z_m$ and by

$$F_m^{\uparrow}(z) = \sigma[T(z_m)]^4 + 4\sigma \int_{T(z_m)}^{T(z)} B(z, z')T'^3\, dT' \tag{3.126}$$

$$F_m^{\downarrow}(z) = F_s^{\downarrow}(z) \tag{3.127}$$

for $z > z_m$.

In discrete form these equations become

$$F_m^{\uparrow}(k) = F_s^{\uparrow}(k) \tag{3.128}$$

$$F_m^{\downarrow}(k) = \sigma(\bar{T}^s_m)^4 - 4\sigma \sum_{l=k+1/2}^{m-1/2} \tfrac{1}{2}[B(k, l - \tfrac{1}{2}) + B(k, l + \tfrac{1}{2})]T_l^3\, \Delta T_l$$

$$F_m^{\downarrow}(m) = \sigma(\bar{T}^s_m)^4 \tag{3.129}$$

for $k \leqslant m$ and

$$F_m^{\uparrow}(k) = \sigma(\bar{T}_m^s)^4 + 4\sigma \sum_{l=m+1/2}^{k-1/2} \tfrac{1}{2}[B(k, l - \tfrac{1}{2}) + B(k, l + \tfrac{1}{2})] T_l^3 \, \Delta T_l \quad (3.130)$$

$$F_m^{\downarrow}(k) = F_s^{\downarrow}(k) \quad (3.131)$$

for $k > m$. To compute the flux from clouds at the surface, we set $\bar{T}_l^s = T_a$.

In the case of partial cloudiness at all layers, we allow clouds at all layers below some specified layer $M$, where $M \leqslant K$. We denote the fractional cloudiness in each layer by $C_m$. Conceptually, the clouds are just above the integer $k$ levels. The total flux at each layer is then the sum of the fluxes resulting from each source attenuated by the partial cloudiness between the layer and each source. The upward fluxes are given by

$$\begin{aligned} F^{\uparrow}(0) &= F_s^{\uparrow}(0) \\ F^{\uparrow}(k) &= \sum_{m=-1}^{k-2} F_m^{\uparrow}(k) C_m \prod_{l=m+1}^{k-1} (1 - C_l) + C_{k-1} F_{k-1}^{\uparrow}(k); \\ & \quad 1 \leqslant k \leqslant M + 1 \\ F^{\uparrow}(k) &= \sum_{m=-1}^{M-1} F_m^{\uparrow}(k) C_m \prod_{l=m+1}^{M} (1 - C_l) + C_M F_M^{\uparrow}(k); \\ & \quad M + 2 \leqslant k \leqslant K \end{aligned} \quad (3.132)$$

where we define $C_{-1} = 1$ and $F_{-1}^{\uparrow}(k) = F_s^{\uparrow}(k)$. We note from the form of (3.132) that the flux at any layer is the sum of fluxes resulting from each source $F_m^{\uparrow} C_m$ attenuated by partial cloudiness $\prod(1 - C_l)$ between the layer and each source. The downward fluxes have a similar form:

$$\begin{aligned} F^{\downarrow}(k) &= \sum_{m=k+1}^{M+1} F_m^{\downarrow}(k) C_m \prod_{l=k}^{m-1} (1 - C_l) + C_k F_k^{\downarrow}(k); \quad 0 \leqslant k \leqslant M \\ F^{\downarrow}(k) &= F_s^{\downarrow}(k); \quad M + 1 \leqslant k \leqslant K \end{aligned} \quad (3.133)$$

where we define $C_{M+1} = 1$ and $F_{M+1}^{\downarrow}(k) = F_s^{\downarrow}(k)$.

Finally, the heating function from atmospheric radiation is calculated from (2.59) which in finite-difference form becomes

$$Q_{\mathrm{RL}}(k) = -(1/\rho)\delta_z(F^{\uparrow} - F^{\downarrow}). \quad (3.134)$$

The long-wave radiation changes very little from one time step to the next and is very time-consuming to compute. Therefore, the heating from the long-wave radiation is computed only every 3 hr at any one grid point and these values are used at each time step in the following 3 hr period. The fluxes at the surface, $F^{\downarrow}(0)$ and $F^{\uparrow}(0)$, are needed every time step to compute the ground temperature; therefore, this partial calculation is performed every time step over non-ocean grid points.

The solar radiation flux at level $k$ is given by

$$F_s(k) = S_0 \cos \zeta \left[ 1 - \sum_{i=1}^{N} \beta_i(\mu_i''(k) \sec \zeta) \right] \tag{3.135}$$

where the solar constant $S_0 = 1.395 \times 10^6$ erg cm$^{-2}$ sec$^{-1}$ and the solar zenith angle is given by

$$\cos \zeta = \sin \phi \sin \delta + \cos \phi \cos \delta \cos \xi \tag{3.136}$$

where $\delta$ is the sun's declination angle and $\xi$ is the sun's hour angle. We take $\xi = -\pi$ at Greenwich at $t = 0$. If $\cos \zeta < 0$, it is night and we set the flux equal to zero.

For water vapor ($i = 1$),

$$\mu_1''(k) = \mu_1(k) \tag{3.137}$$

$$\beta_1 = 0.110(\mu_1''(k) \sec \zeta + 6.31 \times 10^{-4})^{0.30} - 0.0121. \tag{3.138}$$

For carbon dioxide ($i = 2$),

$$\mu_2''(k) = \mu_2(k) \tag{3.139}$$

$$\beta_2 = 2.35 \times 10^{-3}(\mu_2'' \sec \zeta + 1.29 \times 10^{-2})^{0.262} - 7.5 \times 10^{-4} \tag{3.140}$$

and for ozone ($i = 3$)

$$\mu_3''(k) = v_3(k)\mu_3(\text{total}) \tag{3.141}$$

where $v_3$ is the normalized ozone distribution which is tabulated in the model as a function of total ozone and height

$$\beta_3 = 0.0450(\mu_3'' \sec \zeta + 8.34 \times 10^{-4})^{0.377} - 3.11 \times 10^{-3}. \tag{3.142}$$

The heating function for solar radiation is calculated from (2.60). In finite-difference form, the equation becomes

$$Q_S(k) = (1/\rho)\delta_z(F_s). \tag{3.143}$$

The correction due to clouds gives

$$Q_{RS}(k) = Q_S(k) \prod_{m=k+1}^{M} (1 - C_m) \tag{3.144}$$

and the flux at the earth's surface needed for the surface temperature calculation over land is

$$F_{RS} = F_s(1) \prod_{m=1}^{M} (1 - C_m). \tag{3.145}$$

*c. Stable Latent Heat Release.* To compute the amount of water that condenses due to large-scale vertical motion and the corresponding latent heat released, we first forecast the distribution of $\rho q$ or $q$ at time $n + 1$ using (2.9) or (2.10) with $M$ set to 0. The temperature at time $n + 1$ is determined from the forecast values of pressure at time $n + 1$ using the diagnostic relations in Section III, C. These pressures and temperatures determine the saturation specific humidity at time $n + 1$. The saturation partial pressure $e_{st}$ used in (2.63) for $q_{st}$ is computed from Teton's formula, as modified by Murray (1967),

$$e_{st} = 6.11 \times 10^3 \exp\left[\frac{a(T - 273.16)}{T - b}\right] \tag{3.146}$$

where $a = 21.874$ and $b = 7.66$ if $T \leqslant 273.16$ and $a = 17.269$ and $b = 35.86$ if $T > 273.16$.

The excess moisture over the saturation value is assumed to condense giving

$$(M/\rho)2\Delta t = q^{n+1} - \beta q_{st}^{n+1} \tag{3.147}$$

where $\beta$ is the empirical relative humidity factor of 95% which allows precipitation before the entire grid volume becomes saturated and $M$, as defined earlier, is the rate of condensation of water vapor per unit volume. The factor $2\Delta t$ appears in (3.147) since the condensation is assumed to take place during the time $n - 1$ to $n + 1$. If $q \leqslant \beta q_{st}$, or $w < 0$, we set $M = 0$.

Finally, to convert $M$ to the heat rate, we use

$$Q_{\mathrm{CS}} = LM/\rho \tag{3.148}$$

where $L$ is latent heat of condensation.

*d. Convective Adjustment.* The mean temperature (2.73) in the first level in the z-system model used for the convective adjustment calculation is

$$\bar{T}_k = \frac{\rho_{k+1/2} T_{k+1/2}\,\Delta z + \rho_m T_m \sigma}{\rho_{k+1/2}\,\Delta z + \rho_m \sigma}. \tag{3.149}$$

The equation for layers above the $m$ level in the z-system model and at all levels in the s-system model is

$$\bar{T}_k = \frac{\rho_{k+1/2} T_{k+1/2} + \rho_{k-1/2} T_{k-1/2}}{\rho_{k+1/2} + \rho_{k-1/2}}. \tag{3.150}$$

The discrete form for lapse rates is $\delta_z T$ as used earlier in the vertical flux of sensible heat in (3.80). The assumption of constancy of internal energy during the adjustment (2.71) is written in discrete form as

$$\rho_{k+1/2} T^*_{k+1/2} + \rho_{k-1/2} T^*_{k-1/2} = \rho_{k+1/2} T_{k+1/2} + \rho_{k-1/2} T_{k-1/2} \tag{3.151}$$

where the asterisks denote the new temperatures after adjustment. Following (2.72), depending upon whether $w_k \gtrless 0$, we have two equations for the unknowns, namely, the appropriate critical lapse rate (2.72) and (3.151).

This system of equations is solved by starting with the two bottom layers and working up to the top of the model atmosphere using two layers at a time. Following the computation of the adjusted temperatures, pressures are recomputed from

$$p_{k+1} = p_k \exp\left[\frac{-g(z_{k+1/2} - z_{k-1/2})}{RT_k}\right]. \tag{3.152}$$

The surface pressure is assumed not to change in the adjustment process.

*e. Convective Latent Heat Release.* The discrete form of the equations for the Kuo–Krishnamurti scheme is given in Kanamitsu (1971). To save space, we only outline the procedure here.

First, a test is made on the vertical velocity of the air in the boundary layer. If $w_a > 0$, the calculation of the Kuo–Krishnamurti scheme is entered. Otherwise, it is assumed that no cumulus convection occurs. This differs from the approach of Krishnamurti *et al.* (1973), in which they test conditional instability to determine whether to proceed or not.

If the convective heat release is to be computed, the lifting condensation level (LCL) is determined by lifting a particle dry adiabatically from the boundary layer to the level where it condenses. It is then assumed that further ascent from this lifting condensation level follows the moist adiabat. As can be seen from Fig. 9, the LCL determines the cloud base. The vertical velocity is vertically interpolated between $w$ levels of the model to the cloud base and if it is positive, then $Q_{cc}$ is computed from (2.64)–(2.67). If the model temperature exceeds the temperature of the moist adiabat $T_{ma}$, the heating is assumed to be zero. This usually occurs at the cloud top. The shaded portion in Fig. 9 represents cumulus heating. Finally, the moisture transport by cumulus is computed from (2.68) with the assumption that the density of saturated air is the same as that for dry air.

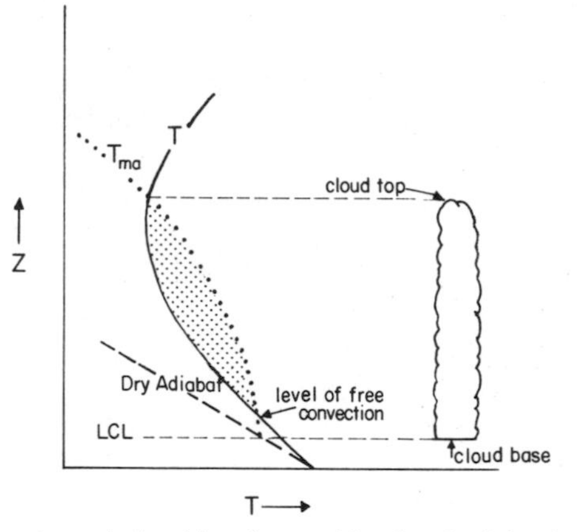


FIG. 9. Diagram showing relationship of quantities involved in the Kuo–Krishnamurti cumulus convection parameterization.

## D. FILTERING

Because both s- and z-system models use centered time and space differences, a small amount of smoothing is needed to prevent the growth of $2\Delta t$ time computational modes and to prevent nonlinear instability. As mentioned earlier, in the z-system model the horizontal diffusion is designed to control both problems. In addition, the z-system model has a nonlinear smoothing operator applied to pressure every time step. The smoothing operator is

$$\mathscr{E} = \frac{5 \times 10^{-4}}{\max |\delta p|} [\delta_\lambda(|\delta_\lambda p|\, \delta_\lambda p) + \delta_\phi(|\delta_\phi p|\, \delta_\phi p)] \tag{3.153}$$

where max $|\delta p|$ is the maximum value of the two values of $|\delta_\lambda p|$ and the two values of $|\delta_\phi p|$ used in (3.153). The factor $5 \times 10^{-4}$ was found by experimentation. In (3.153) the value of $p$ at point $(i, j)$ is at time $n + 1$ while the values at points $(i \pm 1, j)$ and $(i, j \pm 1)$ are at time $n$. After $p^{n+1}$ is calculated from the pressure tendency equation (3.43) or (3.44), it is smoothed according to

$$p^{n+1}(\text{smoothed}) = p^{n+1} + \mathscr{E}. \tag{3.154}$$

The term $\mathscr{E}$ is not defined if max $|\delta p|$ is zero. The smoothing operator is not performed at any mesh point where that is the case.

The a level is, in general, not horizontal and the operator (3.153) is only applicable as defined on a horizontal surface. Consequently, for application to the a level, the operator is modified slightly using $\hat{p}$'s which are defined on a horizontal surface. Thus at the a level at points $(i \pm 1, j)$ and $(i, j \pm 1)$, $\hat{p}$ as defined in (3.38) is used in (3.153) instead of $p_a$.

The s system uses a filter for the pressure field based on that of Shapiro (1970). We make use of the operators

$$(\delta_\lambda)^4\psi = (\psi_{i-2} - 4\psi_{i-1} + 6\psi_i - 4\psi_{i+1} + \psi_{i+2})/(\Delta\lambda)^4 \tag{3.155}$$

and

$$(\delta_\phi)^4\psi = (\psi_{j-2} - 4\psi_{j-1} + 6\psi_j - 4\psi_{j+1} + \psi_{j+2})/(\Delta\phi)^4 \tag{3.156}$$

The smoothing formulas used are

$$\tilde{p}^\lambda = p - \beta(\Delta\lambda/2)^4(\delta_\lambda)^4 p \tag{3.157}$$

and

$$\tilde{p}^\phi = p - \beta(\Delta\phi/2)^4(\delta_\phi)^4 p. \tag{3.158}$$

After the pressure is computed at $n + 1$, the value at $n$ is smoothed using

$$p(\text{smoothed}) = \widetilde{\tilde{p}^{\phi}}^{\lambda} \tag{3.159}$$

We generally use a value of 0.001 for the coefficient $\beta$ to keep the model from developing small-scale noise in the pressure field.

To control the time computational mode associated with centered finite-difference approximations, the s-system model includes the frequency filter originally designed by Robert (1966) and later studied by Asselin (1972). The filter function is

$$\tilde{\psi}^{n} = \psi^{n} + 0.5\nu(\tilde{\psi}^{n-1} - 2\psi^{n} + \psi^{n+1}) \tag{3.160}$$

where the tilde denotes the smoothed variable and $\nu$ is the filter parameter, generally set at 0.05. The filter is applied to recompute the prognostic variables at time $n$ after the values are computed at time $n + 1$.

The discrete grid of the s-system model is uniform in latitude and longitude and hence the distances in the longitudinal direction between grid points become very small near the poles. The resulting small time step required by the linear stability condition is prohibitive for grids of 5° or less with present-day computers. The small time step is circumvented by filtering out the shorter wavelength fast-moving waves of the prognostic variables in the polar regions by Fourier analysis. This method was tested in numerical integrations by Williamson and Browning (1973) for the shallow-water equations, while the stability condition was determined by Williamson (1976a). Examples of stable time steps and the corresponding filtering configuration are given there.

## IV. Application of NCAR Models

The NCAR models have been used for numerical experiments by NCAR and university scientists. These models are coded such that scientists can modify with a minimum of effort the portion of the model they are most interested in without becoming involved with the entire code. A committee of NCAR GCM-related scientists, established to work closely with potential model users to design and execute experiments, recently published a report (GCM Steering Committee, 1975) briefly summarizing the history and uses of the models.

The large number of experiments with the z-system model that have evolved from collaboration of NCAR and university scientists can be separated into two broad categories—climate simulation and sensitivity and numerical weather prediction. References detailing these experiments are summarized in Table II.

TABLE II

NCAR GCM REFERENCES

| Area | References |
|---|---|
| *I. Climate Simulation and Sensitivity* | |
| 1. *Discretization* | |
| Horizontal resolution | Wellck *et al.* (1971) |
| Vertical resolution | Washington and Kasahara (1970) |
| | Kasahara and Washington (1971) |
| Height of model top | Kasahara *et al.* (1973) |
| | Williams (1976) |
| Computer precision | Williamson and Washington (1973) |
| 2. *Diagnostic Studies* | |
| Basic models | Washington and Kasahara (1970) |
| | Kasahara and Washington (1971) |
| | Wellck *et al.* (1971) |
| | Kasahara *et al.* (1973) |
| | Kasahara and Sasamori (1974) |
| | Washington *et al.* (1976) |
| Monsoon studies | Washington and Daggupaty (1975) |
| | Washington (1976) |
| Geostrophic turbulence | Charney (1971) |
| | Wellck *et al.* (1971) |
| Harmonic analysis | Otto-Bliesner (1976) |
| | Williams (1976) |
| Equatorial wave analysis | Tsay (1974) |
| Dispersion of particles | Kao *et al.* (1976) |
| 3. *Parameterization* | |
| Radiation | Washington (1971b) |
| | Fels and Kaplan (1975) |
| Cumulus convection | Washington and Kasahara (1970) |
| | Washington and Daggupaty (1975) |
| | Washington *et al.* (1976) |
| Clear-air turbulence | Schenk (1974) |
| Mountain wave momentum flux | Lilly (1972) |
| Surface hydrology | Washington *et al.* (1976) |
| Planetary boundary layer | Kasahara and Washington (1967) |
| | Deardorff (1972) |
| | Benoit (1976) |
| Ozone transport | London and Park (1973, 1974) |
| 4. *Lower boundary* | |
| Orography | Kasahara (1968) |
| | Kasahara and Washington (1969, 1971) |
| | Kasahara *et al.* (1973) |
| | Kasahara and Sasamori (1974) |

(*continued*)

TABLE II (*Continued*)

| Area | References |
| --- | --- |
| Sea surface temperature anomalies | Houghton *et al.* (1974) |
| | Chervin *et al.* (1976) |
| Ice age | Williams (1974, 1975a, b) |
| | Williams *et al.* (1974) |
| | Barry (1975) |
| | Williams and Barry (1974) |
| Thermal pollution | Washington (1971a, 1972) |
| | Llewellyn and Washington (1977) |
| 5. *Statistical significance* | Washington (1972) |
| | Chervin *et al.* (1974, 1976) |
| | Chervin and Schneider (1976a, b) |
| *II. Numerical Weather Prediction* | |
| 1. *Forecast skill* | |
| Predictability studies with simulated data | Williamson and Kasahara (1971) |
| | Kasahara (1972) |
| | Williamson (1973) |
| Predictability studies with real data | Baumhefner (1970, 1976) |
| | Baumhefner and Downey (1976) |
| Forecast sensitivity | Baumhefner (1971, 1972) |
| | Baumhefner and Julian (1972, 1975) |
| | Williamson and Washington (1973) |
| | Williamson and Browning (1974) |
| Diagnostics | Bettge *et al.* (1976) |
| 2. *Initialization* | |
| Balance equations | Houghton and Washington (1969) |
| | Houghton *et al.* (1971) |
| | Washington and Baumhefner (1975) |
| Normal mode | Dickinson and Williamson (1972) |
| | Williamson and Dickinson (1976) |
| | Kasahara (1976) |
| | Williamson (1976b) |
| Four-dimensional assimilation | Williamson and Kasahara (1971) |
| | Kasahara (1972) |
| | Kasahara and Williamson (1972) |
| | Williamson and Dickinson (1972) |
| | Williamson (1973) |

For basic climatological studies we usually start the model from an arbitrary initial state so as not to bias the model results. This state is given by an isothermal dry atmosphere at rest

$$\begin{aligned} &T = T_0 \qquad p = p_0 \exp(-gz/RT_0) \\ &u = v = q = 0 \end{aligned} \tag{4.1}$$

where $T_0$, initial isothermal temperature, and $p_0$, mean sea-level pressure, are usually taken to be 240 °K and 1013.25 mb, respectively.

Since these calculations start from an isothermal atmosphere, heating differences between land and ocean cause intense large-scale direct circulations (rising motions over warm oceans and sinking motions over cold continents) to develop at the onset. After about 10–15 days, the north–south temperature gradients become so large that the large-scale direct circulation breaks down into eddy motions due to the mechanism known as baroclinic instability. After about 25 days, the model reaches an equilibrium state with fairly constant model climate statistics and the flow patterns become meteorological. This spin-up phase is described in more detail in Washington and Kasahara (1970) and Kasahara and Washington (1971).

To save computer time, many experiments are started from simulated data from one of these basic climatological runs after it reaches its climatological equilibrium. The first section of Table II—Climate Simulation and Sensitivity—lists papers describing experiments starting from both these modes. We have divided the papers into several categories for convenience, although such a division is not unique and there is much overlap.

The first category on discretization examines the effect of the discrete nature of the models. The papers compare 10°, 5°, and $2\frac{1}{2}$° horizontal resolutions, two and six vertical layers in a tropospheric version, and a stratospheric 12-layer version with a six-layer tropospheric version. The last paper considers the effect of computer precision on climate simulations.

The diagnostic studies category compares the model simulation statistics with the corresponding atmospheric statistics to validate the model. The first group includes the primary statistics of the basic 5° and $2\frac{1}{2}$° tropospheric versions and of the vertically extended 5° tropospheric–stratospheric version. The second group of papers deals with regional climatological studies of the African–Asian monsoon. The next group examines model-computed wave number spectra in view of theoretical work in two-dimensional turbulence appropriate for large-scale flows. Zonal harmonic analysis is also used to examine the standing and transient waves in conjunction with Hovmöller diagrams of various wave number bands. An analysis of the model characteristics of equatorial waves such as Kelvin, Rossby-gravity, and Rossby–Haurwitz types is given. The last reference discusses dispersion characteristics of particle tracers placed in the GCM.

The parameterization category reports the effect on simulated climatology of changing various parameterizations of the models. These include tests of a more sophisticated method of infrared radiation calculation than described earlier in this chapter, a discussion of different cumulus convective parameterizations, an attempt to incorporate the effect of clear-air turbulence on the simulations, and the effect of wave momentum flux occurring over mountainous areas. The importance of a more complete surface hydrology calculation and the more sophisticated planetary boundary layer parameterization is also included in this category. The last area shows preliminary attempts to compute the distribution of ozone in the lower stratosphere.

The fourth category involves changes in the lower boundary configuration. The first four papers examine the question of the importance of large-scale forcing of mountains versus the land–ocean thermal contrasts on the large-scale circulation. These are followed by studies of the response of the model to sea surface temperature anomalies in both the Atlantic and Pacific. The next series of papers in this category uses the model to study the last ice age by changing surface albedo, mountain height, ice and snow cover, and ocean temperatures. Finally, the model has been used to study possible climatological effects of mankind-generated thermal pollution.

The last area on statistical significance is rather new in global circulation modeling. The papers represent a beginning methodology for measuring significance of changes observed in a simulation when various parameters, boundary conditions, or external forcing are changed.

The second broad classification of experiments carried out using the GCM involves numerical weather prediction. Observed atmospheric data are used for initial conditions, and we are primarily interested in the detailed evolution of the model and its ability to predict the day-to-day changes correctly for short periods.

The first experiments under forecast skill are designed to determine the theoretical predictability of the models by using simulated data generated by the GCM in place of observed atmospheric data for comparison. The second group examines the ability of the models to forecast the atmosphere from observed data, and the third group investigates the sensitivity of these forecasts to observational errors, forecast domains, and computer precision. The last paper examines the forecast kinetic energy budget.

The second category under numerical weather prediction examines various methods of providing consistent initial data. These include use of the balance equation, removal of the vertical integral of horizontal divergence, and selective elimination of undesirable waves by expansion into the normal modes of the forecast model. The last category, four-dimensional assimilation, uses the models themselves to merge partial observational data taken at many different times and places into a complete data set consistent in both space and time, eventually to provide initial data from which to start forecasts.

## Acknowledgments

In an enterprise as large as the GCM, it is difficult to name all who have contributed to its success. Those to whom appreciation should be extended are our colleagues at NCAR, members of the NCAR Computing Facility who helped with programming and operation of the Control Data Corporation computers, and the many visitors to NCAR whose comments have aided in the development of the models. Major credit must go to A. Kasahara for collaboration in the original design of the NCAR model and to T. Sasamori for designing the infrared and solar radiation aspects of the model. We thank W. Baker, D. Baumhefner, A. Kasahara, B. Otto-Bliesner, R. Sato, T. Schlatter, R. Somerville, and G. Williamson for their comments on early versions of this chapter. We are grateful to A. Modahl for editing, N. Perkey for typing the manuscript, and E. Rosenberg for drafting the figures.

## References

Arakawa, A. (1966). *J. Comput. Phys.* **1**, 119.
Asselin, R. (1972). *Mon. Weather Rev.* **100**, 487.
Barry, R. G. (1975). *Palaeogeogr., Palaeoclimatol., Palaeoecol.* **17**, 123.
Baumhefner, D. P. (1970). *Mon. Weather Rev.* **98**, 92.
Baumhefner, D. P. (1971). *J. Atmos. Sci.* **28**, 42.
Baumhefner, D. P. (1972). *J. Atmos. Sci.* **29**, 768.
Baumhefner, D. P. (1976). *Mon. Weather Rev.* **104**, 1175.
Baumhefner, D. P., and Downey, P. (1976). *Ann. Meteorol.* [N.S.] **11**, 205.
Baumhefner, D. P., and Julian, P. R. (1972). *J. Atmos. Sci.* **29**, 285.
Baumhefner, D. P., and Julian, P. R. (1975). *Mon. Weather Rev.* **103**, 273.
Benoit, R. (1976). "A Comprehensive Parameterization of the Atmospheric Boundary Layer for General Circulation Models." Ph.D. Thesis, Dept. Meteorol., McGill University, Montreal (in preparation).
Bettge, T. W., Smith, P. J., and Baumhefner, D. P. (1976). *Mon. Weather Rev.* **104**, 1242.
Bryan, K. (1969). *Mon. Weather Rev.* **97**, 806.
Budyko, M. I. (1956). "Heat Balance of the Earth's Surface." (Teplovoĭ balans zemnoĭ poverkhnostĭ) (N. A. Stepanova, translator). U.S. Weather Bureau, Washington, D.C.
Charney, J. (1971). *J. Atmos. Sci.* **28**, 1087.
Chervin, R. M., and Schneider, S. H. (1976a). *J. Atmos. Sci.* **33**, 391.
Chervin, R. M., and Schneider, S. H. (1976b). *J. Atmos. Sci.* **33**, 405.
Chervin, R. M., Gates, W. L., and Schneider, S. H. (1974). *J. Atmos. Sci.* **31**, 2216.
Chervin, R. M., Washington, W. M., and Schneider, S. H. (1976). *J. Atmos. Sci.* **33**, 413.
Deardorff, J. W. (1966). *J. Atmos. Sci.* **23**, 503.
Deardorff, J. W. (1972). *Mon. Weather Rev.* **100**, 93.
Dickinson, R. E., and Williamson, D. L. (1972). *J. Atmos. Sci.* **29**, 623.
Fels, S. B., and Kaplan, L. D. (1975). *J. Atmos. Sci.* **32**, 779.
GCM Steering Committee. (1975). "Development and Use of the NCAR GCM," Tech. Note TN/STR-101. National Center for Atmospheric Research, Boulder, Colorado.

Holloway, J. L., Jr., and Manabe, S. (1971). *Mon. Weather Rev.* **99**, 335.
Houghton, D., and Washington, W. M. (1969). *J. Appl. Meteorol.* **8**, 726.
Houghton, D., Kasahara, A., and Washington, W. (1966). *Mon. Weather Rev.* **94**, 141.
Houghton, D., Baumhefner, D. P., and Washington, W. M. (1971). *J. Appl. Meteorol.* **10**, 626.
Houghton, D. D., Kutzbach, J. E., McClintock, M., and Suchman, D. (1974). *J. Atmos. Sci.* **31**, 857.
Kanamitsu, M. (1971). "Students' Summary Reports Fellowship in Scientific Computing," pp. 17–30. National Center for Atmospheric Research, Boulder, Colorado.
Kao, S. K., Chi, C. N., and Washington, W. M. (1976). *J. Atmos. Sci.* **33**, 1042.
Kasahara, A. (1965). *Mon. Weather Rev.* **93**, 27.
Kasahara, A. (1968). *Proc. Symp. Mountain Meteorol., 1967* pp. 193–221.
Kasahara, A. (1972). *Bull. Am. Meteorol. Soc.* **53**, 252.
Kasahara, A. (1974). *Mon. Weather Rev.* **102**, 509.
Kasahara, A. (1976). *Mon. Weather Rev.* **104**, 669.
Kasahara, A., and Sasamori, T. (1974). *J. Atmos. Sci.* **31**, 408.
Kasahara, A., and Washington, W. M. (1967). *Mon. Weather Rev.* **95**, 389.
Kasahara, A., and Washington, W. M. (1969). *Proc. WMO/IUGG Symp. Numerical Weather Prediction, 1968* pp. IV47–IV56.
Kasahara, A., and Washington, W. M. (1971). *J. Atmos. Sci.* **28**, 657.
Kasahara, A., and Williamson, D. (1972). *Tellus* **24**, 100.
Kasahara, A., Sasamori, T., and Washington, W. M. (1973). *J. Atmos. Sci.* **30**, 1229.
Kreiss, H.-O., and Oliger, J. (1972). *Tellus* **24**, 199.
Krishnamurti, T. N., and Moxim, W. J. (1971). *J. Appl. Meteorol.* **10**, 3.
Krishnamurti, T. N., Kanamitsu, M., Ceselski, B., and Mathur, M. (1973). *Tellus* **25**, 523.
Kuo, H. L. (1965). *J. Atmos. Sci.* **22**, 40.
Kurihara, Y. (1965). *Mon. Weather Rev.* **93**, 399.
Leith, C. E. (1965). *In* "Methods of Computational Physics" (B. Alder, ed.), Vol. 4, pp. 1–27. Academic Press, New York.
Lilly, D. K. (1972). *Bull. Am. Meteorol. Soc.* **53**, 17.
Llewellyn, R. A., and Washington, W. M. (1977). "Energy and Climate: Outer Limits to Growth—Energy Consumption and Climate Impact—Regional/Global Estimates," Rep. Nat. Acad. Sci., Washington, D.C. (to appear).
London, J. (1957). "Final Report," Contract AF19(122)-165. Dept. Meteorol. Oceanogr., New York University (ASTIA No. 117227).
London, J., and Park, J. H. (1973). *Pure Appl. Geophys.* 106 and 1611.
London, J., and Park, J. H. (1974). *Can. J. Chem.* **52**, 1599.
Manabe, S. (1969a). *Mon. Weather Rev.* **97**, 739.
Manabe, S. (1969b). *Mon. Weather Rev.* **97**, 775.
Manabe, S., Smagorinsky, J., and Strickler, R. F. (1965). *Mon. Weather Rev.* **93**, 769.
Manabe, S., Smagorinsky, J., Holloway, J. L., Jr., and Stone, H. M. (1970). *Mon. Weather Rev.* **98**, 175.
Mintz, Y. (1964). *Proc. WMO/IUGG Symp. Res. Dev. Aspects Long-Range Forecasting, 1964* pp. 144–167 (reprinted in *W.M.O. Tech. Note* **66**, 141–167).
Murray, F. (1967). *J. Appl. Meteorol.* **6**, 203.
Oliger, J. E., Wellck, R. E., Kasahara, A., and Washington, W. M. (1970). "Description of NCAR Global Circulation Model," Tech. Note TN/STR-56. National Center for Atmospheric Research, Boulder, Colorado.
Otto-Bliesner, B. (1976). "Diagnostic Analyses of the NCAR General Circulation Model Using Hovmöller Diagrams and Harmonic Analysis," NCAR Manuscript No. 0903-76-2. National Center for Atmospheric Research, Boulder, Colorado.
Phillips, N. A. (1957). *J. Meteorol.* **14**, 184.

Posey, J. W., and Clapp, P. F. (1964). *Geofis. Int.* **4**, 33.

Richardson, L. F. (1922). "Weather Prediction by Numerical Process." Cambridge Univ. Press, London and New York (reprinted by Dover, New York, 1965).

Robert, A. J. (1966). *J. Meteorol. Soc. Jpn.* **44**, 237.

Sasamori, T. (1968a). *J. Appl. Meteorol.* **7**, 721.

Sasamori, T. (1968b). *Proc. WMO/IUGG Symp. Radiat., Including Satellite Tech. 1968* pp. 479–488 (reprinted in *W.M.O. Tech. Note* **104**, 79).

Sasamori, T. (1970). *J. Atmos. Sci.* **27**, 1122.

Sasamori, T., London, J., and Hoyt, D. V. (1972). *Meteorol. Monogr.* **13**, 9.

Schenk, H. A. (1974). "A Simulation of the Influence of Clear-Air Turbulence on the Large Scale Circulation of the Atmosphere." Master of Science Thesis, Dept. Meteorol., Pennsylvania State University, University Park.

Shapiro, R. (1970). *Rev. Geophys. Space Phys.* **8**, 359.

Smagorinsky, J. (1963). *Mon. Weather Rev.* **91**, 99.

Smagorinsky, J., Manabe, S., and Holloway, J. L., Jr. (1965). *Mon. Weather Rev.* **93**, 727.

Tsay, C.-Y. (1974). *J. Atmos. Sci.* **31**, 330.

Washington, W. M. (1971a). *In* "Man's Impact on the Climate" (W. H. Matthews, W. W. Kellogg and G. D. Robinson, eds.), Part V, Vol. 18, pp. 265–276. MIT Press, Cambridge, Massachusetts.

Washington, W. M. (1971b). *In* "The Miami Workshop on Remote Sensing, pp. 39–67. U.S. Dept. of Commerce, National Oceanographic and Atmospheric Administration, Boulder, Colorado.

Washington, W. M. (1972). *J. Appl. Meteorol.* **11**, 768.

Washington, W. M. (1974). *In* "Modeling for the First GARP Global Experiment," Rep. No. 14 GARP Publ. Ser. pp. 61–78.

Washington, W. M. (1976). *Mon. Weather Rev.* **104**, 1023.

Washington, W. M., and Baumhefner, D. P. (1975). *J. Appl. Meteorol.* **14**, 114.

Washington, W. M., and Daggupaty, S. M. (1975). *Mon. Weather Rev.* **103**, 105.

Washington, W. M., and Kasahara, A. (1970). *Mon. Weather Rev.* **98**, 559.

Washington, W. M., and Thiel, L. G. (1970). "Digitized Global Monthly Mean Ocean Surface Temperatures," Tech. Note No. 54. National Center for Atmospheric Research, Boulder, Colorado.

Washington, W. M., Otto-Bliesner, B., and Williamson, G. (1976). "January and July Simulation Experiments with the 2.5° Latitude-Longitude Version of the NCAR General Circulation Model," Ms. No. 0903-75-2. National Center for Atmospheric Research, Boulder, Colorado.

Wellck, R. E., Kasahara, A., Washington, W. M., and De Santo, G. (1971). *Mon. Weather Rev.* **99**, 673.

Williams, J. (1974). "Simulation of the Atmospheric Circulation Using the NCAR Global Circulation Model with Present-day and Glacial Period Boundary Conditions." Ph.D. Dissertation, University of Colorado (NCAR Cooperative/INSTAAR Occas. Pap. No. 10).

Williams, J. (1975a). *J. Appl. Meteorol.* **14**, 137.

Williams, J. (1975b). *Proc. WMO/IAMAP Symp. Long-Term Climatic Fluctuations, 1975* pp. 373–380.

Williams, J. (1976). *Mon. Weather Rev.* **104**, 249.

Williams, J., and Barry, R. G. (1974). *In* "Climate of the Arctic" (G. Weller and S. A. Bowling, eds.), pp. 143–149. University of Alaska, Fairbanks.

Williams, J., Barry, R. G., and Washington, W. M. (1974). *J. Appl. Meteorol.* **13**, 305.

Williamson, D. L. (1966). *J. Comput. Phys.* **1**, 51.

Williamson, D. L. (1968). *Tellus* **20**, 642.

Williamson, D. L. (1970). *Mon. Weather Rev.* **98**, 512.

Williamson, D. L. (1973). *J. Atmos. Sci.* **30**, 537.

Williamson, D. L. (1976a). *Mon. Weather Rev.* **104**, 31.
Williamson, D. L. (1976b). *Mon. Weather Rev.* **104**, 195.
Williamson, D. L., and Browning, G. L. (1973). *J. Appl. Meteorol.* **12**, 264.
Williamson, D. L., and Browning, G. L. (1974). *J. Appl. Meteorol.* **13**, 8.
Williamson, D. L., and Dickinson, R. E. (1972). *J. Atmos. Sci.* **29**, 190.
Williamson, D. L., and Dickinson, R. E. (1976). *Mon. Weather Rev.* (in press).
Williamson, D. L., and Kasahara, A. (1971). *J. Atmos. Sci.* **28**, 1313.
Williamson, D. L., and Washington, W. M. (1973). *J. Appl. Meteorol.* **12**, 1254.

# Computational Design of the Basic Dynamical Processes of the UCLA General Circulation Model

AKIO ARAKAWA AND VIVIAN R. LAMB

DEPARTMENT OF ATMOSPHERIC SCIENCES
UNIVERSITY OF CALIFORNIA
LOS ANGELES, CALIFORNIA

I. Outline of the General Circulation Model . . . 174
II. Principles of Mathematical Modeling . . . 176
III. Finite Difference Schemes for Homogeneous Incompressible Flow . . . 179
A. Distribution of Variables over the Grid Points . . . 180
B. Two-Dimensional Nondivergent Flow . . . 190
C. Finite Difference Scheme for the Nonlinear Shallow Water Equations . . . 201
IV. Basic Governing Equations . . . 207
A. The Vertical Coordinate . . . 207
B. The Equation of State . . . 209
C. The Hydrostatic Equation . . . 209
D. The Equation of Continuity . . . 209
E. The Individual Time Derivative and Its Flux Form . . . 211
F. The Momentum Equation . . . 211
G. The Thermodynamic Energy Equation . . . 212
H. The Water Vapor and Ozone Continuity Equations . . . 212
V. The Vertical Difference Scheme of the Model . . . 213
A. Some Integral Properties of the Adiabatic Frictionless Atmosphere . . . 213
B. A Vertical Difference Scheme Which Maintains Integral Properties . . . 218
C. Vertical Propagation of Wave Energy in an Isothermal Atmosphere . . . 229
D. Final Determination of the Vertical Difference Scheme . . . 234
VI. The Horizontal Difference Scheme of the Model . . . 236
A. The Governing Equations in Orthogonal Curvilinear Coordinates . . . 236
B. Horizontal Differencing of the Governing Equations . . . 239
C. Modification of the Horizontal Differencing near the Poles . . . 246
VII. Vertical and Horizontal Differencing of the Water Vapor and Ozone Continuity Equations . . . 251
A. Vertical Differencing . . . 251
B. Horizontal Transport of Water Vapor and Ozone . . . 258
C. Large-Scale Condensation and Precipitation . . . 259
VIII. Time Differencing . . . 260
IX. Summary and Conclusions . . . 262
References . . . 264

## I. Outline of the General Circulation Model

A NEW GENERAL CIRCULATION model, which has an improved finite-difference formulation, greater vertical resolution, and new parameterizations of the subgrid-scale processes, has been developed at UCLA to replace the earlier two- and three-level UCLA general circulation models.

The primary prognostic variables of the model are *horizontal velocity*, *temperature*, and *surface pressure*, governed by the horizontal momentum equation, the thermodynamic energy equation, and the surface pressure tendency equation, respectively. These governing equations, together with the hydrostatic equation, form the system of "quasi-static equations" or so-called "primitive equations."

A number of secondary prognostic variables, with corresponding governing equations, are added to the system to determine the heating and friction. The most important of the secondary prognostic variables is *water vapor*, which is governed by the water vapor continuity equation. *Ozone*, governed by an ozone continuity equation with parameterized sources and sinks, is added as a prognostic variable for use in the radiational heating calculation. The planetary *boundary layer depth* and the magnitudes of the *temperature discontinuity*, *moisture discontinuity*, and *momentum discontinuity* at the top of the boundary layer are made prognostic variables to determine the boundary layer structure. The *ground temperature*, *ground water storage*, and *mass of snow on the ground* are also taken as prognostic variables, governed by the energy and water budget equations of the ground.

The horizontal momentum equation includes the convergence of vertical flux of horizontal momentum due to the boundary layer turbulence and cumulus convection. The thermodynamic energy equation includes a heating term that consists of solar and infrared radiational heating, the convergence of vertical flux of sensible heat due to the boundary layer turbulence and cumulus convection, the release of latent heat due to cumulus-convective and large-scale condensation processes, and cooling due to evaporation of clouds and falling raindrops. The water vapor continuity equation includes the convergence of vertical flux of water vapor due to the boundary layer turbulence and cumulus convection, and both cumulus-convective and large-scale condensation and evaporation. The formulation of horizontal and vertical diffusion due to turbulence in the free atmosphere depends on the version of the model. We plan to introduce, in the near future, a formulation based on the quasi-geostrophic turbulence theory.

To use the general circulation model the following parameters must be prescribed for each grid point: surface characteristics (open ocean, ice-covered ocean, bare land, and land covered by glacial ice); elevation of the land; surface roughness; thickness of the sea ice; and ocean surface temperature.

Figure 1 shows the vertical structure of the model. The lower boundary follows the earth's topography, where the surface pressure is defined. The upper boundary is the 1 mb pressure surface, which is approximately at the height of the stratopause. The atmosphere between the upper and lower boundaries is divided into twelve layers, and the boundaries of these layers follow the coordinate surfaces of a generalized $\sigma$ coordinate (Section IV). From 100 mb upward these coordinate surfaces are also constant pressure surfaces. The lowest four layers have equal depth in pressure $p$, and the uppermost seven layers have equal depth in log $p$.

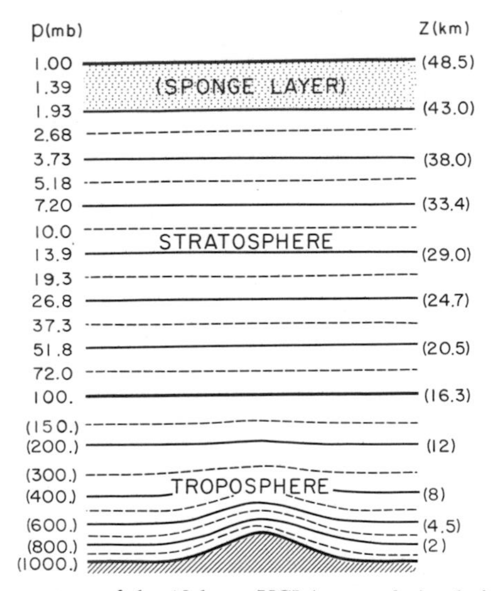


FIG. 1. The vertical structure of the 12-layer UCLA general circulation model. Solid lines define layers, broken lines indicate levels within layers at which prognostic variables are carried. Pressure levels and heights in parentheses are approximate, given for purposes of illustration.

The broken lines in Fig. 1 show the levels at which the prognostic variables of horizontal velocity, temperature, water vapor, and ozone are carried for each of the layers. These levels are approximately centered in $p$ for the layers below 100 mb, and centered in log $p$ for the layers above 100 mb.

The uppermost layer of the model, called the "sponge layer," has a damping term designed to absorb upward propagating wave energy and thus prevent a spurious reflection of wave energy at the upper boundary. The actual formulation of the damping term, however, is still in an experimental stage.

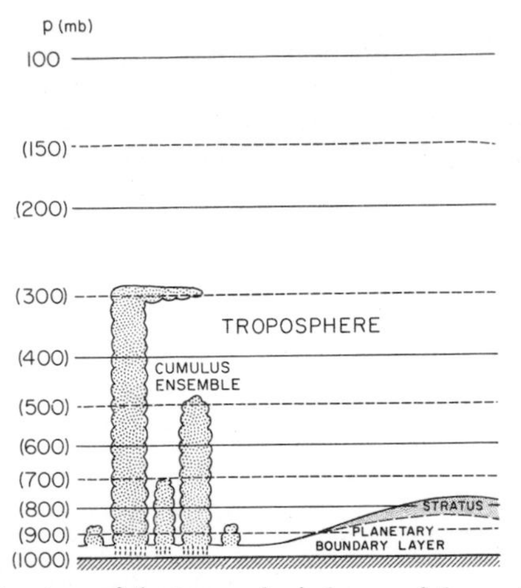


FIG. 2. The vertical structure of the tropospheric layers of the model showing the parameterized planetary boundary layer (light shaded area) and possible cloud types associated with the planetary boundary layer (stippled area).

Figure 2 shows the vertical structure of the lower layers of the model. The light shaded area represents the parameterized planetary boundary layer, which may or may not contain stratus cloud, and out of which there may or may not extend a cumulus cloud ensemble. The boundary layer depth may be less than, equal to, or greater than one or more of the model layers.

Although the model is designed and programmed to have as many as 12 layers, it can be used with fewer layers. A six-layer version of the model, with a vertical structure identical to the lower half of the 12 layer version, is also used at UCLA.

The horizontal coordinates are longitude and latitude; the current grid size is 5° of longitude and 4° of latitude. The convergence of the meridians toward the poles would normally necessitate the use of an extremely short time interval to maintain computational stability. To avoid this requirement, a longitudinal averaging is done of *selected terms* in the prognostic equations near the poles. At present, the finite-difference time step is 6 min, except for the heating, friction, and source and sink terms, for which the time step is 30 min.

## II. Principles of Mathematical Modeling

The space finite difference scheme of the model is designed to maintain many of the important integral constraints of the continuous atmosphere, such as the conservation of total mass; the conservation of total kinetic energy during inertial processes; the conservation of enstrophy (mean square vorticity) during vorticity advection by the nondivergent part of the horizontal velocity; the conservation of the integral constraint on the pressure gradient force; the conservation of total energy during adiabatic and nondissipative processes; and the conservation of total entropy and total potential enthalpy during adiabatic processes.

As the grid size approaches zero, the finite-difference solution obtained with any convergent scheme will approach the true solution and, therefore, in the limit will satisfy the integral constraints. The order of accuracy of a convergent scheme determines how rapidly its solution approaches the true solution as the grid size approaches zero. Although many schemes share the same order of accuracy, the solutions of such schemes generally approach the true solution along different paths in a function space, and with different statistics. One of the basic principles used in the design of the finite difference scheme for the model is the desirability of seeking that finite difference scheme in which the solutions approach the true solution along a path on which the statistics are analogous to those of the true solution. To this end, in the finite difference scheme used in the model, discretized analogs of the integral constraints are maintained, regardless of the grid size, which approach the true integral constraints as the grid size approaches zero.

Maintenance of the integral constraints by the finite difference scheme may not be a critical requirement for short-range numerical weather prediction (over a period of a day or two), because there the concern is with the local accuracy of the solution in space and time, and a formal maintenance of the integral constraints does not necessarily mean a greater accuracy of the solution at a particular place and time. In short-range predictions, the period of integration is usually not long enough for significant changes to occur in the integral properties. The local accuracy in short-range predictions is therefore more or less determined by the grid size and the order of accuracy of the scheme.

In numerical general circulation simulations, however, the governing equations are integrated beyond the physical limit of deterministic prediction, which is of the order of a few weeks. Because the atmosphere is turbulent, in a long-term integration there is no "true" solution in the deterministic sense, and such integrations (including long-range numerical weather prediction from an observed initial state) can only predict the statistical properties of the atmosphere. In a long-term integration, then, it is the accuracy of the statistical properties of the solution that concerns us.

It is shown in Section III that maintaining the conservation of enstrophy as well as of kinetic energy is of great advantage in the control of the statistical properties of nondivergent horizontal flow. It not only prevents nonlinear computational instability, but it also maintains the constraint on the kinetic energy exchange between motions of different size. A false systematic computational cascade of kinetic energy into small-scale motions is prevented, and because there is then relatively little energy in the small-scale motions, the overall error is small. In this way, other statistical properties of the solution, such as conservation of the higher moments of the statistical distribution of vorticity, are approximately maintained.

If the energy in the shortest scale is the result of a spurious computational energy cascade, a decrease of the grid size does not help insofar as the long-term simulation of nonviscous flow is concerned. Such a result is completely different from that which might be expected from the usual analysis of truncation error, which is a measure of the formal difference of the finite-difference equation from the original differential equation. The paradox occurs because a decrease of the grid size allows a further computational cascade of energy into the added part of the spectral domain. After a sufficient period of integration, the cascading energy will again reach and accumulate in the shortest resolvable scale. The overall error will become large again and the prediction of some of the statistical properties will become even worse than with the coarser grid (Section III will show an example of this).

The existence of lateral viscosity can make a false computational cascade of energy less harmful. Since such viscosity is more effective for smaller scales, however, a spurious computational energy cascade into these scales falsely enhances the total amount of energy dissipation.

The second part of Section III describes a finite difference Jacobian that maintains the conservation of enstrophy and kinetic energy and that is suitable for the representation of advection of any quantity in two-dimensional, incompressible flow. The usefulness of this scheme as a guide in the formulation of a finite-difference scheme for the primitive equations rests on the fact that although the motions of the atmospheric general circulation are not exactly horizontal and nondivergent, they are to a good approximation quasi-geostrophic. This type of motion is quasi-nondivergent, as far as horizontal advection is concerned; divergence is important only in the linear, or approximately linear, terms. Thus as far as the consideration of the (nonlinear) advection terms is concerned, the finite-difference scheme for advection by the nondivergent part of the flow is crucial; indeed, a scheme that is inadequate for purely nondivergent motion is almost certainly inadequate for quasi-nondivergent motion.

The other integral constraints maintained by the finite-difference scheme of the UCLA model are not for the prevention of a computational cascade and, therefore, do not directly increase the overall accuracy of the statistics of the solution. The maintenance of these other integral constraints does help make the errors less systematic, however, in terms of the generation, destruction, and conversion of energy, entropy, and angular momentum or vorticity.

Therefore, in a statistical sense, they make the physics of the discrete model more analogous to the physics of the continuous atmosphere.

The following examples may serve to illustrate this point. A small error in the meridional velocity is tolerable if that error is random; but if the error is a systematic one resulting, say, from some latitudinal distribution of false mass sources and sinks, there will be a systematic false generation of the relative angular momentum of the global atmosphere. A small false residual of the line integral of the pressure gradient force, which is an irrotational vector, can drastically affect the angular momentum and vorticity budgets. A systematic small error in the vertical distribution of potential temperature in the troposphere can cause a significant error in the gross static stability, and thereby produce large errors in the motion field.

It is important to note that the integral constraints are maintained regardless of the initial condition, because their maintenance is guaranteed by the form of the finite-difference scheme. Difference schemes that do not have such a formal guarantee may approximately maintain the integral constraints with a particular set of initial conditions, but may not do so with another set of initial conditions. Because the governing equations are nonlinear, we have no way of knowing in advance the integral properties of the solutions obtained with such schemes.

In numerical models of the atmosphere, the energy propagation in physical space, as well as in spectral space, must be properly simulated. In particular, the energy propagation by small-scale dispersive inertia-gravity waves, excited by a local breakdown of the quasi-geostrophic balance, is important in restoring an approximately quasi-geostrophic flow by geostrophic adjustment. Unless the geostrophic adjustment process can operate properly, nothing is gained by maintaining integral constraints on quasi-geostrophic motion. The finite-difference scheme of the model is designed to control the small-scale inertia-gravity waves and the accompanying geostrophic adjustment process.

Computational problems also arise in the simulation of the vertical propagation of wave energy forced from below. The vertical differencing scheme and the location of the levels in the stratosphere are designed to eliminate any false computational internal reflections of the wave energy in a resting isothermal atmosphere.

## III. Finite Difference Schemes for Homogeneous Incompressible Flow

Our governing equations are the primitive equations. Under typical conditions in the atmosphere (low Rossby and Froude numbers), these equations govern two well-separable types of motion. One type is the high-frequency inertia-gravity wave, for which nonlinearity is usually small; the other is low-frequency, quasi-geostrophic motion, for which nonlinearity is usually dominant. It is known that the energy of locally excited inertia-gravity waves disperses away into a wider space, leaving the slowly changing quasi-geostrophic motion behind. This process is called "geostrophic adjustment."

Consequently, there are two main computational problems in the simulation of large-scale motions with the primitive equations. One computational problem is the proper simulation of the geostrophic adjustment. The other is the simulation of the slowly changing quasi-geostrophic (and, therefore, quasi-nondivergent) motion after it has been established by geostrophic adjustment.

This section discusses finite-difference schemes to deal with both of these computational problems for the case of homogeneous incompressible flow. The results of this section will be used in Section VI as a guide for the design of the horizontal finite-difference scheme for the model.

### A. Distribution of Variables over the Grid Points

Winninghoff (1968) found that the simulation of the geostrophic adjustment process with a finite-difference scheme is highly dependent on the manner in which variables are distributed over the grid points. The following discussion is based on his work.

Consider the simplest fluid in which geostrophic adjustment can take place—namely, an incompressible, homogeneous, nonviscous, hydrostatic, rotating fluid with a flat bottom and a free top surface. The basic equations which govern such a fluid are the so-called shallow water equations, given by

$$du/dt - fv + g(\partial h/\partial x) = 0, \tag{1}$$

$$dv/dt + fu + g(\partial h/\partial y) = 0, \tag{2}$$

$$dh/dt + h(\partial u/\partial x + \partial v/\partial y) = 0, \tag{3}$$

where $t$ is time, $x$ and $y$ are the horizontal cartesian coordinates, $u$ and $v$ are the velocity components in the $x$ and $y$ directions, respectively, $h$ is the depth of the fluid, $f$ is a constant coriolis parameter, and $g$ is gravity. The individual time rate of change is defined by

$$\frac{d}{dt} \equiv \frac{\partial}{\partial t} + u\frac{\partial}{\partial x} + v\frac{\partial}{\partial y}. \tag{4}$$

In most of this study a linearized version of these equations is used, which is obtained by replacing $d/dt$ by $\partial/\partial t$, and by replacing $h$ as the factor on $(\partial u/\partial x + \partial v/\partial y)$ in Eq. (3) by $H$, the mean value of $h$. This procedure is justified when the Rossby number is small and the horizontal scale is of the order of the radius of deformation or less.

Consider the five distributions of the dependent variables $h$, $u$, and $v$, on a square grid illustrated in Fig. 3. Each of the following five space finite-difference schemes used with the linearized equations is the simplest second-order scheme for the correspondingly labeled distribution.

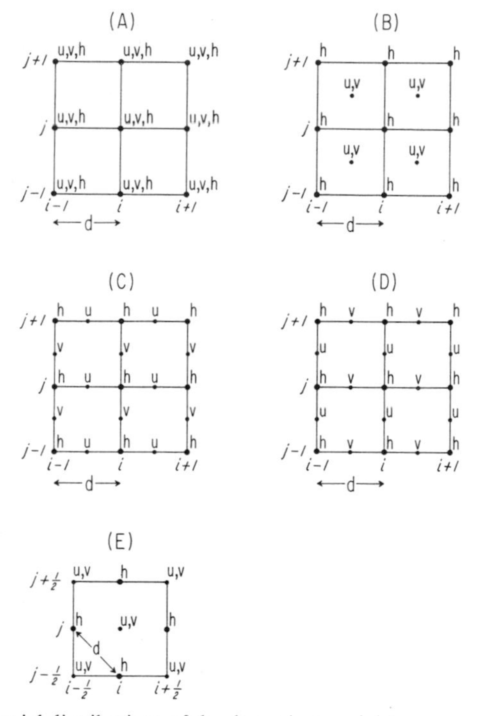


FIG. 3. Spatial distributions of the dependent variables on a square grid.

Scheme A:

$$\partial u/\partial t - fv + (g/d)\overline{(\delta_x h)}^x = 0, \tag{5}$$

$$\partial v/\partial t + fu + (g/d)\overline{(\delta_y h)}^y = 0, \tag{6}$$

$$\partial h/\partial t + (H/d)[\overline{(\delta_x u)}^x + \overline{(\delta_y v)}^y] = 0; \tag{7}$$

Scheme B:

$$\partial u/\partial t - fv + (g/d)(\overline{\delta_x h})^y = 0, \tag{8}$$

$$\partial v/\partial t + fu + (g/d)(\overline{\delta_y h})^x = 0, \tag{9}$$

$$\partial h/\partial t + (H/d)[(\overline{\delta_x u})^y + (\overline{\delta_y v})^x] = 0; \tag{10}$$

Scheme C:

$$\partial u/\partial t - f\overline{v}^{xy} + (g/d)(\delta_x h) = 0, \tag{11}$$

$$\partial v/\partial t + f\overline{u}^{xy} + (g/d)(\delta_y h) = 0, \tag{12}$$

$$\partial h/\partial t + (H/d)[(\delta_x u) + (\delta_y v)] = 0; \tag{13}$$

Scheme D:

$$\partial u/\partial t - f\overline{u}^{xy} + (g/d)(\overline{\delta_x h})^{xy} = 0, \tag{14}$$

$$\partial v/\partial t + f\overline{v}^{xy} + (g/d)(\overline{\delta_y h})^{xy} = 0; \tag{15}$$

$$\partial h/\partial t + (H/d)[(\overline{\delta_x u})^{xy} + (\overline{\delta_y v})^{xy}] = 0; \tag{16}$$

Scheme E:

$$\partial u/\partial t - fv + (g/d^*)(\delta_x h) = 0, \tag{17}$$

$$\partial v/\partial t + fu + (g/d^*)(\delta_y h) = 0, \tag{18}$$

$$\partial h/\partial t + (H/d^*)[(\delta_x u) + (\delta_y v)] = 0; \tag{19}$$

where we define

$$(\delta_x \alpha)_{ij} \equiv \alpha_{i+1/2,\,j} - \alpha_{i-1/2,\,j}, \tag{20}$$

$$(\overline{\alpha}^x)_{ij} \equiv \tfrac{1}{2}(\alpha_{i+1/2,\,j} + \alpha_{i-1/2,\,j}), \tag{21}$$

and where $i$ and $j$ are the indices of the grid points in the $x$ and $y$ directions, respectively. The symbols $(\delta_y \alpha)_{ij}$ and $(\overline{\alpha}^y)_{ij}$ are defined in a similar manner, but with respect to the $y$ direction, and

$$(\overline{\alpha}^{xy})_{ij} \equiv (\overline{\overline{\alpha}^x}^y)_{ij}. \tag{22}$$

For Schemes A through D, $d$ is the grid size shown in Fig. 3. For Scheme E, $d^*$ equals $\sqrt{2}d$; with this choice Scheme E will have the same number of grid points as the other schemes in a given two-dimensional domain.

In this study, all analyses with the linearized equations leave the time-change terms in differential form. If an explicit scheme is to be used for the time differencing, the time interval must be chosen to satisfy the Courant–Friedrich–Lewy type condition for linear computational stability of the wave with the largest possible phase speed, which for the primitive equations of atmospheric motion is the Lamb wave. A time interval so chosen is adequately small for all other waves, including internal gravity waves, and the time discretization error can be ignored in the first approximation.

Consider, first, the following one-dimensional linear equations:

$$\partial u/\partial t - fv + g(\partial h/\partial x) = 0, \tag{23}$$

$$\partial v/\partial t + fu = 0, \tag{24}$$

$$\partial h/\partial t + H(\partial u/\partial x) = 0. \tag{25}$$

Eliminating $v$ and $h$ yields

$$\partial^2 u/\partial t^2 + f^2 u - gH(\partial^2 u/\partial x^2) = 0. \tag{26}$$

If the solution is assumed proportional to $\exp[i(kx - \nu t)]$, then the angular frequency $\nu$ for the inertia-gravity waves is given by

$$(\nu/f)^2 = 1 + gH(k/f)^2, \tag{27}$$

where $k$ is the wave number in the $x$ direction. The frequency of inertia-gravity waves is a monotonically increasing function of the wave number $k$ unless the radius of deformation $\lambda$ defined by $\sqrt{gH}/f$, is zero. The group velocity $\partial \nu/\partial k$ is not zero unless $\lambda = 0$; this nonzero group velocity is very important for the geostrophic adjustment process.

The effect of the space discretization error on the frequency can now be examined. The space distributions of the dependent variables in this one-dimensional case for Schemes A through D are shown in Fig. 4; Scheme E is not shown, since it is equivalent to Scheme A, but with a smaller grid size. For Schemes A through D the following frequencies are obtained:

Scheme A:

$$(\nu/f)^2 = 1 + (\lambda/d)^2 \sin^2(kd), \tag{28}$$

Scheme B:

$$(\nu/f)^2 = 1 + 4(\lambda/d)^2 \sin^2(kd/2), \tag{29}$$

Scheme C:

$$(\nu/f)^2 = \cos^2(kd/2) + 4(\lambda/d)^2 \sin^2(kd/2), \tag{30}$$

Scheme D:

$$(\nu/f)^2 = \cos^2(kd/2) + (\lambda/d)^2 \sin^2(kd). \tag{31}$$

In all cases, the nondimensional frequency $\nu/f$ depends on the two parameters $kd$ and $\lambda/d$.

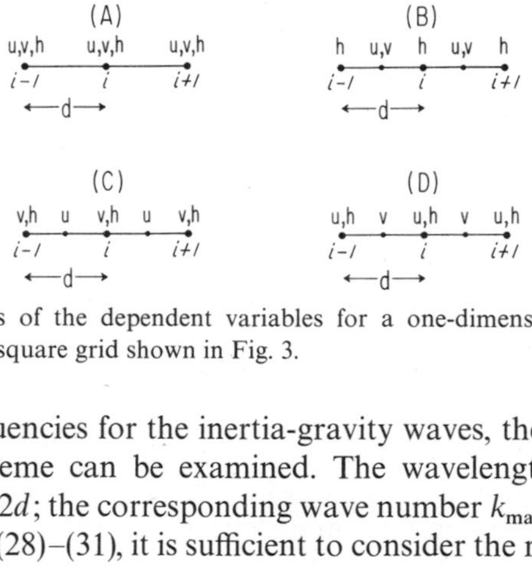

FIG. 4. Distributions of the dependent variables for a one-dimensional grid which correspond to those for a square grid shown in Fig. 3.

With these frequencies for the inertia-gravity waves, the dispersion properties of each scheme can be examined. The wavelength of the shortest resolvable wave is $2d$; the corresponding wave number $k_{max}$ is $\pi/d$. Therefore, in examining Eqs. (28)–(31), it is sufficient to consider the range $0 < kd < \pi$.

*Scheme A.* The frequency reaches its maximum at $kd = \pi/2$, which means that the group velocity at $kd = \pi/2$ is zero. When inertia-gravity waves at about this wave number are excited somewhere in the domain (by nonlinearity, heating, etc.), the wave energy stays there. In this scheme, a wave with $kd = \pi$ behaves like a pure inertia oscillation.

*Scheme B.* For nonzero $\lambda$ the frequency is monotonically increasing in the range $0 < kd < \pi$.

*Scheme C.* The frequency is monotonically increasing for $\lambda/d > \frac{1}{2}$ and monotonically decreasing for $\lambda/d < \frac{1}{2}$. For $\lambda/d = \frac{1}{2}$, $\nu^2 = f^2$ and the group velocity is zero for all $k$.

*Scheme D.* The frequency reaches a maximum at $(\lambda/d)^2 \cos(kd) = \frac{1}{4}$. Moreover, $kd = \pi$ is a stationary wave.

These results for the one-dimensional case show that Scheme B is the most satisfactory. However, when $\lambda/d$ is sufficiently larger than 1/2, Scheme C is as good as Scheme B. To illustrate, Fig. 5 shows a comparison of the dependence of $|\nu|/f$ on $kd/\pi$ for the case $\lambda/d = 2$.

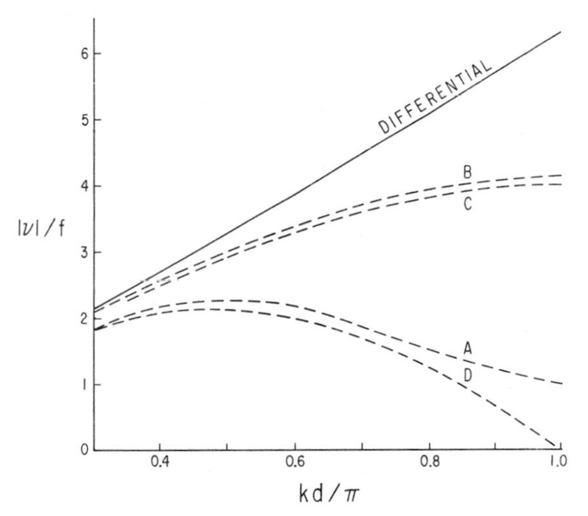


FIG. 5. The dependence of the (nondimensional) frequency on the (nondimensional) wave number for the shallow water equations for the case $\lambda/d = 2$. Solid line corresponds to the differential case, given by Eq. (27); dashed lines represent the difference Schemes A–D, given by Eqs. (28)–(31).

Cahn (1945) gave the solution of an initial value problem for which Eqs. (23)–(25) are the governing equations. At the initial time, he let $h =$ constant, $u = 0$, $v = V_0$ in the domain from $x = -a$ to $x = a$, and $v = 0$ outside of this domain.

A form of the solution $u(x, t)$ for these same initial conditions, suitable for use in a comparison of the differential and difference formulations, is obtained by first expressing $u(x, t)$ in Fourier integral form:

$$u(x, t) = \frac{1}{2\pi} \operatorname{Re} \int_{-\infty}^{\infty} e^{ikx} u^*(k, t)\, dk, \tag{32}$$

where

$$u^*(k, t) = \int_{-\infty}^{\infty} e^{-ikx} u(x, t)\, dx, \tag{33}$$

and $k$ is the wave number in the $x$ direction. The function $u^*(k, t)$ then satisfies the following equation

$$\frac{\partial^2 u^*(k, t)}{\partial t^2} + (f^2 + k^2 gH)u^*(k, t) = 0, \tag{34}$$

which has the general solution

$$u^*(k, t) = A(k) \cos (\nu t) + B(k) \sin (\nu t), \tag{35}$$

where

$$\nu^2 = f^2(1 + \lambda^2 k^2). \tag{36}$$

To determine $A(k)$, Eqs. (35) and (33) are applied at $t = 0$ to give

$$A(k) = u^*(k, 0) = \int_{-\infty}^{\infty} e^{-ikx} u(x, 0)\, dx = 0. \tag{37}$$

Moreover, Eqs. (35) and (33) give

$$\frac{\partial u^*(k, t)}{\partial t} = \nu[-A(k) \sin (\nu t) + B(k) \cos (\nu t)] \tag{38}$$

and

$$\frac{\partial u^*(k, t)}{\partial t} = \int_{-\infty}^{\infty} e^{-ikx} \frac{\partial u(x, t)}{\partial t}\, dx. \tag{39}$$

Applying Eqs. (38) and (39) at $t = 0$ gives an expression for $B(k)$,

$$B(k) = \frac{1}{\nu}\left(\frac{\partial u^*(k, t)}{\partial t}\right)_{t=0} = \frac{1}{\nu}\int_{-\infty}^{\infty} e^{-ikx}\left(\frac{\partial u(x, t)}{\partial t}\right)_{t=0} dx. \tag{40}$$

From the initial conditions and Eq. (23), we have

$$\left(\frac{\partial u(x, t)}{\partial t}\right)_{t=0} = \begin{cases} f V_0 & \text{for } |x| \leqslant a \\ 0 & \text{for } |x| > a. \end{cases} \tag{41}$$

Therefore, from Eq. (40),

$$B(k) = \frac{1}{v}\int_{-a}^{a} e^{-ikx} f V_0\, dx = -\frac{f V_0 e^{-ikx}}{vik}\bigg|_{x=-a}^{x=a} = \frac{2f V_0}{kv}\sin(ak). \quad (42)$$

Finally, Eq. (32) gives, with Eqs. (35) and (42), the desired solution

$$u(x, t) = \frac{faV_0}{\pi}\,\mathrm{Re}\int_{-\infty}^{\infty} \frac{\sin(ak)}{ak}\frac{\sin(vt)}{v} e^{ikx}\, dx, \quad (43)$$

or

$$u(x, t) = \frac{faV_0}{\pi}\int_{-\infty}^{\infty} \frac{\sin(ak)}{ak}\frac{\sin(vt)}{v}\cos kx\, dk. \quad (44)$$

Expressions for $h$ were obtained using Eq. (44) with the equation of continuity (25) in the differential case and with the finite-difference analogs of the equation of continuity for each of the Schemes A–D. In the differential case, $v$ was given by Eq. (36), while with finite-difference Schemes A–D the frequency $v$ was given instead by Eqs. (28)–(31), respectively.* The integral in these expressions for $h$ was evaluated numerically using Simpson's rule with 600 intervals in $k$ from 0 to $\pi/a$. The solutions for $h$ were calculated, with $f = 10^{-4}\ \mathrm{sec}^{-1}$, for constant $x$ for values of $t$ up to 40 hr at 15-min intervals, and for constant $t$ over a range of $x$.

Some results of these calculations, with $a/d = 1$ and $\lambda/d = 2$, are shown in Figs. 6 and 7. Figure 6 shows the *time* variation of $h$ at $x = a$ for the differential case, which approximates the solution obtained by Cahn, and for each of the difference schemes. Figure 7 gives the *space* variation of $h$ in the differential case and for each of the schemes at $t = 80$ hr. As expected, Schemes B and C simulate the geostrophic adjustment better than the other schemes.

However, in the two-dimensional case there is a difficulty with Scheme B. Figure 8 shows $|v|/f$ for each of Schemes A through E, as a function of $kd/\pi$ and $ld/\pi$, where $k$ and $l$ are the wave numbers in the $x$ and $y$ directions; again $\lambda/d = 2$. For comparison, $|v|/f$ for the differential case is shown in Fig. 9. The chain lines in Fig. 8 show the maximum $|v|/f$ for each of a range

*Note added in proof*. Professor Arthur L. Schoenstadt, Department of Mathematics, United States Naval Postgraduate School, Monterey, has pointed out in a personal communication that $f$ in Eq. (44) must also be modified for Schemes C and D. Figures 6 and 7 are based on his corrected expressions.

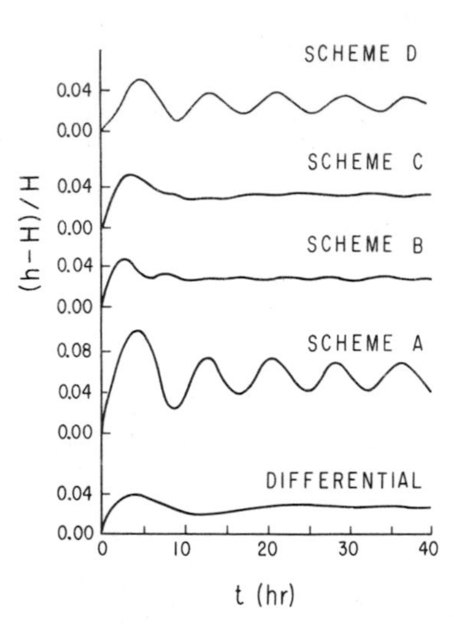


FIG. 6. Time variation of the (nondimensional) height perturbation at $x = a$ for the initial value problem posed by Cahn (1945): comparison of results for the differential case and for difference Schemes A–D.

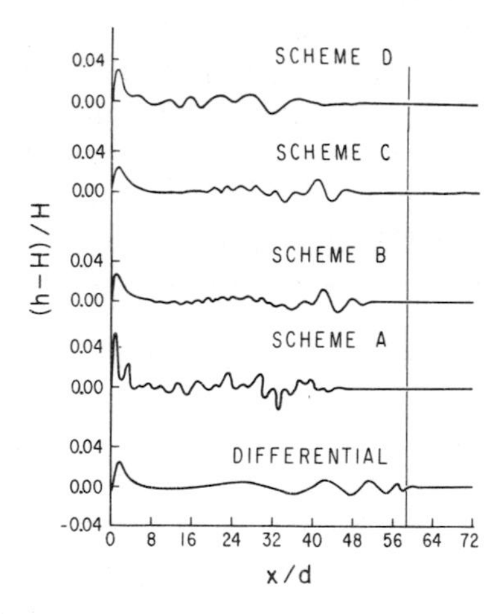


FIG. 7. The spatial variation of the (nondimensional) height perturbation at $t = 80$ hours for the same initial value problem: comparison of results for the differential case and for Schemes A–D. The thin vertical line at $x/d \doteqdot 59$ indicates the theoretical limit of influence.

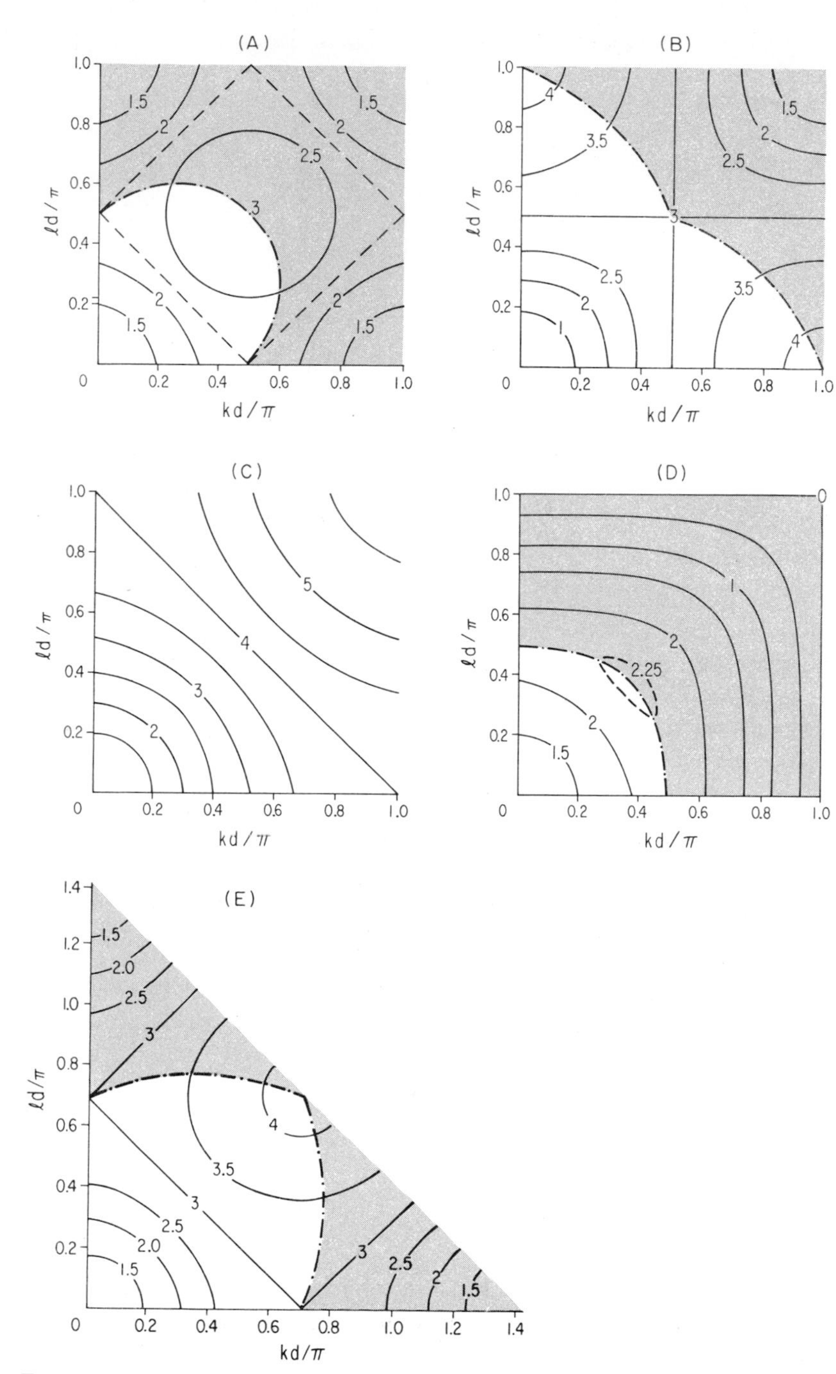


FIG. 8. Contours of the (nondimensional) frequency $|\nu|/f$ for Schemes A–E, as a function of the (nondimensional) horizontal wave numbers for the shallow water equations, for fixed $\lambda/d = 2$. Chain lines show the position of maximum values of the function for a range of the ratio $l/k$.

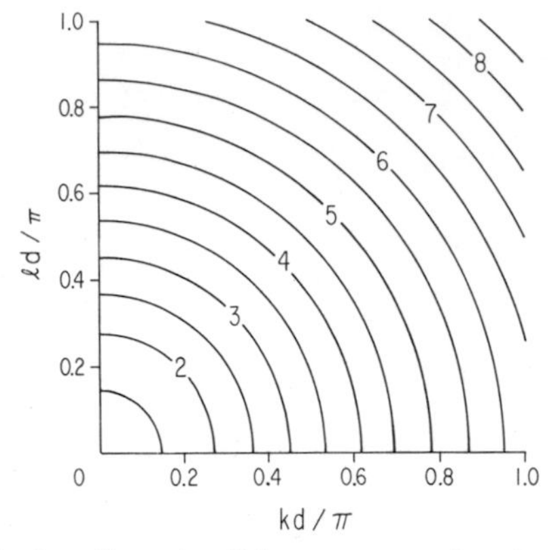


FIG. 9. Contours of the (nondimensional) frequency as a function of the (nondimensional) horizontal wave numbers for the differential shallow water equation for $\lambda/d = 2$, presented for comparison with Fig. 8.

of values of the ratio $l/k$. Note that there is no such maximum for Scheme C or the differential case.

In conclusion, the simulation of geostrophic adjustment is best with Scheme C, except for abnormal situations in which $\lambda/d$ is less than or close to 1.

## B. Two-Dimensional Nondivergent Flow

The next consideration must be the simulation of the slowly changing quasi-geostrophic (and, therefore, quasi-nondivergent) motion after it is established by the geostrophic adjustment process.

Consider, first, a flow which is purely horizontal and nondivergent, governed by the vorticity equation

$$\partial\zeta/\partial t + \mathbf{v} \cdot \nabla\zeta = 0, \tag{45}$$

where

$$\mathbf{v} = \mathbf{k} \times \nabla\psi, \qquad \zeta = \mathbf{k} \cdot \nabla \times \mathbf{v} = \nabla^2\psi, \tag{46}$$

and $\psi$ is the stream function, $\nabla$ is the two-dimensional del operator, and $\mathbf{k}$ is the unit vector normal to the plane of motion. Equation (45) can also be written as

$$\partial\zeta/\partial t = J(\zeta, \psi), \tag{47}$$

where $J$ is the Jacobian operator, defined by

$$J(\zeta, \psi) = (\partial\zeta/\partial x)(\partial\psi/\partial y) - (\partial\zeta/\partial y)(\partial\psi/\partial x). \tag{48}$$

There are the following integral constraints, among others, on the Jacobian:

$$\overline{J(\zeta, \psi)} = 0, \tag{49}$$
$$\overline{\zeta J(\zeta, \psi)} = 0, \tag{50}$$
$$\overline{\psi J(\zeta, \psi)} = 0, \tag{51}$$

where the bar denotes the average over the domain, along the boundary of which $\psi$ is constant. From these integral constraints it is seen that the mean vorticity $\overline{\zeta}$, the enstrophy (one half of the mean square vorticity) $\frac{1}{2}\overline{\zeta^2}$, and the mean kinetic energy $\frac{1}{2}\overline{(\nabla\psi)^2}$ are conserved with time. Conservation of these quantities during the advection process poses important constraints on the statistical properties of two-dimensional incompressible flow, as pointed out by Fjørtoft (1953). In particular, the average wave number $k$ defined by

$$k^2 = \overline{(\nabla^2\psi)^2}/\overline{(\nabla\psi)^2}, \tag{52}$$

is conserved with time, so that no systematic cascade of energy into shorter waves can occur.

If the statistical properties are to be simulated numerically, a finite-difference scheme must be used that approximately conserves these quadratic quantities. Avoiding computational instability in the nonlinear sense is necessary but not sufficient for this purpose. Two examples of stable schemes that have a false energy cascade into shorter waves will be shown later.

It should be noted that if Eq. (47) is applied to a one-dimensional problem, the nonlinearity will be lost. Therefore, the tests of a finite-difference scheme for incompressible flow must be made with two-dimensional problems.

The finite-difference approximation for Eq. (47) may be written in a relatively general form as

$$\zeta_{ij}^{n+1} - \zeta_{ij}^{\ n} = \Delta t \mathbb{J}_{ij}(\zeta^*, \psi^*), \tag{53}$$

where $\zeta_{ij}^{\ n} = (\nabla_{ij}^{\ 2}\psi)^n$ is a finite-difference approximation of $\zeta = \nabla^2\psi$ at the grid point $x = id$, $y = jd$, and at time $t = n\,\Delta t$. Here, $d$ is the grid size, $\Delta t$ is the time interval, and $\nabla_{ij}^{\ 2}$ and $\mathbb{J}_{ij}$ are finite difference approximations for the operators $\nabla^2$ and $J$ at the grid point $x = id$, $y = jd$. Hereafter, the subscripts $i, j$ will be omitted unless they are necessary for clarity.

There are a number of time-difference schemes corresponding to different choices of $\zeta^*$ and $\psi^*$. For example, $\zeta^*$ may be equal to $\zeta^{n+1/2}$, as in the leapfrog scheme; or $\zeta^*$ may be a linear combination of $\zeta^n$ and $\zeta^{n+1}$ such as

$$\zeta^* = \tfrac{1}{2}(\zeta^n + \zeta^{n+1}), \tag{54}$$

which is an implicit scheme of the Crank–Nicholson type. As another example, $\zeta^*$ may be a provisional value of $\zeta$, predicted by

$$\zeta^* = S\zeta^n + \alpha\,\Delta t \mathbb{J}^*(\zeta^n, \psi^n), \tag{55}$$

where $S$ and $\alpha$ may be equal to 1, as in the Matsuno scheme, or $S$ may be a smoothing operator and $\alpha = \frac{1}{2}$, as in the two-step Lax–Wendroff scheme. Here $\mathbb{J}^*$ is not necessarily the same as $\mathbb{J}$.

The change of enstrophy is obtained from Eq. (53), as

$$\tfrac{1}{2}[\overline{(\zeta^{n+1})^2} - \overline{(\zeta^n)^2}] = \Delta t\overline{[(\zeta^{n+1} + \zeta^n)/2]\mathbb{J}(\zeta^*, \psi^*)}, \tag{56}$$

where the bar denotes an average over all grid points in the domain considered. Equation (56) can be rewritten as

$$\begin{aligned}\tfrac{1}{2}[\overline{(\zeta^{n+1})^2} - \overline{(\zeta^n)^2}] &= \overline{\{[(\zeta^{n+1} + \zeta^n)/2] - \zeta^*\}(\zeta^{n+1} - \zeta^n)} \\ &\quad + \overline{\Delta t \zeta^* \mathbb{J}(\zeta^*, \psi^*)}.\end{aligned} \tag{57}$$

To conserve enstrophy, $\zeta^*$ and the form of $\mathbb{J}$ must be chosen in such a way that the right-hand side of Eq. (57) vanishes. The first term on the right vanishes if $\zeta^*$ is chosen as $(\zeta^{n+1} + \zeta^n)/2$. The second term vanishes if the finite-difference Jacobian $\mathbb{J}$ maintains the integral constraint given by Eq. (50) for the differential Jacobian $J$. Similarly, it can be shown that a properly defined kinetic energy is conserved if $\psi^*$ is chosen as $(\psi^{n+1} + \psi^n)/2$ and $\mathbb{J}$ maintains the integral constraint given by Eq. (51).

Consider the grid shown in Fig. 10. There are three basic second-order, finite-difference Jacobians:

$$\begin{aligned}\mathbb{J}_1 &= \Delta_x\zeta\,\Delta_y\psi - \Delta_y\zeta\,\Delta_x\psi, \\ \mathbb{J}_2 &= \Delta_y(\psi\,\Delta_x\zeta) - \Delta_x(\psi\,\Delta_y\zeta), \\ \mathbb{J}_3 &= \Delta_x(\zeta\,\Delta_y\psi) - \Delta_y(\zeta\,\Delta_x\psi),\end{aligned} \tag{58}$$

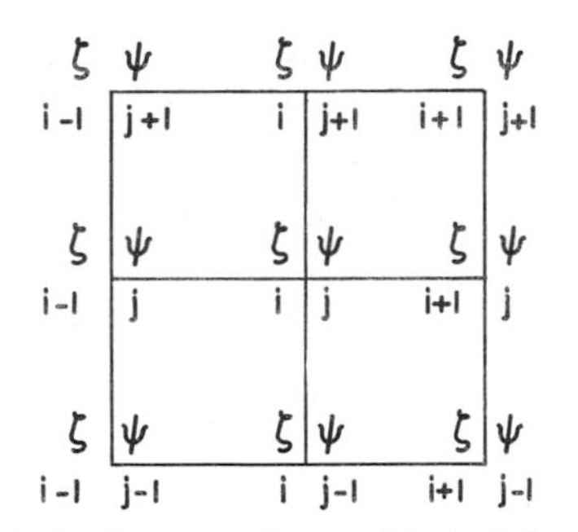


FIG. 10. Grid showing indexing for $\zeta$, $\psi$ points used in the finite-difference Jacobian schemes of Eq. (58).

where $(\Delta_x \alpha)$ is defined by $(\alpha_{i+1,j} - \alpha_{i-1,j})/2d$, and $\Delta_y \alpha$ is defined similarly with respect to $y$. It was shown by Arakawa (1966) that the Jacobian $\mathbb{J}$ given by

$$\mathbb{J} = \alpha \mathbb{J}_1 + \gamma \mathbb{J}_2 + \beta \mathbb{J}_3, \qquad \alpha + \gamma + \beta = 1, \tag{59}$$

conserves mean square vorticity if $\alpha = \beta$ and conserves energy if $\alpha = \gamma$. Examples of Jacobians which have the form of (59) are

$$\begin{aligned} \mathbb{J}_4 &= \tfrac{1}{2}(\mathbb{J}_1 + \mathbb{J}_2), \\ \mathbb{J}_5 &= \tfrac{1}{2}(\mathbb{J}_2 + \mathbb{J}_3), \\ \mathbb{J}_6 &= \tfrac{1}{2}(\mathbb{J}_3 + \mathbb{J}_1), \\ \mathbb{J}_7 &= \tfrac{1}{3}(\mathbb{J}_1 + \mathbb{J}_2 + \mathbb{J}_3). \end{aligned} \tag{60}$$

A schematic representation of the $\zeta$ and $\psi$ points used in constructing the seven finite-difference Jacobians introduced above is given in Fig. 11.

$\mathbb{J}_7$ is the Jacobian proposed by Arakawa (1966) as conserving both enstrophy and energy. $\mathbb{J}_2$ and $\mathbb{J}_6$ conserve enstrophy, but not energy. $\mathbb{J}_3$ and $\mathbb{J}_4$ conserve energy, but not enstrophy. All five schemes mentioned thus far are stable. $\mathbb{J}_1$ does not conserve either quantity, and an analysis similar to that by Phillips (1959), but with the implicit scheme (54), shows that it is unstable. $\mathbb{J}_5$, also, does not conserve either quantity, but experience with numerical tests shows that the instability is very weak, if it exists at all. This is not surprising, since $2\mathbb{J}_5 = 3\mathbb{J}_7 - \mathbb{J}_1$; because $\mathbb{J}_7$ is a quadratic-conserving scheme the time rates of change of the mean quadratic quantities using $\mathbb{J}_5$, for given $\zeta$ and $\psi$, have opposite sign to the time rates of change of the mean quadratic quantities using $\mathbb{J}_1$.

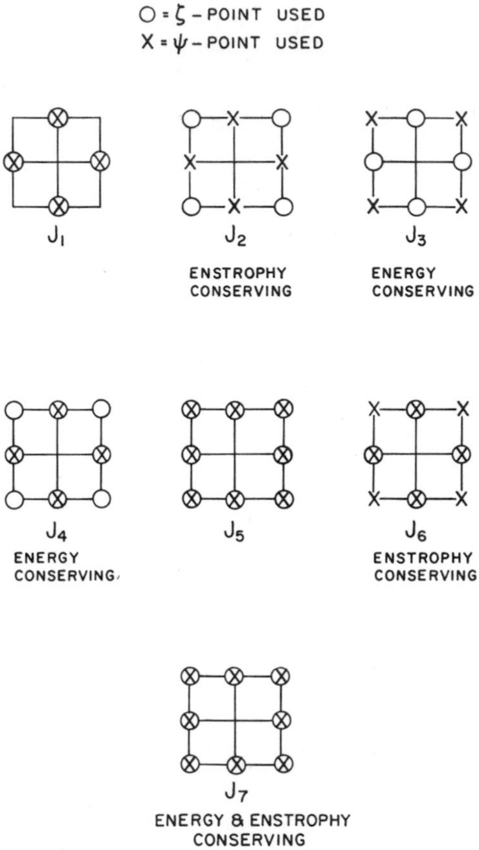

FIG. 11. Schematic representation of $\zeta$ and $\psi$ points used in constructing the finite-difference Jacobians defined by Eqs. (58) and (60).

$\mathbb{J}_7$ is the best second-order scheme because of its formal guarantee for maintaining the integral constraints on the quadratic quantities. $\mathbb{J}_7$ is also just as accurate as any other second-order scheme. A further increase in accuracy can be obtained by going to higher order schemes. The more accurate fourth-order scheme that has the same integral constraints as $\mathbb{J}_7$ was also given by Arakawa (1966).

Numerical tests have been made with the above seven Jacobians. In these tests, the initial condition was given by

$$\psi = \Psi \sin(\pi i/8)[\cos(\pi j/8) + 0.1 \cos(\pi j/4)], \tag{61}$$

and $\Delta t$ was chosen such that $\Delta t/d^2 = 0.7$. The leapfrog scheme was used instead of the implicit scheme. In order to eliminate the gradual separation of the solutions at even and odd time steps that occurs in the leapfrog scheme, a two-level scheme was inserted every 240 time steps. The simplest five-point Laplacian was used. Figures 12 and 13 show the time change of enstrophy and energy obtained with the seven Jacobians. The expected conservation properties are observed, even though the implicit scheme was not used. The energy conserving schemes $\mathbb{J}_3$ and $\mathbb{J}_4$ show considerable increase of enstrophy. On the other hand, the enstrophy conserving schemes $\mathbb{J}_2$ and $\mathbb{J}_6$ approximately conserve energy in spite of the lack of a formal guarantee. This is reasonable because the enstrophy is more sensitive to shorter waves for which the truncation errors are large. $\mathbb{J}_5$ approximately conserves both quantities, again in spite of the lack of formal guarantees.

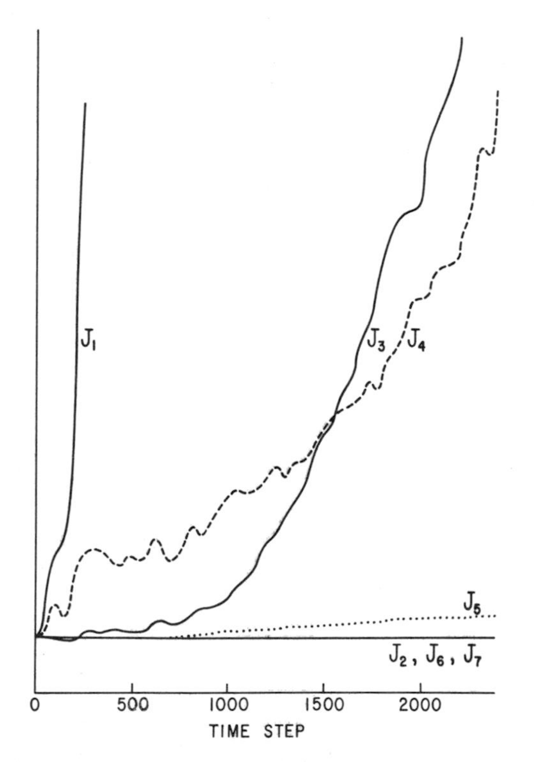


FIG. 12. Comparison of the time variation of the mean square vorticity (units arbitrary) during a numerical integration with the seven finite-difference Jacobians under consideration. (Arakawa, 1970). Reprinted with permission of the publisher American Mathematical Society from *SIAM-AMS Proceedings*. Copyright © 1970, Vol. 2, Fig. 5, p. 35.

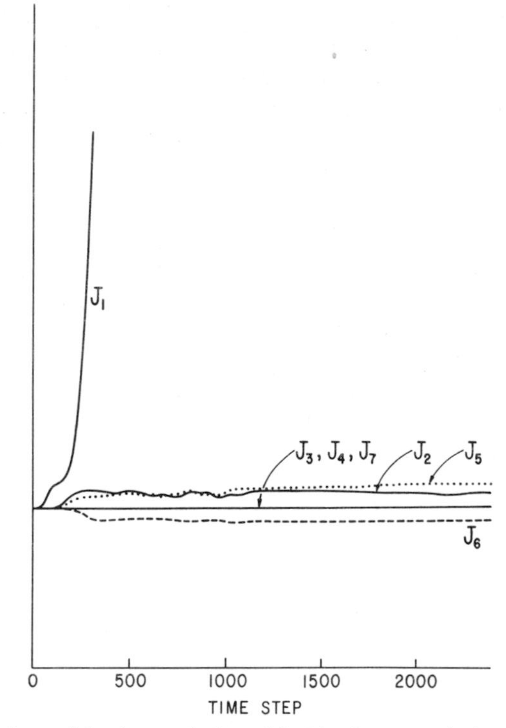


FIG. 13. Comparison of the time variation of the kinetic energy during a numerical integration with the seven finite-difference Jacobians under consideration (Arakawa, 1970). Reprinted with permission of the publisher American Mathematical Society from *SIAM-AMS Proceedings*. Copyright © 1970, Vol. 2, Fig. 6, p. 36.

$\mathbb{J}_7$ conserves both quantities, with only negligible errors arising from the leapfrog scheme. $\mathbb{J}_5$, like $\mathbb{J}_1$ and $\mathbb{J}_7$, maintains the property of the Jacobian $J(\zeta, \psi) = -J(\psi, \zeta)$.

Figure 14 shows the spectral distribution of kinetic energy obtained by the energy and enstrophy conserving scheme $\mathbb{J}_7$ and by the energy conserving scheme $\mathbb{J}_3$ at the end of the calculations. The small arrow shows the wave number for $\sin(\pi i/8) \cos(\pi j/8)$, which contained almost all of the energy at the initial time. Although the total energy was approximately conserved with $\mathbb{J}_3$ there was a considerable spurious energy cascade into the high wave numbers, whereas with $\mathbb{J}_7$ more energy went into a lower wave number than into the higher wave numbers, in agreement with the conservation of the average wave number as given by Eq. (52).

Whether the increase of the enstrophy is important in the simulation of large-scale atmospheric motion will depend on the viscosity used with the

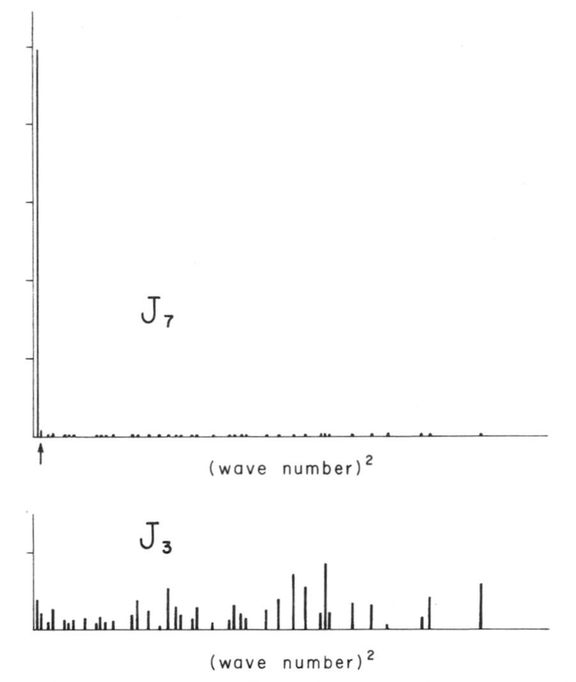


FIG. 14. A comparison of the spectral distribution of kinetic energy, obtained with $\mathbb{J}_3$ and $\mathbb{J}_7$, after a numerical integration of 2400 time steps. Arrow shows the wave number that contained most of the energy at the initial time.

complete equation. A relatively small amount of viscosity may be sufficient to keep the enstrophy quasi-constant in time. However, the viscosity will also remove energy, and as a result the average wave number, defined by Eq. (52), will falsely increase with time.

In Section II it was pointed out that when a scheme that produces a strong computational cascade is used, a decrease in grid size does not mean an increase in overall accuracy as far as long-term numerical integrations are concerned. Figure 15 shows such an example. With an identical initial condition, experiments have been made using $\mathbb{J}_3$ with three different grid sizes. The nondimensional parameter $\Psi\Delta t/d^2$ is kept the same for the three experiments. A two-level scheme was inserted every 120 time steps to suppress separation of the solution due to the leapfrog scheme. The figure shows a more rapid increase of enstrophy with the smaller grid sizes. Since the kinetic energy is practically conserved in all three experiments, a larger enstrophy means a smaller average scale of the motion. These results show that the convergence of the scheme, in the nonlinear sense, must be seriously questioned.

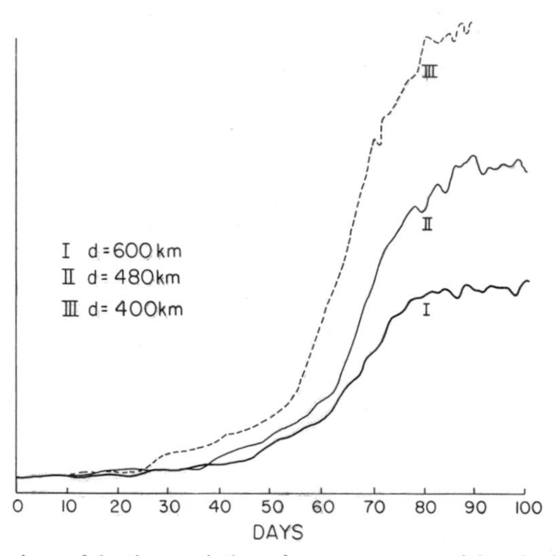


FIG. 15. A comparison of the time variation of mean square vorticity obtained by numerical integrations using $\mathbb{J}_3$ for three different grid sizes.

With the grid shown in Fig. 16, $\mathbb{J}_7$ may be written as

$$-\mathbb{J}_7(\zeta, \psi) = \frac{2}{3}\frac{1}{d^2}\{\delta_x(-\overline{\delta_y\psi}^{xy}\overline{\zeta}^x) + \delta_y(\overline{\delta_x\psi}^{xy}\overline{\zeta}^y)\}$$
$$+ \frac{1}{3}\frac{1}{2d^2}\{\delta_{x'}(-\delta_{y'}\psi\overline{\zeta}^{x'}) + \delta_{y'}(\delta_{x'}\psi\overline{\zeta}^{y'})\}, \tag{62}$$

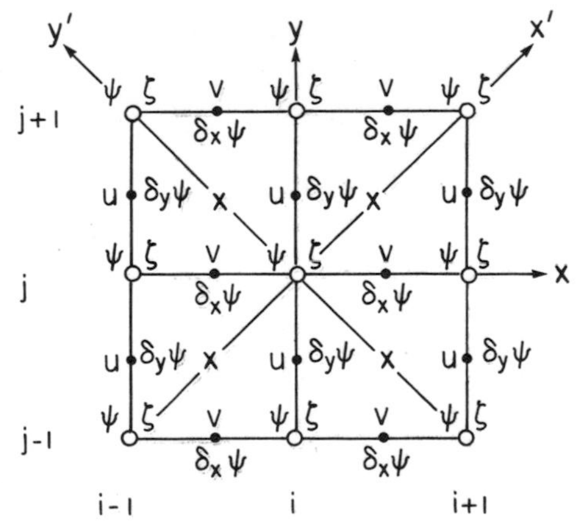


FIG. 16. Grid showing points of definition of the dependent variables and axes of definition for mean and difference operators used in the differencing of the momentum advection terms.

where $\delta_x \alpha$ and $\bar{\alpha}^x$ are defined as

$$(\delta_x \alpha)_{i+1/2,\, j} = \alpha_{i+1,\, j} - \alpha_{i,\, j} \tag{63}$$

and

$$(\bar{\alpha}^x)_{i+1/2,\, j} = \tfrac{1}{2}(\alpha_{i+1,\, j} + \alpha_{i,\, j}). \tag{64}$$

The symbols $\delta_y \alpha$ and $\bar{\alpha}^y$ are defined in a similar manner, but with respect to the $y$ direction, and $\bar{\alpha}^{xy} = \bar{\bar{\alpha}}^{x^{y}}$. The symbols $\delta_{x'}$, $\delta_{y'}$, $\bar{\alpha}^{x'}$, and $\bar{\alpha}^{y'}$ follow the same definitions, but in the $x'$ and $y'$ directions and with the spacing $\sqrt{2}d$. It can easily be shown that

$$\delta_{x'}\psi = \overline{\delta_x \psi}^y + \overline{\delta_y \psi}^x, \qquad \delta_{y'}\psi = -\overline{\delta_x \psi}^y + \overline{\delta_y \psi}^x. \tag{65}$$

Since it is the momentum equation and not the vorticity equation that is used in the model, the next problem is to find a finite-difference scheme for the advection term in the momentum equation. The guiding assumption in our present approach is that a scheme that is inadequate for purely nondivergent motion is almost certainly also inadequate for the quasi-nondivergent motion typical of large-scale atmospheric motions. Thus the first constraint on a finite-difference scheme for the momentum equation is that it become equivalent to $\partial\zeta/\partial t = \mathbb{J}_7(\zeta, \psi)$ when the flow is horizontal and nondivergent.

The vorticity can be expressed as

$$\begin{aligned}\zeta_{ij} = (\nabla^2\psi)_{ij} &\equiv \frac{1}{d}\left(\frac{\psi_{i+1,\, j} - \psi_{ij}}{d} - \frac{\psi_{ij} - \psi_{i-1,\, j}}{d} + \frac{\psi_{i,\, j+1} - \psi_{ij}}{d} - \frac{\psi_{ij} - \psi_{i,\, j-1}}{d}\right) \\ &= \frac{1}{d^2}(\psi_{i+1,\, j} + \psi_{i-1,\, j} + \psi_{i,\, j+1} + \psi_{i,\, j-1} - 4\psi_{ij}).\end{aligned} \tag{66}$$

For the grid points shown in Fig. 16, $u$ and $v$ are defined by

$$u_{i,\, j+1/2} \equiv -\frac{(\delta_y \psi)_{i,\, j+1/2}}{d}, \qquad v_{i+1/2,\, j} \equiv \frac{(\delta_x \psi)_{i+1/2,\, j}}{d}. \tag{67}$$

Then the vorticity given by Eq. (66) is

$$\zeta_{ij} = (1/d)[(\delta_x v)_{ij} - (\delta_y u)_{ij}], \tag{68}$$

and the vorticity equation may be written as

$$(\partial/\partial t)[(\delta_x v)_{ij} - (\delta_y u)_{ij}] = \mathbb{J}_{ij}(\delta_x v - \delta_y u, \psi). \tag{69}$$

Here the symbol $\mathbb{J}$ is used for $\mathbb{J}_7$.

Consider $\mathbb{J}_{i,\,j+1/2}(u, \bar{\psi}^y)$. From a property of the Jacobian, which is maintained by $\mathbb{J}_7$,

$$\mathbb{J}_{i,\,j+1/2}(u, \bar{\psi}^y) = \mathbb{J}_{i,\,j+1/2}(u, \bar{\psi}^y + \tfrac{1}{2}ud), \tag{70}$$

and

$$\mathbb{J}_{i,\,j-1/2}(u, \bar{\psi}^y) = \mathbb{J}_{i,\,j-1/2}(u, \bar{\psi}^y - \tfrac{1}{2}ud). \tag{71}$$

Note that $(\bar{\psi}^y + \frac{1}{2}ud)_{i,\,j+1/2} = (\bar{\psi}^y - \frac{1}{2}ud)_{i,\,j-1/2} = \psi_{ij}$ for *arbitrary* $i$, $j$. Using (70) and (71),

$$\begin{aligned}[\delta_y \mathbb{J}(u, \bar{\psi}^y)]_{ij} &\equiv \mathbb{J}_{i,\,j+1/2}(u, \bar{\psi}^y) - \mathbb{J}_{i,\,j-1/2}(u, \bar{\psi}^y), \\ &= \mathbb{J}_{ij}(\delta_y u, \psi).\end{aligned} \tag{72}$$

Similarly,

$$[\delta_x \mathbb{J}(v, \bar{\psi}^x)]_{ij} = \mathbb{J}_{ij}(\delta_x v, \psi). \tag{73}$$

Equations (72) and (73) are analogs, respectively, of

$$(\partial/\partial y)J(u, \psi) = J(\partial u/\partial y, \psi), \qquad (\partial/\partial x)J(v, \psi) = J(\partial v/\partial x, \psi).$$

From Eqs. (72), (73), and (68),

$$\begin{aligned}[\delta_x \mathbb{J}(v, \bar{\psi}^x)]_{ij} - [\delta_y \mathbb{J}(u, \bar{\psi}^y)]_{ij} &= \mathbb{J}_{ij}(\delta_x v - \delta_y u, \psi) \\ &= d\mathbb{J}_{ij}(\zeta, \psi).\end{aligned} \tag{74}$$

The conclusion is that

$\mathbb{J}(u, \bar{\psi}^y)$ for $-\mathbf{v} \cdot \nabla u$ at the $u$ points
$\mathbb{J}(v, \bar{\psi}^x)$ for $-\mathbf{v} \cdot \nabla v$ at the $v$ points

are consistent with

$\mathbb{J}(\zeta, \psi)$ for $-\mathbf{v} \cdot \nabla \zeta$ at the $\psi$ points.

Equations (62) and (65), with $\zeta$ replaced by $u$ and $\psi$ by $\overline{\psi}^y$, give the form

$$-\mathbb{J}(u, \overline{\psi}^y) = \frac{2}{3d^2}\left[\delta_x\left(-\overline{\delta_y\overline{\psi}}^{yxy}\,\overline{u}^x\right) + \delta_y\left(\overline{\delta_x\overline{\psi}}^{yxy}\,\overline{u}^y\right)\right]$$

$$+ \frac{1}{6d^2}\left[\delta_{x'}\left\{\left(\overline{\delta_x\overline{\psi}}^{yy} - \overline{\delta_y\overline{\psi}}^{yx}\right)\overline{u}^{x'}\right\} + \delta_{y'}\left\{\left(\overline{\delta_x\overline{\psi}}^{yy} + \overline{\delta_y\overline{\psi}}^{yx}\right)\overline{u}^{y'}\right\}\right]. \qquad (75)$$

Define $u^*$ and $v^*$ by

$$u^* = -(1/d)\overline{\delta_y\overline{\psi}}^x, \qquad v^* = (1/d)\overline{\delta_x\overline{\psi}}^y. \qquad (76)$$

Then (75), which is the divergence of $u$-momentum transport, becomes

$$\frac{2}{3d}\left[\delta_x\left(\overline{u^*}^{yy}\,\overline{u}^x\right) + \delta_y\left(\overline{v^*}^{xy}\,\overline{u}^y\right)\right] + \frac{1}{6d}\left[\delta_{x'}\left(\overline{v^* + u^*}^{y}\,\overline{u}^{x'}\right) + \delta_{y'}\left(\overline{v^* - u^*}^{y}\,\overline{u}^{y'}\right)\right]. \qquad (77)$$

Similarly, the divergence of $v$-momentum transport becomes

$$\frac{2}{3d}\left[\delta_x\left(\overline{u^*}^{xy}\,\overline{v}^x\right) + \delta_y\left(\overline{v^*}^{xx}\,\overline{v}^y\right)\right] + \frac{1}{6d}\left[\delta_{x'}\left(\overline{v^* + u^*}^{x}\,\overline{v}^{x'}\right) + \delta_{y'}\left(\overline{v^* - u^*}^{x}\,\overline{v}^{y'}\right)\right]. \qquad (78)$$

In Fig. 16, the distribution of $u$ and $v$ is staggered as in Schemes C and D of the last subsection. Results of the last subsection indicate, however, that Scheme C is definitely better than Scheme D in view of the geostrophic adjustment and therefore the $x$ points rather than the $\psi$ points in Fig. 16 carry pressure and temperature.

## C. Finite Difference Scheme for the Nonlinear Shallow Water Equations

For use with the advection term of the momentum equations in the general circulation model, the finite-difference expressions (77) and (78) derived for the case of horizontal nondivergent flow must be generalized to the case of divergent flow. In this subsection, the principles guiding such a generalization will be illustrated through the derivation of a finite-difference scheme suitable for integration of the nonlinear shallow water equations on a square grid with the variables staggered as in the C scheme. The analogous development for the momentum equations governing three-dimensional motion in curvilinear coordinates is presented in Section VI.

The governing differential equations are Eqs. (1)–(3), restated below for convenience:

$$\frac{\partial u}{\partial t} + u\frac{\partial u}{\partial x} + v\frac{\partial u}{\partial y} - fv + g\frac{\partial h}{\partial x} = 0, \tag{79}$$

$$\frac{\partial v}{\partial t} + u\frac{\partial v}{\partial x} + v\frac{\partial v}{\partial y} + fu + g\frac{\partial h}{\partial y} = 0, \tag{80}$$

$$\frac{\partial h}{\partial t} + \frac{\partial (hu)}{\partial x} + \frac{\partial (hv)}{\partial y} = 0. \tag{81}$$

Combining Eq. (81) with Eqs. (79) and (80) gives another useful form of the momentum equations,

$$\frac{\partial (uh)}{\partial t} + \frac{\partial (huu)}{\partial x} + \frac{\partial (hvu)}{\partial y} - fhv + gh\frac{\partial h}{\partial x} = 0, \tag{82}$$

$$\frac{\partial (vh)}{\partial t} + \frac{\partial (huv)}{\partial x} + \frac{\partial (hvv)}{\partial y} + fhu + gh\frac{\partial h}{\partial y} = 0. \tag{83}$$

Multiplying Eq. (79) by $u$ and Eq. (80) by $v$ and combining the results with Eq. (81) yields the equations for the time change of kinetic energy,

$$\frac{\partial}{\partial t}(h\tfrac{1}{2}u^2) + \frac{\partial [hu\tfrac{1}{2}u^2]}{\partial x} + \frac{\partial [hv\tfrac{1}{2}u^2]}{\partial y} - fhuv + ghu\frac{\partial h}{\partial x} = 0, \tag{84}$$

$$\frac{\partial}{\partial t}(h\tfrac{1}{2}v^2) + \frac{\partial [hu\tfrac{1}{2}v^2]}{\partial x} + \frac{\partial [hv\tfrac{1}{2}v^2]}{\partial y} + fhuv + ghv\frac{\partial h}{\partial y} = 0. \tag{85}$$

Multiplying Eq. (81) by $gh$ gives the equation for the change of potential energy,

$$\frac{\partial}{\partial t}\left(\frac{gh^2}{2}\right) + gh\left[\frac{\partial}{\partial x}(hu) + \frac{\partial}{\partial y}(hv)\right] = 0, \tag{86}$$

or

$$\frac{\partial}{\partial t}\left(\frac{gh^2}{2}\right) + \frac{\partial}{\partial x}(gh^2u) + \frac{\partial}{\partial y}(gh^2v) - gh\left[u\frac{\partial h}{\partial x} + v\frac{\partial h}{\partial y}\right] = 0. \tag{87}$$

The coriolis force of course makes no contribution to the change of total kinetic energy. Also, the summation of the last terms in Eqs. (84), (85), and (87) is zero. These points, which lead to conservation of the total energy, are utilized in the construction of the finite-difference scheme.

The differencing for the continuity equation is chosen on the basis of simplicity. At $h$ points, Eq. (81) can be represented as

$$\frac{\partial}{\partial t} h_{i,j} + \frac{1}{d^2}[F_{i+1/2,j} - F_{i-1/2,j} + G_{i,j+1/2} - G_{i,j-1/2}] = 0, \tag{88}$$

where the mass fluxes

$$\begin{aligned} F_{i+1/2,j} &\equiv d[\overline{h}^x u]_{i+1/2,j} \\ G_{i,j+1/2} &\equiv d[\overline{h}^y v]_{i,j+1/2} \end{aligned} \tag{89}$$

are defined at $u$ and $v$ points, respectively. The time change terms are left in differential form throughout this section.

The first requirement on the finite-difference scheme is that it conserve total kinetic energy during inertial processes. To this end, considering first the $u$ momentum equation (82), the terms

$$\frac{\partial}{\partial t}(uh) + \frac{\partial}{\partial x}(huu) + \frac{\partial}{\partial y}(hvu)$$

can be represented by the following form, which automatically guarantees proper conservation of integrated zonal momentum:

$$\begin{aligned} &\frac{\partial}{\partial t}(H^{(u)}u)_{i,j} + \frac{1}{d^2}[\delta_x(\mathscr{F}^{(u)}\overline{u}^x) + \delta_y(\mathscr{G}^{(u)}\overline{u}^y) \\ &\qquad + \delta_{x'}(\tilde{\mathscr{F}}^{(u)}\overline{u}^{x'}) + \delta_{y'}(\tilde{\mathscr{G}}^{(u)}\overline{u}^{y'})]_{i,j} \end{aligned} \tag{90}$$

where $H^{(u)}$ and $\mathscr{F}^{(u)}$, $\mathscr{G}^{(u)}$, $\tilde{\mathscr{F}}^{(u)}$, $\tilde{\mathscr{G}}^{(u)}$ are as yet undefined (see Fig. 17 for the points of definition of the new mass flux symbols). For simplicity, the convention of using the indices $(i, j)$ for the variable whose prognostic equation is under consideration is followed. If these new terms are chosen in such a way that they satisfy

$$\frac{\partial}{\partial t} H^{(u)}_{i,j} + \frac{1}{d^2}[\delta_x\mathscr{F}^{(u)} + \delta_y\mathscr{G}^{(u)} + \delta_{x'}\tilde{\mathscr{F}}^{(u)} + \delta_{y'}\tilde{\mathscr{G}}^{(u)}]_{i,j} = 0, \tag{91}$$

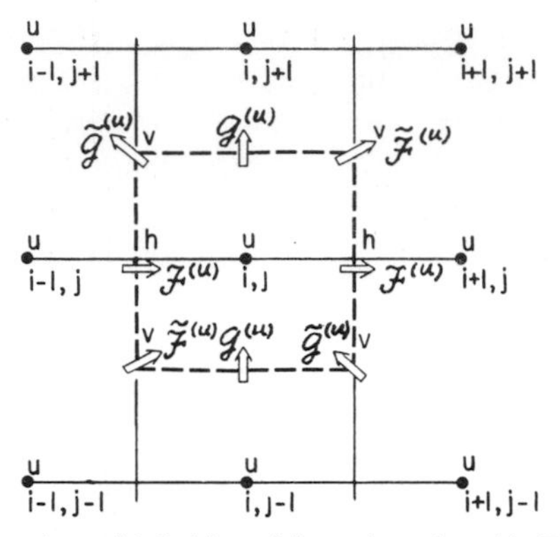


FIG. 17. Grid showing points of definition of fluxes introduced in Eq. (90) in the differencing of the advection terms of the $u$-momentum equation.

then by subtracting Eq. (91) multiplied by $u_{ij}$, (90) can be shown equivalent to

$$H^{(u)}_{i,j}\frac{\partial u_{i,j}}{\partial t} + \frac{1}{d^2}\left[\overline{\mathscr{F}^{(u)}\,\delta_x u}^{x} + \overline{\mathscr{G}^{(u)}\,\delta_y u}^{y} + \overline{\tilde{\mathscr{F}}^{(u)}\,\delta_{x'} u}^{x'} + \overline{\tilde{\mathscr{G}}^{(u)}\,\delta_{y'} u}^{y'}\right]_{i,j}. \tag{92}$$

Multiplying (92) by $u_{ij}$ and combining with Eq. (91) multiplied by $\frac{1}{2}u^2_{i,j}$, a finite-difference analog to the first three terms of Eq. (84) is obtained:

$$\begin{aligned}\frac{\partial}{\partial t}(H^{(u)}\tfrac{1}{2}u^2)_{ij} + \frac{1}{2d^2}\big[&\mathscr{F}^{(u)}_{i+1/2,\,j}u_{i,\,j}u_{i+1,\,j} - \mathscr{F}^{(u)}_{i-1/2,\,j}u_{i-1,\,j}u_{i,\,j}\\
&+ \mathscr{G}^{(u)}_{i,\,j+1/2}u_{i,\,j}u_{i,\,j+1} - \mathscr{G}^{(u)}_{i,\,j-1/2}u_{i,\,j-1}u_{i,\,j}\\
&+ \tilde{\mathscr{F}}^{(u)}_{i+1/2,\,j+1/2}u_{i,\,j}u_{i+1,\,j+1} - \tilde{\mathscr{F}}^{(u)}_{i-1/2,\,j-1/2}u_{i-1,\,j-1}u_{i,\,j}\\
&+ \tilde{\mathscr{G}}^{(u)}_{i-1/2,\,j+1/2}u_{i,\,j}u_{i-1,\,j+1} - \tilde{\mathscr{G}}^{(u)}_{i+1/2,\,j-1/2}u_{i+1,\,j-1}u_{i,\,j}\big].\end{aligned} \tag{93}$$

In (93), each of the kinetic energy flux terms reappears at a neighboring point but with the opposite sign. Thus, regardless of the subsequent definition of $H^{(u)}$, $\mathscr{F}^{(u)}$, $\mathscr{G}^{(u)}$, $\tilde{\mathscr{F}}^{(u)}$ and $\tilde{\mathscr{G}}^{(u)}$, the choice of form (90) and the constraint (91) together ensure that the total kinetic energy over the domain does not falsely increase or decrease.

The additional requirement on the difference scheme is that enstrophy be conserved during advection by the nondivergent part of the horizontal velocity. This will be guaranteed if the finite-difference scheme for the momentum advection terms reduces to (77) for the case of nondivergent flow.

If new symbols, based on Eq. (89), are defined at $h$ points by

$$F^* \equiv \overline{F}^x \qquad G^* \equiv \overline{G}^y, \tag{94}$$

it is seen that in the case of nondivergent motion, $F^*$ and $G^*$ are equivalent, respectively, to (a constant) $hd$ times $u^*$ and $v^*$ given by Eq. (76). The flux terms in (90) then reduce to (77) for this case if

$$\begin{aligned}
\mathscr{F}^{(u)}_{i+1/2,\,j} &= \tfrac{2}{3}(\overline{F^*}^{yy})_{i+1/2,\,j} \\
\mathscr{G}^{(u)}_{i,\,j+1/2} &= \tfrac{2}{3}(\overline{G^*}^{yx})_{i,\,j+1/2} \\
\tilde{\mathscr{F}}^{(u)}_{i+1/2,\,j+1/2} &= \tfrac{1}{6}(\overline{G^* + F^*}^{y})_{i+1/2,\,j+1/2} \\
\tilde{\mathscr{G}}^{(u)}_{i-1/2,\,j+1/2} &= \tfrac{1}{6}(\overline{G^* - F^*}^{y})_{i-1/2,\,j+1/2}.
\end{aligned} \tag{95}$$

It should be noted that this generalization of (77) is not unique.

The definition of $H^{(u)}$ is now determined by the requirement (91). Making use of Eqs. (95) and (94), Eq. (91) can be written in the form

$$\begin{aligned}
\frac{\partial}{\partial t} H^{(u)}_{i,\,j} + \frac{1}{d^2}\frac{1}{8}[&(\delta_x F + \delta_y G)_{i+1/2,\,j+1} + (\delta_x F + \delta_y G)_{i-1/2,\,j+1} \\
&+ (\delta_x F + \delta_y G)_{i+1/2,\,j-1} + (\delta_x F + \delta_y G)_{i-1/2,\,j-1} \\
&+ 2(\delta_x F + \delta_y G)_{i+1/2,\,j} + 2(\delta_x F + \delta_y G)_{i-1/2,\,j}] = 0.
\end{aligned} \tag{96}$$

From the continuity equation (88) it is then clear that Eq. (96) is satisfied only if

$$H^{(u)}_{i,\,j} = (\overline{h}^{xyy})_{i,\,j}. \tag{97}$$

An analogous development for the first terms of the $v$ momentum equation (83) yields the form

$$\begin{aligned}
\frac{\partial}{\partial t}(H^{(v)}v)_{i,\,j} + \frac{1}{d^2}[&\delta_x(\mathscr{F}^{(v)}\overline{v}^x) + \delta_y(\mathscr{G}^{(v)}\overline{v}^y) \\
&+ \delta_{x'}(\tilde{\mathscr{F}}^{(v)}\overline{v}^{x'}) + \delta_{y'}(\tilde{\mathscr{G}}^{(v)}\overline{v}^{y'})]_{i,\,j},
\end{aligned} \tag{98}$$

which guarantees both conservation of kinetic energy, integrated over the domain, under inertial processes and conservation of enstrophy for the case of nondivergent flow with the definitions

$$\begin{aligned}
\mathscr{F}^{(v)}_{i+1/2,\,j} &= \tfrac{2}{3}(\overline{F^{*}}^{xy})_{i+1/2,\,j} \\
\mathscr{G}^{(v)}_{i,\,j+1/2} &= \tfrac{2}{3}(\overline{G^{*}}^{xx})_{i,\,j+1/2} \\
\tilde{\mathscr{F}}^{(v)}_{i+1/2,\,j+1/2} &= \tfrac{1}{6}(\overline{G^{*} + F^{*x}})_{i+1/2,\,j+1/2} \\
\tilde{\mathscr{G}}^{(v)}_{i-1/2,\,j+1/2} &= \tfrac{1}{6}(\overline{G^{*} - F^{*x}})_{i-1/2,\,j+1/2}
\end{aligned} \tag{99}$$

and

$$H^{(v)}_{i,\,j} = (\overline{h}^{xxy})_{i,\,j}. \tag{100}$$

The coriolis term $-fhv$ in Eq. (82) is represented at the $u$ point $(i, j)$ by

$$-f_j(\overline{h\overline{v}^{y}}^{x})_{i,\,j}; \tag{101}$$

and the term $+fhu$ in Eq. (83) at the $v$ point $(i + 1/2, j + 1/2)$ is represented by

$$(\overline{fh\overline{u}^{x}}^{y})_{i+1/2,\,j+1/2}. \tag{102}$$

Here the coriolis parameter $f_j$ is defined at latitudes where $h$ is carried.

The rate of increase in the kinetic energy of the $u$ component at the point $(i, j)$ due to the coriolis force is obtained by multiplying (101) by $u_{i,\,j}$. The contribution to this increase from the $v$ point $(i + 1/2, j + 1/2)$ involves the portion

$$-\tfrac{1}{4}f_j h_{i+1/2,\,j} v_{i+1/2,\,j+1/2} u_{i,\,j}. \tag{103}$$

Similarly, the rate of increase in the kinetic energy of the $v$ component at the point $(i + 1/2, j + 1/2)$ is given by (102) times $v_{i+1/2,\,j+1/2}$; and the fraction due to the $u$ point $(i, j)$ involves the term

$$+\tfrac{1}{4}f_j h_{i+1/2,\,j} u_{i,\,j} v_{i+1/2,\,j+1/2}. \tag{104}$$

Note that (103) and (104) exactly cancel so that total kinetic energy is not influenced by these terms.

Finally, the pressure gradient terms, which convert potential into kinetic energy, can be examined. At the $u$ point $(i, j)$, the term $gh(\partial h/\partial x)$ in Eq. (82) is represented as

$$g[\overline{h}^{x}\,\delta_x h]_{ij}; \tag{105}$$

and at the $v$ point $(i + 1/2, j + 1/2)$, $gh(\partial h/\partial y)$ in Eq. (83) is represented as

$$g[\overline{h}^{y}\, \delta_y h]_{i+1/2,\, j+1/2}. \tag{106}$$

An argument completely analogous to that utilized in the discussion of the coriolis terms can be advanced to show that this finite-difference form of the pressure gradient terms does not cause any false production of total energy.

## IV. Basic Governing Equations

### A. The Vertical Coordinate

The vertical coordinate used in the model is a combination of the $\sigma$ coordinate (Phillips, 1957) for the lower part of the atmosphere, and the pressure coordinate for the upper part of the atmosphere.

Let $p$ be the pressure; $p_T$, the pressure at the top of the model atmosphere, taken as a constant; and $p_S$, the pressure at the earth's surface, which varies with the horizontal coordinates and time. A constant pressure $p_I$ is chosen which lies between $p_T$ and a lower bound of $p_S$, and the vertical coordinate $\sigma$ is then defined by

$$\sigma \equiv \frac{p - p_I}{\pi}, \tag{107}$$

where

$$\pi = \begin{cases} \pi_U \equiv p_I - p_T & \text{for} \quad p_T \leqslant p < p_I, \\ \pi_L \equiv p_S - p_I & \text{for} \quad p_I < p \leqslant p_S. \end{cases} \tag{108}$$

Note that $\pi_U$ is constant, whereas $\pi_L$ is a function of the horizontal coordinates and time. It follows from Eqs. (107) and (108) that

$$\begin{aligned} \sigma &= -1 && \text{for} \quad p = p_T, \\ \sigma &= 0 && \text{for} \quad p = p_I, \\ \sigma &= 1 && \text{for} \quad p = p_S. \end{aligned} \tag{109}$$

Figure 18 shows surfaces of constant $\sigma$ in a vertical cross section. The lower boundary, which follows the earth's topography, is a coordinate surface; and the isobaric surfaces for $p_T \leqslant p \leqslant p_I$ are coordinate surfaces. When $p_I = p_T$, this vertical coordinate system reduces to the $\sigma$ coordinate of earlier versions of the UCLA General Circulation Model (Mintz, 1965, 1968; Arakawa, 1972); and when $p_I = p_T = 0$, it reduces to the original $\sigma$ coordinate of Phillips (1957).

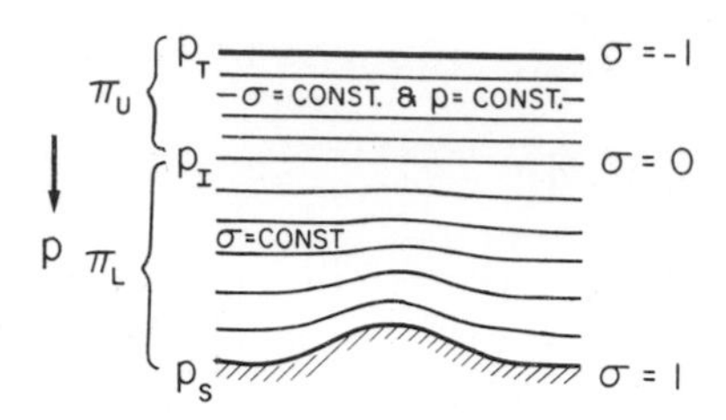


FIG. 18. Definition of the layers of the model in terms of the vertical $\sigma$ coordinate.

Since $\pi$ is either $\pi_U$, which is a constant, or $\pi_L$, which is a function only of the horizontal coordinates and time, (107) gives

$$\delta p = \pi\, \delta\sigma, \tag{110}$$

where $\delta$ denotes the differential under constant horizontal coordinates and time. $\pi\, \delta\sigma/g$ is the mass per unit horizontal area in a layer of depth $\delta\sigma$, where $g$ is the acceleration of gravity.

From Eq. (107), the individual time derivative of pressure is given by

$$\omega \equiv dp/dt = \pi\dot{\sigma} + \sigma[(\partial\pi/\partial t) + \mathbf{v}\cdot\nabla\pi], \tag{111}$$

where $\dot{\sigma} \equiv d\sigma/dt$, $\mathbf{v}$ is the horizontal velocity, and $\nabla$ is the horizontal gradient operator. Note that $\partial\pi/\partial t + \mathbf{v}\cdot\nabla\pi = 0$ for $\sigma < 0$ and, therefore,

$$\omega = \pi\dot{\sigma} \qquad \text{for} \quad \sigma \leqslant 0. \tag{112}$$

At the top of the model atmosphere, Eq. (112) gives $(\pi\dot{\sigma})_{\sigma=-1} = (\omega)_{p=p_T}$. It is assumed that $(\omega)_{p=p_T} = 0$, and thus

$$(\pi\dot{\sigma})_{\sigma=-1} = 0. \tag{113}$$

The earth's surface is a material surface as well as a coordinate surface. The kinematical boundary condition there is simply $\dot{\sigma} = 0$, so that

$$(\pi\dot{\sigma})_{\sigma=1} = 0. \tag{114}$$

Finally, the continuity of $\omega$ at $\sigma = 0$ requires

$$(\pi\dot{\sigma})_{\sigma=0-} = (\pi\dot{\sigma})_{\sigma=0+} = \omega_I, \tag{115}$$

where $\omega_I \equiv (\omega)_{p=p_I}$.

## B. The Equation of State

The model atmosphere is assumed to be a perfect gas, so that

$$\alpha = RT/p, \tag{116}$$

where $\alpha$ is the specific volume, $T$ is the temperature, and $R$ is the gas constant. For simplicity, the difference of the gas constant from that of dry air (which determines the difference between the virtual temperature and the temperature) is neglected except in the parameterizations of subgrid scale turbulence and cumulus convection.

## C. The Hydrostatic Equation

With Eq. (110), the hydrostatic equation $\delta\Phi = -\alpha\delta p$ becomes

$$\delta\Phi = -\pi\alpha\, \delta\sigma, \tag{117}$$

where $\Phi$ is the geopotential $gz$ and $z$ is height.

The following alternate forms of the hydrostatic equation can be derived from Eq. (117) and will be useful:

$$\delta(\Phi\sigma) = -(\pi\sigma\alpha - \Phi)\, \delta\sigma, \tag{118}$$

$$\delta\Phi = -RT\, \delta \ln p, \tag{119}$$

$$= -c_p\theta\, \delta(p/p_0)^\kappa \tag{120}$$

$$= c_p \frac{d \ln \theta}{d(1/\theta)}\, \delta \left(\frac{p}{p_0}\right)^\kappa, \tag{121}$$

$$\delta(c_p T + \Phi) = \left(\frac{p}{p_0}\right)^\kappa c_p\, \delta\theta, \tag{122}$$

where $c_p$ is the specific heat at constant pressure, $\kappa \equiv R/c_p$, and $\theta$ is the potential temperature, $T(p_0/p)^\kappa$, where $p_0$ is a standard pressure.

## D. The Equation of Continuity

In the pressure coordinate system, the equation of continuity takes the form

$$\nabla_p \cdot \mathbf{v} + (\partial\omega/\partial p) = 0. \tag{123}$$

Gradients in the pressure and $\sigma$-coordinate systems are related by

$$\nabla_{\rm p} = \nabla_\sigma + (\nabla_{\rm p}\sigma)(\partial/\partial\sigma), \tag{124}$$

Using the gradient $\nabla_{\rm p}$ of Eq. (107), namely

$$\pi \nabla_{\rm p}\sigma + \sigma \nabla\pi = 0,$$

Eq. (124) becomes

$$\nabla_{\rm p} = \nabla_\sigma - \sigma/\pi \, \nabla\pi \, \partial/\partial\sigma. \tag{125}$$

Note that $\nabla_{\rm p} = \nabla_\sigma$ for $\sigma \leqslant 0$, because $\pi$ is constant for $\sigma \leqslant 0$.

Using Eq. (125) for $\nabla_{\rm p} \cdot \mathbf{v}$ and using Eqs. (111) and (110) for $\partial\omega/\partial p$, Eq. (123) gives

$$\left[\nabla_\sigma \cdot \mathbf{v} - \frac{\sigma}{\pi}\nabla\pi \cdot \frac{\partial \mathbf{v}}{\partial\sigma}\right] + \frac{\partial}{\pi\,\partial\sigma}\left[\pi\dot\sigma + \sigma\left(\frac{\partial}{\partial t} + \mathbf{v}\cdot\nabla\right)\pi\right] = 0,$$

and finally

$$(\partial\pi/\partial t) + \nabla_\sigma \cdot (\pi\mathbf{v}) + (\partial/\partial\sigma)(\pi\dot\sigma) = 0. \tag{126}$$

The equation of continuity (126) is used to compute both $\pi\dot\sigma$ and $\partial\pi_{\rm L}/\partial t = \partial p_{\rm S}/\partial t$. Integrating Eq. (126) with respect to $\sigma$, from $-1$ to $\sigma$, and using Eq. (113) gives

$$\int_{-1}^{\sigma} \frac{\partial\pi}{\partial t}\,d\sigma + \pi\dot\sigma = -\int_{-1}^{\sigma} \nabla\cdot(\pi\mathbf{v})\,d\sigma. \tag{127}$$

Since $\partial\pi/\partial t = \partial\pi_{\rm U}/\partial t = 0$ for $\sigma < 0$ and $\partial\pi/\partial t = \partial\pi_{\rm L}/\partial t$ for $\sigma > 0$, which is constant in $\sigma$,

$$\pi\dot\sigma = -\int_{-1}^{\sigma} \nabla\cdot(\pi\mathbf{v})\,d\sigma \qquad \text{for} \quad \sigma < 0, \tag{128}$$

$$\sigma\frac{\partial\pi_{\rm L}}{\partial t} + \pi\dot\sigma = -\int_{-1}^{\sigma} \nabla\cdot(\pi\mathbf{v})\,d\sigma \qquad \text{for} \quad \sigma > 0. \tag{129}$$

From Eq. (129) applied at $\sigma = 1$, where $\pi\dot\sigma = 0$,

$$\frac{\partial\pi_{\rm L}}{\partial t} = \frac{\partial p_{\rm S}}{\partial t} = -\int_{-1}^{1} \nabla\cdot(\pi\mathbf{v})\,d\sigma. \tag{130}$$

Substituting $\partial\pi_{\rm L}/\partial t$ from Eq. (130) into Eq. (129) gives $\pi\dot\sigma$ for $\sigma > 0$.

## E. The Individual Time Derivative and Its Flux Form

With the $\sigma$ coordinate, the individual time derivative $d/dt$ is expressed as

$$d/dt = (\partial/\partial t)_\sigma + \mathbf{v} \cdot \nabla_\sigma + \dot{\sigma}(\partial/\partial\sigma). \tag{131}$$

With $A$ an arbitrary scalar, (131) gives

$$dA/dt = [(\partial/\partial t)_\sigma + \mathbf{v} \cdot \nabla_\sigma]A + \dot{\sigma}(\partial/\partial\sigma)A, \tag{132}$$

which is the advective form for $dA/dt$. Use of the continuity equation (126) then gives the flux form

$$\pi(dA/dt) = (\partial/\partial t)_\sigma(\pi A) + \nabla_\sigma \cdot (\pi\mathbf{v}A) + (\partial/\partial\sigma)(\pi\dot{\sigma}A). \tag{133}$$

## F. The Momentum Equation

The pressure gradient force is given by $-\nabla_p\Phi$. Applying (125) to $\Phi$ gives

$$\nabla_p\Phi = \nabla_\sigma\Phi - (\sigma/\pi)\,\nabla\pi(\partial\Phi/\partial\sigma), \tag{134}$$

which with substitution from Eq. (117) becomes

$$\nabla_p\Phi = \nabla_\sigma\Phi + \sigma\alpha\,\nabla\pi. \tag{135}$$

For $\sigma < 0$, $\nabla_p\Phi = \nabla_\sigma\Phi$. For $\sigma > 0$, the pressure gradient force consists of two terms, as shown by Eq. (135). Where the slope of the earth's surface is steep, the individual terms are large but are approximately in opposite directions. In the particular case where $\nabla_p\Phi = 0$, complete compensation occurs.

The horizontal component of the equation of motion becomes

$$d\mathbf{v}/dt + f\mathbf{k} \times \mathbf{v} + \nabla_\sigma\Phi + \sigma\alpha\,\nabla\pi = \mathbf{F}, \tag{136}$$

where $\mathbf{F}$ is the horizontal frictional force and $d\mathbf{v}/dt$ is the horizontal acceleration. Note that

$$\pi(\nabla_\sigma\Phi + \sigma\alpha\,\nabla\pi) = \nabla_\sigma(\pi\Phi) + (\sigma\pi\alpha - \Phi)\,\nabla\pi, \tag{137}$$

which gives us another form of the equation of motion

$$\pi(d\mathbf{v}/dt + f\mathbf{k} \times \mathbf{v}) + \nabla_\sigma(\pi\Phi) + (\sigma\pi\alpha - \Phi)\,\nabla\pi = \pi\mathbf{F}, \tag{138}$$

or, using Eq. (118) in Eq. (138),

$$\pi(d\mathbf{v}/dt) + f\mathbf{k} \times \pi\mathbf{v} + \nabla_\sigma(\pi\Phi) - (\partial(\Phi\sigma)/\partial\sigma)\,\nabla\pi = \pi\mathbf{F}. \tag{139}$$

## G. The Thermodynamic Energy Equation

The specific entropy is $c_p \ln \theta = \text{const}$, and the first law of thermodynamics is

$$d/dt\; c_p \ln \theta = Q/T, \tag{140}$$

where $Q$ is the heating rate per unit mass. The flux form which corresponds to Eq. (140) is

$$\frac{\partial}{\partial t}(\pi c_p \ln \theta) + \nabla \cdot (\pi \mathbf{v} c_p \ln \theta) + \frac{\partial}{\partial \sigma}(\pi \dot{\sigma} c_p \ln \theta) = \pi \frac{Q}{T}. \tag{141}$$

The first law of thermodynamics can also be written as

$$c_p(dT/dt) = \omega\alpha + Q, \tag{142}$$

where $c_p T$ is the specific enthalpy and

$$\omega \equiv dp/dt = \pi\dot{\sigma} + \sigma(\partial/\partial t + \mathbf{v} \cdot \nabla)\pi,$$

as given by (111). The corresponding flux form is

$$\frac{\partial}{\partial t}(\pi c_p T) + \nabla_\sigma \cdot (\pi \mathbf{v} c_p T) + \frac{\partial}{\partial \sigma}(\pi \dot{\sigma} c_p T) = \pi(\omega\alpha + Q). \tag{143}$$

## H. The Water Vapor and Ozone Continuity Equations

Let $q$ be the mixing ratio of either water vapor or ozone. The continuity equation for either variable is expressed by

$$dq/dt = S, \tag{144}$$

where $S$ is the source term. The corresponding flux form is

$$(\partial/\partial t)(\pi q) + \nabla_\sigma \cdot (\pi \mathbf{v} q) + (\partial/\partial \sigma)(\pi \dot{\sigma} q) = \pi S. \tag{145}$$

## V. The Vertical Difference Scheme of the Model

### A. Some Integral Properties of the Adiabatic Frictionless Atmosphere

The following integral properties of the governing equations, or of selected terms in these equations, are useful in designing the vertical finite difference scheme.

#### 1. *Mass Conservatism*

Equation (130) gives

$$\frac{\partial p_S}{\partial t} = -\nabla \cdot \int_{-1}^{1} \pi \mathbf{v} \, d\sigma. \tag{146}$$

The area integral of Eq. (146) over the entire globe makes the divergence term vanish, which means that the total mass of the model atmosphere is conserved.

#### 2. *Vertically Integrated Horizontal Pressure Gradient Force*

With the $p$ coordinate, the horizontal pressure gradient force per unit mass is $-\nabla_p \Phi$. Vertical integration with respect to mass gives

$$\begin{aligned} -\frac{1}{g}\int_{p_T}^{p_S} \nabla_p \Phi \, dp &= -\frac{1}{g}\left[\nabla \int_{p_T}^{p_S} \Phi \, dp - \Phi_S \nabla p_S\right] \\ &= -\frac{1}{g}\left[\nabla \int_{p_T}^{p_S} (\Phi - \Phi_S) \, dp + (p_S - p_T) \nabla \Phi_S\right], \end{aligned} \tag{147}$$

where $\Phi_S \equiv g z_S$, and $z_S$ is the height of the earth's surface. The first term in brackets in Eq. (147) is a gradient vector, and a line integral of its tangential component taken along an arbitrary closed curve on the sphere always vanishes. Only the second term contributes to such a line integral and therefore only when there is a nonhorizontal boundary surface can there be any acceleration of the circulation (any "spin-up" or "spin down" of the vertically integrated atmosphere) by the pressure gradient force.

With the $\sigma$ coordinate, the horizontal pressure gradient force per unit $\delta\sigma$ is given by

$$-1/g\{\nabla_\sigma(\pi\Phi) - [\partial(\Phi\sigma)/\partial\sigma]\, \nabla\pi\} \tag{148}$$

[see Eq. (139)]. Vertical integration with respect to $\sigma$ gives

$$-\frac{1}{g}\int_{-1}^{1}\left[\nabla_\sigma(\pi\Phi) - \frac{\partial(\Phi\sigma)}{\partial\sigma}\nabla\pi\right]d\sigma = -\frac{1}{g}\left[\nabla\int_{-1}^{1}\pi\Phi\, d\sigma - \Phi_S\nabla\pi\right], \quad (149)$$

where the fact that $\nabla\pi = 0$ for $\sigma < 0$ has been used. From Eqs. (110) and (108) it is easy to show that Eq. (149) is equivalent to Eq. (147).

### 3. *Conservation of Total Energy*

The equation of motion (136) readily gives the kinetic energy equation

$$\pi(d/dt)\tfrac{1}{2}\mathbf{v}^2 = -\pi\mathbf{v}\cdot(\nabla_\sigma\Phi + \sigma\alpha\,\nabla\pi) + \pi\mathbf{v}\cdot F. \quad (150)$$

The left-hand side of Eq. (150) can be written in the flux form given by Eq. (133) with $A = \frac{1}{2}\mathbf{v}^2$ as follows

$$\left(\frac{\partial}{\partial t}\right)_\sigma(\pi\tfrac{1}{2}\mathbf{v}^2) + \nabla_\sigma\cdot(\pi\mathbf{v}\tfrac{1}{2}\mathbf{v}^2) + \frac{\partial}{\partial\sigma}(\pi\dot{\sigma}\tfrac{1}{2}\mathbf{v}^2)$$
$$= -\pi\mathbf{v}\cdot[\nabla_\sigma\Phi + \sigma\alpha\,\nabla\pi] + \pi\mathbf{v}\cdot F. \quad (151)$$

The rate of kinetic energy generation by the pressure gradient force per unit $\delta\sigma/g$ is thus $-\pi\mathbf{v}\cdot[\nabla_\sigma\Phi + \sigma\alpha\,\nabla\pi]$. Using Eqs. (126), (117), (118), and (111), this becomes

$$\begin{aligned}
-\pi\mathbf{v}\cdot[\nabla_\sigma\Phi + \sigma\alpha\,\nabla\pi] &= -\nabla_\sigma\cdot(\pi\mathbf{v}\Phi) + \Phi\,\nabla_\sigma\cdot(\pi\mathbf{v}) - \sigma\pi\alpha\mathbf{v}\cdot\nabla\pi \\
&= -\nabla_\sigma\cdot(\pi\mathbf{v}\Phi) - \Phi\left[\frac{\partial}{\partial\sigma}(\pi\dot{\sigma}) + \frac{\partial\pi}{\partial t}\right] - \sigma\pi\alpha\mathbf{v}\cdot\nabla\pi \\
&= -\nabla_\sigma\cdot(\pi\mathbf{v}\Phi) - \frac{\partial}{\partial\sigma}(\pi\dot{\sigma}\Phi) + \pi\dot{\sigma}\frac{\partial\Phi}{\partial\sigma} - \Phi\frac{\partial\pi}{\partial t} - \sigma\pi\alpha\mathbf{v}\cdot\nabla\pi \\
&= -\nabla_\sigma\cdot(\pi\mathbf{v}\Phi) - \frac{\partial}{\partial\sigma}(\pi\dot{\sigma}\Phi) + (\sigma\pi\alpha - \Phi)\frac{\partial\pi}{\partial t} \\
&\quad - \pi\left[\sigma\left(\frac{\partial\pi}{\partial t} + \mathbf{v}\cdot\nabla\pi\right) + \pi\dot{\sigma}\right]\alpha \\
&= -\nabla_\sigma\cdot(\pi\mathbf{v}\Phi) - \frac{\partial}{\partial\sigma}\left(\pi\dot{\sigma}\Phi + \Phi\sigma\frac{\partial\pi}{\partial t}\right) - \pi\omega\alpha,
\end{aligned} \quad (152)$$

so that

$$\nabla_\sigma \cdot (\pi \mathbf{v}\Phi) + \frac{\partial}{\partial\sigma}\left[\left(\sigma\frac{\partial\pi}{\partial t} + \pi\dot{\sigma}\right)\Phi\right] = \pi\mathbf{v} \cdot [\nabla_\sigma\Phi + \sigma\alpha\,\nabla\pi] - \pi\omega\alpha. \tag{153}$$

The first law of thermodynamics as given by Eq. (143) is

$$\frac{\partial}{\partial t}(\pi c_p T) + \nabla_\sigma \cdot (\pi\mathbf{v}c_p T) + \frac{\partial}{\partial\sigma}(\pi\dot{\sigma}c_p T) = \pi Q + \pi\omega\alpha. \tag{154}$$

Taking the sum of Eqs. (151), (153), and (154), and integrating with respect to $\sigma$ from $-1$ to 1 gives

$$\frac{\partial}{\partial t}\left[p_S\Phi_S + \int_{-1}^{1}\pi(\tfrac{1}{2}\mathbf{v}^2 + c_p T)\,d\sigma\right] + \nabla \cdot \int_{-1}^{1}\pi\mathbf{v}(\tfrac{1}{2}\mathbf{v}^2 + c_p T + \Phi)\,d\sigma$$
$$= \int_{-1}^{1}\pi(\mathbf{v} \cdot F + Q)\,d\sigma. \tag{155}$$

Here $\partial\pi/\partial t = 0$ at $\sigma = -1$, $\partial\pi/\partial t = \partial p_S/\partial t$ at $\sigma = 1$, $\partial\Phi_S/\partial t = 0$ and Eqs. (113) and (114) have been used. The area integral of Eq. (155) over the entire globe makes the contribution of the divergence term vanish, and total energy is thus conserved when $F = 0$ and $Q = 0$.

4. *Conservation of Total Potential Enthalpy and Total Entropy*

Under adiabatic processes the potential temperature $\theta$ and therefore any function of the potential temperature $f(\theta)$ are conserved with respect to an air parcel. The flux form which corresponds to $df(\theta)/dt = 0$ is given by Eq. (133), with $A$ replaced by $f(\theta)$; that is,

$$(\partial/\partial t)_\sigma[\pi f(\theta)] + \nabla_\sigma \cdot [\pi\mathbf{v}f(\theta)] + (\partial/\partial\sigma)[\pi\dot{\sigma}f(\theta)] = 0. \tag{156}$$

Integrating Eq. (156) with respect to $\sigma$ from $-1$ to 1 gives

$$\frac{\partial}{\partial t}\int_{-1}^{1}\pi f(\theta)\,d\sigma + \nabla \cdot \int_{-1}^{1}\pi\mathbf{v}f(\theta)\,d\sigma = 0, \tag{157}$$

where $f(\theta)$ can be any arbitrary function of $\theta$ whose global integral with respect to mass exists. Because the divergence term in Eq. (157) vanishes when the area integral is taken over the entire globe, the global integral of $f(\theta)$ with respect to mass is conserved under adiabatic processes. Choosing $f(\theta) = c_p\theta$ gives conservation of the total potential enthalpy, and choosing $f(\theta) = c_p \ln \theta + \text{const}$ gives conservation of the total entropy.

The conservation of these quantities can be interpreted from a different point of view. For simplicity, consider motion in a stably stratified atmosphere. Under adiabatic processes, air parcels that carry potential temperatures larger than $\theta_0$ stay above the isentropic surface $\theta = \theta_0$, and air parcels that carry potential temperatures smaller than $\theta_0$ stay below the isentropic surface $\theta = \theta_0$; therefore the total mass of air above the isentropic surface is constant. This holds even when the isentropic surface intersects the ground, as does the surface $\theta = \theta_1$ in Fig. 19. In this respect, the earth's surface can be regarded as a continuation of the isentropic surface, as shown by the heavy line in the figure. Then, for quasi-static motion, the horizontal average of the pressure on each isentropic surface $\overline{p}(\theta)$ does not change with time. (This constraint was used by Lorenz (1955) in deriving an expression for available potential energy.) Because $(1/g)\, d\overline{p}\,(\theta)/d\theta$ is the mass of air per unit horizontal area and per unit increment of $\theta$ in the vertical, $d\overline{p}\,(\theta)/d\theta$ is termed the "mass density function in $\theta$ space." Since $\overline{p}(\theta)$ is constant in time, the mass density function is also constant in time. Figure 20 shows the shape of the function for a typical situation. The reciprocal of the density function is closely related to the static stability (but not exactly related, unless the isentropic surfaces coincide with the isobaric surfaces).

The global integral of $f(\theta)$ with respect to mass, where $f(\theta)$ is any function for which the integral exists, can be related to the mass density function

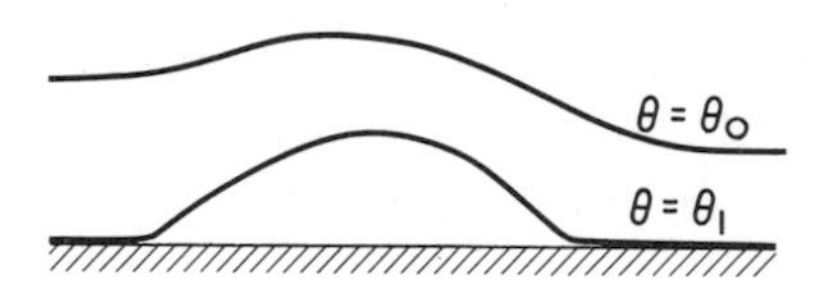


FIG. 19. Isentropic surfaces, one of which intersects the earth's surface.

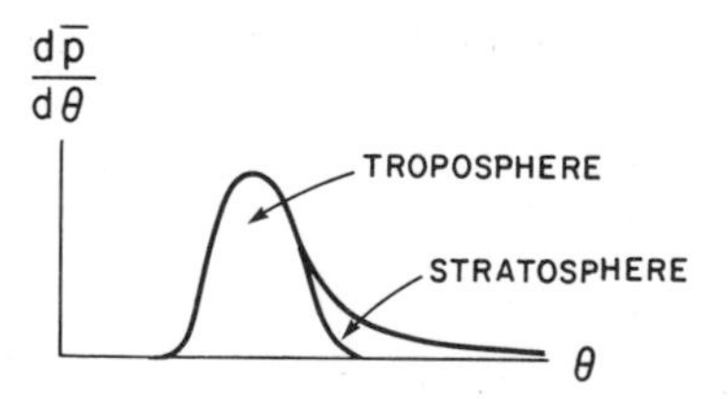


FIG. 20. Schematic representation of a typical distribution of the mass density function in $\theta$ space.

as follows

$$\overline{\int_0^{p_S} f(\theta)\,dp} = \overline{\int_{\theta_S}^{\infty} f(\theta)\frac{\partial p}{\partial \theta}\,d\theta}$$

$$= \overline{\int_{\theta_{min}}^{\infty} f(\theta)\frac{\partial p}{\partial \theta}\,d\theta}$$

$$= \int_{\theta_{min}}^{\infty} f(\theta)\frac{d\overline{p}}{d\theta}\,d\theta, \tag{158}$$

where for simplicity it has been assumed that $p_T = 0$ and therefore $\theta_T = \infty$. Here, the bar over the integral denotes the horizontal average; $p_S$ and $\theta_S$ are, respectively, $p$ and $\theta$ at the earth's surface, and $\theta_{min} = \min(\theta_S)$. In changing the lower limit of the integral from $\theta_S$ to $\theta_{min}$, $\partial p/\partial\theta = 0$ for $\theta_S > \theta > \theta_{min}$ has been used. Thus conservation of the global integral of $f(\theta)$ with respect to mass is equivalent to a constraint on the density function. For example, when $f(\theta) = \theta^n$, the integral gives the $n$th moment of the density function.

In order to fully constrain the density function, it is generally necessary to specify an infinite sequence of moments or an integral transform such as the momentum generating function or the characteristic function. In a discrete system, however, such a full constraint on the density function is not possible unless the isentropic surfaces are taken as coordinate surfaces. In the next subsection it is shown that reasonably simple vertical difference schemes can exactly conserve global integrals with respect to mass of only two independent functions of $\theta$, say $f(\theta)$ and $g(\theta)$; that is, only two independent constraints on the density function can be formally satisified. Consequently, some false distortion of the density function by discretization errors cannot be avoided in numerical simulation. It is to be expected, however, that certain features of the density function can be maintained by proper choice of $f(\theta)$ and $g(\theta)$.

The vertical difference scheme for the first law of thermodynamics in the current UCLA general circulation model has been derived with $f(\theta) = \theta$ and $g(\theta) = \ln\theta$ as the two functions. This choice is based on the following physical reasoning. Choosing $f(\theta) = \theta$ guarantees conservation of the first moment of the density function and, therefore, guarantees conservation of mean potential enthalpy, which is of physical importance. Lorenz (1960) showed that if we define a gross static stability $S$ by

$$S = \left(\frac{\overline{p}_S}{p_0}\right)^{\kappa}\frac{E}{1+\kappa} - (P + I), \tag{159}$$

where $E$ and $(P + I)$ are, respectively, the potential enthalpy and the *total* potential energy of the whole atmosphere and $\bar{p}_S$ is the mean surface pressure, $S$ becomes a weighted vertical integral of the static stability. Then, because $d(P + I)/dt = -dK/dt$, when $Q = 0$ and $F = 0$, where $K$ is the kinetic energy of the whole atmosphere, $dE/dt = 0$ guarantees

$$dS/dt = dK/dt. \tag{160}$$

Thus when potential enthalpy is conserved, energy conversion from total potential energy to kinetic energy, which requires rising of warmer air and sinking of colder air, stabilizes the atmosphere.

The earlier UCLA general circulation models used $g(\theta) = \theta^2$. That choice, together with $f(\theta) = \theta$, guaranteed conservation of the second moment about the mean of the density function and, therefore, guaranteed conservation of the variance of the potential temperature. That was a reasonable choice for the earlier versions of the model, for they covered only the troposphere and the potential temperature distribution in the troposphere does not deviate greatly from a Gaussian distribution. That choice also guaranteed the approximate conservation of the total entropy because

$$(\ln \theta)_{\mathrm{m}} \fallingdotseq \ln \theta_{\mathrm{m}} + [(\theta'/\theta_{\mathrm{m}})^2]_{\mathrm{m}}, \tag{161}$$

for small $\theta'/\theta_{\mathrm{m}}$, where the subscript m denotes the mean and $\theta' \equiv \theta - \theta_{\mathrm{m}}$.

However, the potential temperature distribution in the coupled troposphere-stratosphere system is highly skewed (see Fig. 20); and conservation of the second moment is not necessarily an effective constraint on the density function near its maximum, because the very large potential temperatures in the stratosphere make a dominant contribution to the second moment. With the present choice of $g(\theta) = \ln \theta$, instead of conservation of the variance, there is conservation of $(\ln \theta)_{\mathrm{m}} - \theta_{\mathrm{m}}$, which is a measure of the broadening of the density function near its maximum [see Eq. (161)]. In addition, $g(\theta) = \ln \theta$ guarantees the conservation of total entropy, which is a quantity of physical importance. Furthermore, as is shown in the next subsection, the finite-difference hydrostatic equation that is energetically consistent with this choice of $g(\theta)$ is very accurate for a wide range of vertical profiles of temperature.

### B. A Vertical Difference Scheme Which Maintains Integral Properties

In this subsection the vertical differencing of all the basic equations except the water vapor and ozone continuity equations is presented. The vertical differencing is designed to maintain finite-difference analogs of the integral constraints discussed in the last subsection.

1. *The Vertical Index*

The model atmosphere is divided into $K$ layers by $K$-1 levels of constant $\sigma$. The layers are identified with odd $k$ and carry the velocity $\mathbf{v}$, the temperature $T$, the water vapor mixing ratio $q$, and the ozone mixing ratio $O_3$. The levels which divide the layers are identified with even $k$ and carry $\pi\dot{\sigma}$. The upper boundary $p = p_T$, the level $p = p_I$, and the lower boundary $p = p_S$ are identified with $k = 0$, $k = k_I$, and $k = K + 1$, respectively (see Fig. 21). Define, for odd $k$,

$$\Delta\sigma_k \equiv \sigma_{k+1} - \sigma_{k-1}; \tag{162}$$

then

$$\sum_{k=1}^{k_I - 1}{}' \Delta\sigma_k = 1 \quad \text{and} \quad \sum_{k=k_I+1}^{K}{}' \Delta\sigma_k - 1, \tag{163}$$

where $\sum'$ represents a summation over odd $k$.

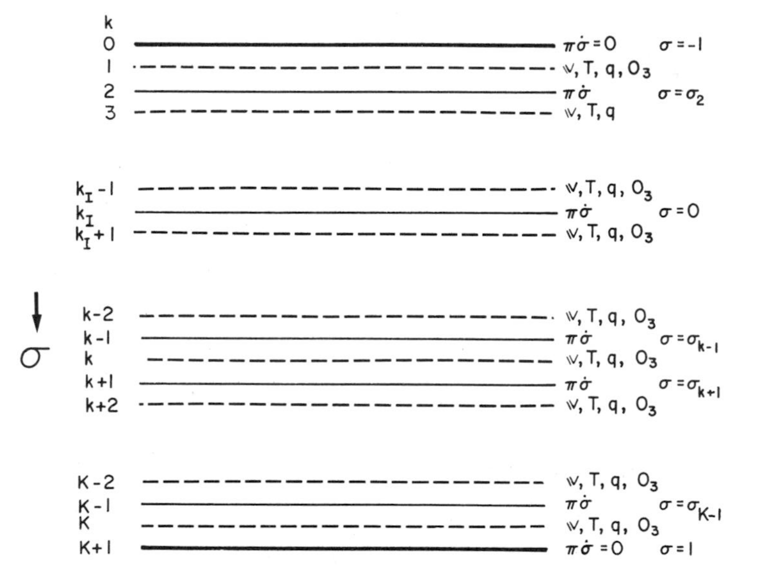


FIG. 21. The vertical structure of the model, showing distribution of the prognostic variables; solid lines (even $k$) indicate the levels dividing the layers; dashed lines (odd $k$) indicate levels within layers at which prognostic variables are carried (exact position discussed in Section V).

## 2. *The Equation of Continuity*

The continuity equation is written in the form

$$\frac{\partial \pi_k}{\partial t} + \nabla \cdot (\pi_k \mathbf{v}_k) + \frac{1}{\Delta \sigma_k} [(\pi \dot{\sigma})_{k+1} - (\pi \dot{\sigma})_{k-1}] = 0, \tag{164}$$

where $k$ is odd. We have

$$\pi_k = \begin{cases} \pi_{\mathrm{U}} & \text{for} \quad k < k_{\mathrm{I}} \\ \pi_{\mathrm{L}} & \text{for} \quad k > k_{\mathrm{I}}. \end{cases} \tag{165}$$

With $\partial \pi_k / \partial t = 0$ for $k < k_{\mathrm{I}}$, $\partial \pi_k / \partial t = \partial \pi_{\mathrm{L}} / \partial t$ for $k > k_{\mathrm{I}}$, and $(\pi \dot{\sigma})_0 = (\pi \dot{\sigma})_{K+1} = 0$, $\sum_{k=1}^{K}$ (164) $\Delta \sigma_k$ gives

$$\frac{\partial \pi_{\mathrm{L}}}{\partial t} = -\sum_{k=1}^{K}{}' \nabla \cdot (\pi_k \mathbf{v}_k) \Delta \sigma_k, \tag{166}$$

which is an analog of Eq. (130). Because $\partial \pi_{\mathrm{L}} / \partial t = \partial p_{\mathrm{S}} / \partial t$, and the area integral of the right-hand side of Eq. (166) over the entire globe vanishes, total mass conservation is maintained with this vertical differencing for the continuity equation.

The quantity $(\pi \dot{\sigma})_{k+1}$ is given by

$$\begin{aligned} (\pi \dot{\sigma})_{k+1} &= -\sum_{k=1}^{k}{}' \nabla \cdot (\pi_k \mathbf{v}_k) \Delta \sigma_k & \text{for} \quad k < k_{\mathrm{I}}, \\ (\pi \dot{\sigma})_{k+1} &= -\sum_{k=1}^{k}{}' \nabla \cdot (\pi_k \mathbf{v}_k) \Delta \sigma_k - \sigma_{k+1} \frac{\partial \pi_{\mathrm{L}}}{\partial t} & \text{for} \quad k > k_{\mathrm{I}}, \end{aligned} \tag{167}$$

which are analogs of Eqs. (128) and (129).

## 3. *Flux Forms*

For any variable $A$ carried by the layers, the flux form analogous to Eq. (133) can be written as

$$\frac{\partial}{\partial t} (\pi_k A_k) + \nabla \cdot (\pi_k \mathbf{v}_k A_k) + \frac{1}{\Delta \sigma_k} [(\pi \dot{\sigma})_{k+1} \hat{A}_{k+1} - (\pi \dot{\sigma})_{k-1} \hat{A}_{k-1}]. \tag{168}$$

where the variable $\hat{A}$, defined at the levels between layers, is obtained by some manner of interpolation from $A$. Whatever the form of the interpolation, however, Eq. (168) guarantees that the analog of the global integral of $A$ with respect to mass is conserved as far as advective processes are concerned, because $\sum'^{K}_{k=1}$ (168) $\Delta\sigma_k$ gives

$$\frac{\partial}{\partial t}\sum_{k=1}^{K}{}' \pi_k A_k \, \Delta\sigma_k + \sum_{k=1}^{K}{}' \nabla \cdot (\pi_k \mathbf{v}_k A_k)\, \Delta\sigma_k,$$

and the second term vanishes when the area integral over the entire globe is taken.

Equations (164) and (168) give the expression

$$\left(\pi \frac{dA}{dt}\right)_k = \pi_k \left(\frac{\partial}{\partial t} + \mathbf{v}_k \cdot \nabla\right) A_k + \frac{1}{\Delta\sigma_k}[(\pi\dot{\sigma})_{k+1}(\hat{A}_{k+1} - A_k) + (\pi\dot{\sigma})_{k-1}(A_k - \hat{A}_{k-1})], \quad (169)$$

which when divided by $\pi_k$ gives the advective form for $dA/dt$ which is consistent with the flux form in Eq. (168).

So far, the choice of $\hat{A}$ is completely arbitrary, provided that the choice does not violate the consistency of the scheme with the original differential equation. It is possible, then, to satisfy an additional requirement.

Let us require also that the finite-difference analog of the global integral of $F(A)$ with respect to mass be conserved. Let $F_k \equiv F(A_k)$ and $F'_k \equiv dF(A_k)/dA_k$. Then (169) multiplied by $F'_k$ gives

$$\pi_k \left(\frac{\partial}{\partial t} + \mathbf{v}_k \cdot \nabla\right) F_k + \frac{1}{\Delta\sigma_k}[(\pi\dot{\sigma})_{k+1} F'_k(\hat{A}_{k+1} - A_k) + (\pi\dot{\sigma})_{k-1} F'_k(A_k - \hat{A}_{k-1})]. \quad (170)$$

Using the equation of continuity, (170) can be rewritten as

$$\frac{\partial}{\partial t}(\pi_k F_k) + \nabla \cdot (\pi_k \mathbf{v}_k F_k) + \frac{1}{\Delta\sigma_k}[(\pi\dot{\sigma})_{k+1}\{F'_k(\hat{A}_{k+1} - A_k) + F_k\} - (\pi\dot{\sigma})_{k-1}\{-F'_k(A_k - \hat{A}_{k-1}) + F_k\}]. \quad (171)$$

In order that (171) be in flux form, it is necessary that

$$\hat{F}_{k+1} = F'_k(\hat{A}_{k+1} - A_k) + F_k, \tag{172}$$

$$\hat{F}_{k-1} = -F'_k(A_k - \hat{A}_{k-1}) + F_k. \tag{173}$$

Replacing $k$ in Eq. (173) by $k + 2$ and eliminating $\hat{F}_{k+1}$ with Eq. (172) gives

$$\hat{A}_{k+1} = \frac{(F'_{k+2}A_{k+2} - F_{k+2}) - (F'_k A_k - F_k)}{F'_{k+2} - F'_k}. \tag{174}$$

This may be interpreted as a finite-difference analog to the identity

$$A \equiv \frac{d(F'A - F)}{dF'}. \tag{175}$$

When $F(A) = A^2$, for example, Eq. (174) gives

$$\hat{A}_{k+1} = \tfrac{1}{2}(A_k + A_{k+2}). \tag{176}$$

That this constraint on $\hat{A}_{k+1}$ leads to conservation of the global integral of $A^2$ with respect to mass was first pointed out by Lorenz (1960).

4. *Vertically Integrated Horizontal Pressure Gradient Force*

In order to maintain the property of the vertically integrated horizontal pressure gradient force discussed in Section V, A, 2, it is convenient to start from the form given in Eq. (139). The terms $\nabla_\sigma(\pi\Phi) - \partial/\partial\sigma(\Phi\sigma)\,\nabla\pi$ are written for odd $k$ as

$$\nabla(\pi_k\Phi_k) - \frac{1}{\Delta\sigma_k}(\hat{\Phi}_{k+1}\sigma_{k+1} - \hat{\Phi}_{k-1}\sigma_{k-1})\,\nabla\pi_k. \tag{177}$$

Again, the caret is a reminder that a variable is evaluated at the levels, that is, at even $k$. The analog to Eq. (149) is

$$-\frac{1}{g}\sum_{k=1}^{K}{}'\,(180)\,\Delta\sigma_k = \frac{1}{g}\left[\nabla\left(\sum_{k=1}^{K}{}'\,\pi_k(\Phi_k - \hat{\Phi}_S)\,\Delta\sigma_k\right) + (p_S - p_T)\,\nabla\hat{\Phi}_S\right]. \tag{178}$$

In this way the integral property is maintained.

The terms in (177) are equivalent to

$$\pi_k \nabla\Phi_k + [\Phi_k - (1/\Delta\sigma_k)(\hat{\Phi}_{k+1}\sigma_{k+1} - \hat{\Phi}_{k-1}\sigma_{k-1})] \nabla\pi_k. \tag{179}$$

If we let

$$\pi_k(\sigma\alpha)_k \equiv \Phi_k - (1/\Delta\sigma_k)(\hat{\Phi}_{k+1}\sigma_{k+1} - \hat{\Phi}_{k-1}\sigma_{k-1}), \tag{180}$$

(179) can be written as

$$\pi_k[\nabla\Phi_k + (\sigma\alpha)_k \nabla\pi_k], \tag{181}$$

which is the analog to $\pi(\nabla\Phi + \sigma\alpha \nabla\pi)$, another form of the horizontal pressure gradient force. Equation (180) provides an analog to Eq. (118), one form of the hydrostatic equation. However, because $\hat{\Phi}$ is not yet specified, Eq. (180) must be considered at this stage only a definition of the symbol $(\sigma\alpha)_k$.

5. *The Kinetic Energy Equation*

Following (169), the acceleration term is written as

$$\left(\pi \frac{d\mathbf{v}}{dt}\right)_k = \pi_k \left[\frac{\partial}{\partial t} + (\mathbf{v}_k \cdot \nabla)\right] \mathbf{v}_k + \frac{1}{\Delta\sigma_k}[(\pi\dot{\sigma})_{k+1}(\hat{\mathbf{v}}_{k+1} - \mathbf{v}_k) + (\pi\dot{\sigma})_{k-1}(\mathbf{v}_k - \hat{\mathbf{v}}_{k-1})]. \tag{182}$$

To have a flux form for $\mathbf{v}_k \cdot (\pi\, d\mathbf{v}/dt)_k$, Eq. (176) is used with $A = \mathbf{v}$; that is

$$\hat{\mathbf{v}}_{k+1} = \tfrac{1}{2}(\mathbf{v}_k + \mathbf{v}_{k+2}). \tag{183}$$

This guarantees the conservation of total kinetic energy, insofar as vertical advection is concerned. The finite-difference expression for the kinetic energy in a vertical column per unit horizontal area is

$$\frac{1}{g}\sum_{k=1}^{K}{}' \tfrac{1}{2}\mathbf{v}_k{}^2(\pi\,\Delta\sigma)_k. \tag{184}$$

To obtain the kinetic energy generation in finite-difference form, the procedure used in deriving Eq. (152) is followed:

$$
\begin{aligned}
-\pi_k \mathbf{v}_k \cdot [\nabla\Phi_k + (\sigma\alpha)_k \nabla\pi_k]
&= -\nabla_k \cdot (\pi \mathbf{v}_k \Phi_k) - \Phi_k \left[ \frac{1}{\Delta\sigma_k} \{(\pi\dot{\sigma})_{k+1} - (\pi\dot{\sigma})_{k-1}\} + \frac{\partial \pi_k}{\partial t} \right] \\
&\quad - \pi_k(\sigma\alpha)_k \mathbf{v}_k \cdot \nabla\pi_k \\
&= -\nabla \cdot (\pi_k \mathbf{v}_k \Phi_k) - \frac{1}{\Delta\sigma_k} \{(\pi\dot{\sigma})_{k+1}\hat{\Phi}_{k+1} - (\pi\dot{\sigma})_{k-1}\hat{\Phi}_{k-1}\} \\
&\quad + \frac{1}{\Delta\sigma_k} [(\pi\dot{\sigma})_{k+1}(\hat{\Phi}_{k+1} - \Phi_k) + (\pi\dot{\sigma})_{k-1}(\Phi_k - \hat{\Phi}_{k-1})] \\
&\quad - \Phi_k \frac{\partial \pi_k}{\partial t} - \pi_k(\sigma\alpha)_k \mathbf{v}_k \cdot \nabla\pi_k \\
&= -\nabla \cdot (\pi_k \mathbf{v}_k \Phi_k) - \frac{1}{\Delta\sigma_k} \{(\pi\dot{\sigma})_{k+1}\hat{\Phi}_{k+1} - (\pi\dot{\sigma})_{k-1}\hat{\Phi}_{k-1}\} \\
&\quad + \{\pi_k(\sigma\alpha)_k - \Phi_k\} \frac{\partial \pi_k}{\partial t} - \pi_k \left[ (\sigma\alpha)_k \left( \frac{\partial}{\partial t} + \mathbf{v}_k \cdot \nabla \right) \pi_k - \frac{1}{\pi_k \Delta\sigma_k} \right. \\
&\quad \left. \times \{(\pi\dot{\sigma})_{k+1}(\hat{\Phi}_{k+1} - \Phi_k) + (\pi\dot{\sigma})_{k-1}(\Phi_k - \hat{\Phi}_{k-1})\} \right] \\
&= -\nabla \cdot (\pi_k \mathbf{v}_k \Phi_k) - \frac{1}{\Delta\sigma_k} \left[ \left\{ (\pi\dot{\sigma})_{k+1} + \sigma_{k+1} \frac{\partial \pi_k}{\partial t} \right\} \hat{\Phi}_{k+1} \right. \\
&\quad \left. - \left\{ (\pi\dot{\sigma})_{k-1} + \sigma_{k-1} \frac{\partial \pi_k}{\partial t} \right\} \hat{\Phi}_{k-1} \right] - \pi_k(\omega\alpha)_k. \qquad (185)
\end{aligned}
$$

Here $(\omega\alpha)_k$ is defined by

$$
\begin{aligned}
(\omega\alpha)_k \equiv (\sigma\alpha)_k \left( \frac{\partial}{\partial t} + \mathbf{v}_k \cdot \nabla \right) \pi_k \\
- \frac{1}{\pi_k \Delta\sigma_k} \{(\pi\dot{\sigma})_{k+1}(\hat{\Phi}_{k+1} - \Phi_k) + (\pi\dot{\sigma})_{k-1}(\Phi_k - \hat{\Phi}_{k-1})\}. \qquad (186)
\end{aligned}
$$

At this stage, Eq. (186) is the definition of the symbol $(\omega\alpha)_k$.

From a finite-difference scheme for the first law of thermodynamics, another expression for $(\omega\alpha)_k$ will be derived. With Eq. (186), this will determine a form for $(\sigma\alpha)_k$ that, with (180), will fix the discrete form of the hydrostatic equation.

### 6. *Thermodynamic Energy Equation*

In this subsection, a vertical differencing of the thermodynamic energy equation is presented that maintains conservation of total potential enthalpy and total entropy under adiabatic processes.

To conserve an analog of the global integral of the potential temperature $\theta \equiv T(p_0/p)^\kappa$ with respect to mass, the form given by (168) is used with $A = \theta$. Then,

$$\frac{\partial}{\partial t}(\pi_k\theta_k) + \nabla\cdot(\pi_k\mathbf{v}_k\theta_k) + \frac{1}{\Delta\sigma_k}[(\pi\dot{\sigma})_{k+1}\hat{\theta}_{k+1} - (\pi\dot{\sigma})_{k-1}\hat{\theta}_{k-1}] = 0, \quad (187)$$

where the heating term is omitted for convenience. Here

$$\theta_k \equiv T_k/P_k \quad (188)$$

and $P_k$ is an analog to $(p/p_0)^\kappa$ for the layer $k$. The actual form for $P_k$ used in the model will be described later. Here it is sufficient to assume that $P_k$ is a function of $\pi_k$, $\sigma_{k-1}$, and $\sigma_{k+1}$ only.

The earlier versions of the UCLA general circulation model used $\hat{\theta}_{k+1} = \frac{1}{2}(\theta_k + \theta_{k+2})$ following Eq. (176). The present model, however, requires conservation of an analog of the global integral of ln $\theta$ with respect to mass. Equation (174) with $A = \theta$ and $F(A) = \ln\theta$ gives

$$\hat{\theta}_{k+1} = \frac{\ln\theta_k - \ln\theta_{k+2}}{1/\theta_{k+2} - 1/\theta_k}. \quad (189)$$

The corresponding advective form is given by

$$\pi_k\left(\frac{\partial}{\partial t} + \mathbf{v}_k\cdot\nabla\right)\theta_k + \frac{1}{\Delta\sigma_k}[(\pi\dot{\sigma})_{k+1}(\hat{\theta}_{k+1} - \theta_k) + (\pi\dot{\sigma})_{k-1}(\theta_k - \hat{\theta}_{k-1})] = 0. \quad (190)$$

Substituting Eq. (188) into Eq. (190) gives

$$\pi_k\left(\frac{\partial}{\partial t} + \mathbf{v}_k\cdot\nabla\right)T_k - \pi_k\frac{T_k}{P_k}\frac{\partial P_k}{\partial\pi_k}\left(\frac{\partial}{\partial t} + \mathbf{v}_k\cdot\nabla\right)\pi_k$$
$$+ \frac{1}{\Delta\sigma_k}[(\pi\dot{\sigma})_{k+1}(P_k\hat{\theta}_{k+1} - T_k) + (\pi\dot{\sigma})_{k-1}(T_k - P_k\hat{\theta}_{k-1})] = 0, \quad (191)$$

or, introducing $\hat{T}$ to make the left-hand side an analog of $\pi d(c_p T)/dt$,

$$\pi_k\left(\frac{\partial}{\partial t}+\mathbf{v}_k\cdot\nabla\right)c_p T_k+\frac{1}{\Delta\sigma_k}[(\pi\dot{\sigma})_{k+1}c_p(\hat{T}_{k+1}-T_k)+(\pi\dot{\sigma})_{k-1}c_p(T_k-\hat{T}_{k-1})]$$

$$=\pi_k\frac{c_p T_k}{P_k}\frac{\partial P_k}{\partial\pi_k}\left(\frac{\partial}{\partial t}+\mathbf{v}_k\cdot\nabla\right)\pi_k+\frac{1}{\Delta\sigma_k}[(\pi\dot{\sigma})_{k+1}c_p(\hat{T}_{k+1}-P_k\theta_{k+1})$$

$$+(\pi\dot{\sigma})_{k-1}c_p(P_k\hat{\theta}_{k-1}-\hat{T}_{k-1})]. \tag{192}$$

The dependence of $\hat{T}$ on the odd index temperatures need not be specified at this point. The left-hand side of Eq. (192) may be written in flux form, as

$$\frac{\partial}{\partial t}(\pi c_p T_k)+\nabla\cdot(\pi\mathbf{v}_k c_p T_k)+\frac{c_p}{\Delta\sigma_k}(\pi\dot{\sigma}_{k+1}\hat{T}_{k+1}-\pi\dot{\sigma}_{k-1}\hat{T}_{k-1}). \tag{193}$$

### 7. *Total Energy Conservation and the Hydrostatic Equation*

In order that the total energy be conserved under an adiabatic, frictionless process, the right-hand side of Eq. (192) must agree with $\pi_k(\omega\alpha)_k$, where $(\omega\alpha)_k$ is defined by Eq. (186). For $k < k_I$, $\pi_k = \pi_U = \text{const}$ and therefore $(\partial/\partial t + \mathbf{v}_k\cdot\nabla)\pi_k = 0$. For $k > k_I$, $(\partial/\partial t + \mathbf{v}_k\cdot\nabla)\pi_k$ is generally not zero, so that it is necessary to require

$$(\sigma\alpha)_k=\frac{c_p T_k}{P_k}\frac{\partial P_k}{\partial\pi_k}\quad\text{for}\quad k>k_I. \tag{194}$$

Comparison with Eq. (180) which also defines $(\sigma\alpha)_k$ gives

$$\Phi_k-\frac{1}{\Delta\sigma_k}(\hat{\Phi}_{k+1}\sigma_{k+1}-\hat{\Phi}_{k-1}\sigma_{k-1})=\pi_L\frac{c_p T_k}{P_k}\frac{\partial P_k}{\partial\pi_k}\quad\text{for}\quad k>k_I. \tag{195}$$

This is the form of the hydrostatic equation that corresponds to Eq. (118). It must also be required for all odd $k$ that

$$c_p(\hat{T}_{k+1}-P_k\hat{\theta}_{k+1})=\Phi_k-\hat{\Phi}_{k+1} \tag{196}$$

and

$$c_p(P_k\hat{\theta}_{k-1}-\hat{T}_{k-1})=\hat{\Phi}_{k-1}-\Phi_k. \tag{197}$$

Rearranging the terms,

$$(c_p \hat{T}_{k+1} + \hat{\Phi}_{k+1}) - (c_p T_k + \Phi_k) = P_k c_p(\hat{\theta}_{k+1} - \theta_k), \tag{198}$$

and

$$(c_p T_k + \Phi_k) - (c_p \hat{T}_{k-1} + \hat{\Phi}_{k-1}) = P_k c_p(\theta_k - \hat{\theta}_{k-1}), \tag{199}$$

where $\hat{\theta}_{k+1}$ (and therefore $\hat{\theta}_{k-1}$) is given by Eq. (189). Equations (198) and (199) are analogs of the form of the hydrostatic equation given by Eq. (122).

Replacing $k$ in Eq. (199) by $k + 2$ and adding it to Eq. (198) gives

$$\begin{aligned}(c_p T_{k+2} + \Phi_{k+2}) - (c_p T_k + \Phi_k) \\ = c_p[P_{k+2}(\theta_{k+2} - \hat{\theta}_{k+1}) + P_k(\hat{\theta}_{k+1} - \theta_k)],\end{aligned} \tag{200}$$

or, using Eq. (188),

$$\Phi_{k+2} - \Phi_k = -c_p(P_{k+2} - P_k)\hat{\theta}_{k+1}. \tag{201}$$

Equation (200) is an analog of Eq. (122) and Eq. (201) is an analog of Eq. (120). Using Eq. (189),

$$\Phi_{k+2} - \Phi_k = c_p \frac{\ln \theta_{k+2} - \ln \theta_k}{(1/\theta_{k+2}) - (1/\theta_k)} (P_{k+2} - P_k). \tag{202}$$

Equation (202) is a finite-difference approximation of Eq. (121):

$$\delta\Phi = c_p[d \ln \theta / d(1/\theta)]\delta(p/p_0)^\kappa$$

or of

$$\delta\Phi = c_p[\partial(p/p_0)^\kappa / \partial(1/\theta)]\delta \ln \theta.$$

Equation (202) is used to compute $\Phi_k$ for odd $k$. To do so, it is necessary to know $\Phi$ at a single odd $k$, say $k = K$; and Eq. (195) can be used for this purpose. From Eq. (195),

$$\sum_{k=k_I+1}^{K}{}' \Phi_k \Delta\sigma_k - \Phi_S = \sum_{k=k_I+1}^{K}{}' \pi_L \frac{c_p T_k}{P_k} \frac{\partial P_k}{\partial \pi_k} \Delta\sigma_k. \tag{203}$$

However, $\sum_{k=k_I+1}^{\prime K} \Phi_k \, \Delta\sigma_k$ can be written as

$$\sum_{k=k_I+1}^{K}{}' \Phi_k \, \Delta\sigma_k = \sum_{k=k_I+1}^{K}{}' \Phi_k(\sigma_{k+1} - \sigma_{k-1}) \tag{204}$$

$$= \Phi_K + \sum_{k=k_I+1}^{K-2}{}' \sigma_{k+1}(\Phi_k - \Phi_{k+2}).$$

Equations (204) and (203) then give

$$\Phi_K = \Phi_S + \sum_{k=k_I+1}^{K}{}' \pi \frac{c_p T_k}{P_k} \frac{\partial P_k}{\partial \pi_k} - \sum_{k=k_I+1}^{K-2}{}' \sigma_{k+1}(\Phi_k - \Phi_{k+2}) \tag{205}$$

8. *Summary of Subsections 5–8*

A vertical difference scheme has now been constructed that maintains the property of the vertically integrated horizontal pressure gradient force, total energy conservation under adiabatic and frictionless processes, and conservation of $\theta$ and $\ln \theta$, integrated over the entire mass under adiabatic processes. The function $P_k$, however, which is an analog to $(p/p_0)^\kappa$ for the layer $k$, remains to be determined.

*a. Pressure Gradient Force.* From Eqs. (180) and (194), expression (177) becomes

$$\nabla(\pi_k \Phi_k) + \left(\pi_k \frac{c_p T_k}{P_k} \frac{\partial P_k}{\partial \pi_k} - \Phi_k\right) \nabla \pi_k, \tag{206}$$

where

$$\pi_k = \begin{cases} \pi_U = p_I - p_T, & \text{for} \quad k < k_I \\ \pi_L = p_S - p_I, & \text{for} \quad k > k_I. \end{cases}$$

*b. The Hydrostatic Equation.* Equations (205) and (201) give

$$\Phi_K = \Phi_S + \sum_{k=k_I+1}^{K}{}' \pi_L \frac{c_p T_k}{P_k} \frac{\partial P_k}{\partial \pi_k} - \sum_{k=k_I+1}^{K-2}{}' \sigma_{k+1} c_p \hat{\theta}_{k+1}(P_{k+2} - P_k), \tag{207}$$

$$\Phi_k - \Phi_{k+2} = c_p(P_{k+2} - P_k)\hat{\theta}_{k+1}, \tag{207$'$}$$

where

$$\hat{\theta}_{k+1} = \frac{\ln \theta_k - \ln \theta_{k+2}}{(1/\theta_{k+2}) - (1/\theta_k)}. \tag{208}$$

*c. The Thermodynamic Energy Equation.* Using Eq. (193) for the left-hand side of Eq. (192), rearranging terms, dividing by $c_p$, and restoring the the heating term gives

$$\frac{\partial}{\partial t}(\pi_k T_k) + \nabla \cdot (\pi_k \mathbf{v}_k T_k) + \frac{1}{\Delta\sigma_k}[(\pi\dot{\sigma})_{k+1}(P_k\hat{\theta}_{k+1}) - (\pi\dot{\sigma})_{k-1}(P_k\hat{\theta}_{k-1})]$$

$$= \pi_k \frac{T_k}{P_k}\frac{\partial \pi_k}{\partial \pi_k}\left(\frac{\partial}{\partial t} + \mathbf{v}_k \cdot \nabla\right)\pi_k + \pi_k Q_k/c_p. \tag{209}$$

## C. Vertical Propagation of Wave Energy in an Isothermal Atmosphere

In this subsection, the effect of the vertical differencing scheme in current use in the model on the vertical propagation of wave energy in an isothermal atmosphere is examined. The material presented here is based on part of a forthcoming paper by Tokioka. His study provided the foundation for our choice of the depth of the layers and the function $P_k$ in the stratosphere.

### 1. *The Vertical Structure Equation—Continuous Case*

The quasi-static system of equations, linearized with respect to perturbations on a resting, isothermal basic state, may be written with the pressure coordinate as

$$\frac{\partial u}{\partial t} - (2\Omega \sin \varphi)v + \frac{1}{a \cos \varphi}\frac{\partial \phi}{\partial \lambda} = 0, \tag{210}$$

$$\frac{\partial v}{\partial t} + (2\Omega \sin \varphi)u + \frac{1}{a}\frac{\partial \phi}{\partial \varphi} = 0, \tag{211}$$

$$\frac{\partial T}{\partial t} - \omega \frac{1}{c_p}\frac{RT_0}{p} = 0, \tag{212}$$

$$\partial\phi/\partial p = -RT/p, \tag{213}$$

$$\frac{\partial u}{a \cos \varphi\, \partial \lambda} + \frac{\partial (v \cos \varphi)}{a \cos \varphi\, \partial \varphi} + \frac{\partial \omega}{\partial p} = 0, \tag{214}$$

where $\lambda$ and $\varphi$ are longitude and latitude, $u$ and $v$ are the eastward and northward components of the perturbation velocity, $\phi$ is the perturbation geopotential, $\omega$ is the perturbation $p$ velocity, $T$ is the perturbation temperature, $a$ is the radius of the earth, $\Omega$ is the angular speed of rotation of the earth, and $T_0$ is the constant temperature of the basic state.

Let us consider a solution of the form

$$\begin{pmatrix} u \\ v \\ \phi \\ T \end{pmatrix} = \mathrm{Re} \begin{pmatrix} \hat{u} \\ \hat{v} \\ \hat{\phi} \\ \hat{T} \end{pmatrix} \exp[i(s\lambda + \sigma t)], \tag{215}$$

where $s$ is the longitudinal wave number, assumed positive, and $\sigma$ is the angular frequency. A positive $\sigma$ then represents a westward-moving wave and a negative $\sigma$ represents an eastward-moving wave. Using Eq. (215), Eqs. (210)–(214) become

$$i\sigma\hat{u} - (2\Omega \sin \varphi)\hat{v} + \frac{is}{a \cos \varphi}\hat{\phi} = 0 \tag{216}$$

$$i\sigma\hat{v} + (2\Omega \sin \varphi)\hat{u} + (1/a)(\partial\hat{\phi}/\partial\varphi) = 0, \tag{217}$$

$$i\sigma\hat{T} - (RT_0/c_p p)\hat{\omega} = 0 \tag{218}$$

$$\partial\hat{\phi}/\partial p = -(R/p)\hat{T}, \tag{219}$$

$$is\hat{u} + \frac{\partial(\hat{v} \cos \varphi)}{\partial\varphi} + a \cos \varphi \frac{\partial\hat{\omega}}{\partial p} = 0. \tag{220}$$

Following the theory of the atmospheric tides, $\hat{u}$ and $\hat{v}$ are eliminated from Eqs. (216), (217), and (220), giving

$$\mathscr{L}(i\sigma\hat{\phi}) = 4a^2\Omega^2(\partial\hat{\omega}/\partial p), \tag{221}$$

where the differential operator $\mathscr{L}$ is given by

$$\mathscr{L} \equiv -\frac{\partial}{\partial\mu}\left(\frac{1-\mu^2}{f^2-\mu^2}\frac{\partial}{\partial\mu}\right) + \frac{1}{f^2-\mu^2}\left(\frac{s}{\mu}\frac{f^2+\mu^2}{f^2-\mu^2} + \frac{s^2}{1-\mu^2}\right),$$

and

$$\mu \equiv \sin \varphi \qquad \text{and} \qquad f \equiv \sigma/2\Omega.$$

Equations (218) and (219), on the other hand, give

$$\partial/\partial p(i\sigma\hat{\phi}) = -(R^2T_0/c_p p^2)\hat{\omega}. \tag{222}$$

Eliminating $\hat{\phi}$ between Eqs. (221) and (222) gives a single equation for $\hat{\omega}$,

$$\mathscr{L}(\hat{\omega}) + (c_p p^2/R^2 T_0) 4a^2\Omega^2(\partial^2\hat{\omega}/\partial p^2) = 0. \tag{223}$$

Let

$$\hat{\omega} = F(\mu)W(p). \tag{224}$$

Then Eq. (223) can be separated into horizontal and vertical structure equations, given respectively by

$$\mathscr{L}F = \varepsilon F \tag{225}$$

and

$$d^2 W/dp^2 = -(\kappa H_0/h)(1/p^2)W, \tag{226}$$

where $\varepsilon$ is the separation constant, $h$ is the equivalent depth defined by $\varepsilon \equiv 4\Omega^2 a^2/gh$, $H_0 \equiv RT_0/g$ is the scale height of the isothermal atmosphere, and $\kappa \equiv R/c_p$.

Transformation of the dependent variable in the vertical structure equation (226) from $W$ to $\tilde{W} \equiv (p/p_0)^{-1/2} W$ gives

$$d^2\tilde{W}/d\zeta^2 = -n^2\tilde{W}, \tag{227}$$

where $\zeta$ is $-\ln(p/p_0)$, the height scaled by $H_0$; $p_0$ is a standard pressure; and $n$ is defined by

$$n \equiv (\kappa(H_0/h) - \tfrac{1}{4})^{1/2}. \tag{228}$$

The quantity $n$ gives the vertical wave number and therefore a measure of the index of refraction for vertical wave energy propagation. For a given equivalent depth, the vertical wave number $n$ is constant in height. When $n$ is real, the waves are oscillatory in height (internal waves), and transfer wave energy vertically; $n$ is real for the range $0 < h < 4\,\kappa H_0$, that is, for $\varepsilon > (\Omega a)^2/g\kappa H_0$ ($\sim 10$ for $T \sim 270\ ^\circ\mathrm{K}$).

The thin line in Fig. 22 shows $n$ as a function of the parameter $\varepsilon$. Here $T_0 = 270\ ^\circ\mathrm{K}$ and therefore $H_0 = 7.91$ km. The vertical wavelength is approximately $(49.7/n)$ km.

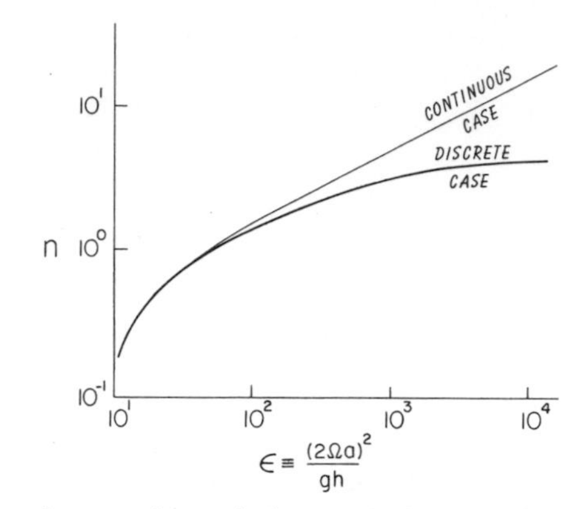


FIG. 22. Comparison between the vertical wave number $n$ in the differential case, defined by Eq. (228), and that given by the vertical difference analog of the vertical structure equation (243).

### 2. *The Vertical Structure Equation—Discrete Case*

That the vertical wave number is constant in height for a given equivalent depth means that the index of refraction is constant in height, so that no internal reflection of wave energy takes place. This important property of an isothermal atmosphere is not necessarily maintained in a discrete model, where vertical differencing is employed. It will be shown here that the vertical differencing described in Section V, B maintains that property when the depths of the layers are equal in log $p$ and

$$P_k = [(p_{k-1}p_{k+1})^{1/2}/p_0]^{\kappa}. \tag{229}$$

Using the vertical index $k$ of Section V, B, the discrete versions of Eqs. (216)–(220) may be written as

$$i\sigma\hat{u}_k - 2\Omega\sin\varphi\hat{v}_k + \frac{is}{a\cos\varphi}\hat{\phi}_k = 0, \tag{230}$$

$$i\sigma\hat{v}_k + 2\Omega\sin\varphi\hat{u}_k + (1/a)(\partial\hat{\phi}_k/\partial\varphi) = 0, \tag{231}$$

$$i\sigma\hat{T}_k - (T_0/\Delta p_k)(Q_k{}^1\hat{\omega}_{k-1} + S_k{}^1\hat{\omega}_{k+1}) = 0, \tag{232}$$

$$\hat{\phi}_k - \hat{\phi}_{k+2} = c_p(S_k{}^2\hat{T}_k + Q_{k+2}^2\hat{T}_{k+2}), \tag{233}$$

$$\hat{u}_k + \frac{\partial(\hat{v}_k\cos\varphi)}{\partial\varphi} + a\cos\varphi\frac{(\hat{\omega}_{k+1} - \hat{\omega}_{k-1})}{\Delta p_k} = 0, \tag{234}$$

where $S_k{}^1$, $S_k{}^2$, $Q_k{}^1$, and $Q_k{}^2$ are coefficients which depend on the vertical differencing of the thermodynamic energy equation and the hydrostatic equation.

With the vertical differencing given by Eqs. (209) and (201), the coefficients are defined as follows:

$$Q_k{}^1 = (P_k/T_0)(\bar{\hat{\theta}}_{k-1} - \bar{\theta}_k), \tag{235}$$

$$S_k{}^1 = (P_k/T_0)(\bar{\theta}_k - \bar{\hat{\theta}}_{k+1}), \tag{236}$$

$$S_k{}^2 = (1/P_k)(P_{k+2} - P_k)(\partial\bar{\hat{\theta}}_{k+1}/\partial\bar{\theta}_k), \tag{237}$$

$$Q_{k+2}^2 = (1/P_{k+2})(P_{k+2} - P_k)(\partial\bar{\hat{\theta}}_{k+1}/\partial\bar{\theta}_{k+2}), \tag{238}$$

where the overbar denotes the basic state. For an isothermal basic state, with the definition of $\hat{\theta}_{k+1}$ given by Eq. (208), these coefficients become

$$Q_k{}^1 = Q_k{}^2 = -\left[1 + \frac{\ln (P_{k-2}/P_k)}{1 - (P_{k-2}/P_k)}\right] \tag{239}$$

$$S_k{}^1 = S_k{}^2 = 1 + \frac{\ln (P_{k+2}/P_k)}{1 - (P_{k+2}/P_k)} \tag{240}$$

The equations corresponding to Eqs. (221) and (222) are then

$$\mathscr{L}(i\sigma\hat{\phi}_k) = 4a^2\Omega^2[(\hat{\omega}_{k+1} - \hat{\omega}_{k-1})/\Delta p_k], \tag{241}$$

$$\begin{aligned} i\sigma(\hat{\phi}_k - \hat{\phi}_{k+2}) = C_{\mathrm{p}}T_0[&(S_k/\Delta p_k)(Q_k\omega_{k-1} + S_k\omega_{k+1}) \\ &+ (Q_{k+2}/\Delta p_{k+2})(Q_{k+2}\omega_{k+1} + S_{k+2}\omega_{k+3})], \end{aligned} \tag{242}$$

where the superscripts of $S$ and $Q$ can now be omitted. Equations (224), (241), and (242) give a finite-difference analog of the vertical structure equation (226):

$$\begin{aligned} \frac{W_{k-1} - W_{k+1}}{\Delta p_k} - \frac{W_{k+1} - W_{k+3}}{\Delta p_{k+2}} = -\frac{C_{\mathrm{p}}T_0}{gh}\bigg[&\frac{S_k}{\Delta p_k}(Q_k W_{k-1} + S_k W_{k+1}) \\ &+ \frac{Q_{k+2}}{\Delta p_{k+2}}(Q_{k+2}W_{k+1} + S_{k+2}W_{k+3})\bigg]. \end{aligned} \tag{243}$$

When the even levels are chosen such that the intervals are equal in log $p$,

$$p_{k+1}/p_{k-1} \equiv e^d \tag{244}$$

for any odd $k$, where the constant $d$ is the depth in $\zeta$ of each layer in the model. With the choice of $P_k$ given by Eq. (229), $P_{k-2}/P_k$ is then constant and, therefore, the coefficients $S_k$ and $Q_k$ are constant. Equations (239), (240), (244), and (229) give

$$Q_k \equiv Q \equiv \frac{e^{-\kappa} - 1 - \kappa d}{1 - e^{-\kappa d}}, \tag{245}$$

and

$$S_k \equiv S \equiv \frac{e^{\kappa d} - 1 - \kappa d}{e^{\kappa d} - 1}. \tag{246}$$

In addition,

$$\frac{p_{k+1}}{\Delta p_k} = \frac{p_{k+1}}{p_{k+1} - p_{k-1}} = \frac{1}{1 - e^{-d}}, \tag{247}$$

and

$$\frac{p_{k+1}}{\Delta p_{k+2}} = \frac{p_{k+1}}{p_{k+3} - p_{k+1}} = \frac{1}{e^d - 1}, \tag{248}$$

which are also constant. Multiplication of Eq. (243) by $p_{k+1}$ and use of Eqs. (247) and (248) gives a constant coefficient difference equation for $W$, whose solution is formally identical to the solution of Eq. (227). As a result, the vertical wave number $n$ is constant in height for an isothermal atmosphere, just as it is in the continuous case. Spurious computational reflection of wave energy due to the discretization is thus prevented as far as a resting isothermal atmosphere is concerned. For these reasons, an equal interval in log $p$ between even levels and the function $P_k$ given by Eq. (229) have been chosen for the stratospheric part of the model (see Fig. 1). A value of $d = 0.657$ is used, that is, $e^d = 1.93$.

The heavy line in Fig. 22 shows the index of refraction obtained from the discrete model for the same isothermal atmosphere as in the continuous case. Although $n$ is constant in height, there is some unavoidable error as $n$ approaches $\pi/d$, the highest resolvable wave number.

### D. Final Determination of the Vertical Difference Scheme

Thus far, each layer of the model and its corresponding representative temperature $T_k$, potential temperature $\theta_k$, and geopotential $\phi_k$, have been identified by an odd value of the index $k$. The variables $\theta_k$, $\phi_k$, and $T_k$ are uniquely related once the function $P_k$ is defined. For a time integration, it is unnecessary to specify the particular *levels* within the layers at which these variables are carried. However, when it is necessary to compare the model results with observations or when it is necessary for the purpose of actual numerical weather predictions to initialize the model from observations, levels must be chosen, somewhere near the center of each layer, to which the values of $T_k$, $\theta_k$, and $\phi_k$ can be assigned. The same odd index $k$ will be used to identify such levels.

Once the function $P_k$ is specified, it is logical to define $p_k$ by

$$(p_k/p_0)^\kappa = P_k. \tag{249}$$

The odd levels $p = p_k$ determined by Eqs. (229) and (249) are constant pressure levels (and, therefore, constant $\sigma$ levels) centered in log $p$. Then, with $\theta_k = T_k(p_0/p_k)^\kappa$, no discretization error exists in the definition of the potential temperature.

The discrete hydrostatic equation, however, given by Eqs. (207) and (207′), is generally subject to discretization errors. As Phillips (1974) pointed out, these errors can be intolerably large unless the function $P_k$ is properly chosen. The function $P_k$ has already been defined for the stratosphere ($k < k_{\mathrm{I}}$) by Eq. (229), based on considerations of vertical energy propagation. This choice turns out to be satisfactory from the point of view of the accuracy of the discretized hydrostatic equation as well. The difference $\Phi_k - \Phi_{k+2}$ given by Eqs. (207′) and (208) is exact for an atmosphere that is isothermal between levels $k$ and $k + 2$.

For the troposphere, however, the function $P_k$ has not yet been defined. The earlier UCLA general circulation models, including the earlier version of the 12-level model, used Eq. (249) with $P_k = [\frac{1}{2}(p_{k-1} + p_{k+1})/p_0]^\kappa$. Phillips has pointed out that with such a choice, calculation of $\theta_k$ from given $\Phi_k$ (which is a necessary procedure for initialization of the model for numerical weather prediction from an observed initial geopotential field) shows a large amplitude oscillation in $\theta_k$ from level to level. Phillips showed that $P_k$ given by

$$P_k = \frac{1}{p_0{}^\kappa}\frac{1}{1 + \kappa}\frac{p_{k+1}^{1+\kappa} - p_{k-1}^{1+\kappa}}{p_{k+1} - p_{k-1}} \tag{250}$$

drastically reduced this deficiency. Tokioka shows in a forthcoming paper that use of Eq. (250) does give the exact value of $\theta_k$ from $\Phi_k$ when the atmosphere is isentropic. He also showed that the optimum choice of $P_k$ for a polytropic atmosphere, for which $T_k(p_0/p_k)^a$ is constant in height, is

$$P_k = \frac{1}{p_0{}^{\kappa}}\left[\frac{1}{1+a}\frac{(p_{k+1}^{1+a} - p_{k-1}^{1+a})}{p_{k+1} - p_{k-1}}\right]^{\kappa}. \tag{251}$$

The current version of the model uses $a = 0.205$, which approximately gives the normally observed stratification.

## VI. The Horizontal Difference Scheme of the Model

### A. The Governing Equations in Orthogonal Curvilinear Coordinates

Let the orthogonal curvilinear coordinates be $\xi$ and $\eta$. Let the actual distances corresponding to $d\xi$ and $d\eta$ be $(ds)_\xi$ and $(ds)_\eta$, respectively, and define the metric factors $m$, $n$ such that

$$(ds)_\xi = (1/m)\, d\xi, \tag{252}$$

and

$$(ds)_\eta = (1/n)\, d\eta. \tag{253}$$

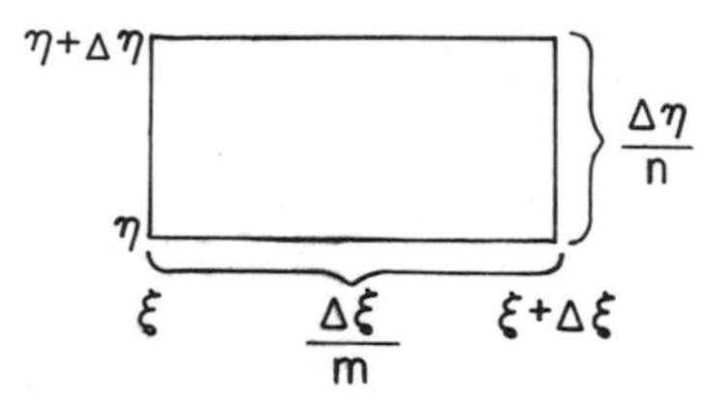


FIG. 23. A rectangular area element in the plane of the orthogonal curvilinear coordinates $(\zeta, \eta)$.

For the rectangular area element in the $\xi - \eta$ plane shown in Fig. 23, the actual lengths of the sides are $\Delta\xi/m$ and $\Delta\eta/n$, and the enclosed area is thus $\Delta\xi\, \Delta\eta/mn$. Let the component of $\mathbf{v}$ in $\xi$ be $u$ and the component of $\mathbf{v}$ in $\eta$ be $v$.

#### 1. *The Equation of Continuity*

The divergence is

$$\frac{\delta_\xi[u(\Delta\eta/n)] + \delta_\eta[v(\Delta\xi/m)]}{(\Delta\xi/m)(\Delta\eta/n)}, \tag{254}$$

where $\delta_\xi$ and $\delta_\eta$ are increments in the $\xi$ and $\eta$ directions, respectively. In the limit as $\Delta\xi, \Delta\eta \to 0$, the divergence can be written

$$\nabla_\sigma \cdot \mathbf{v} = mn\left[\frac{\partial}{\partial\xi}\left(\frac{u}{n}\right) + \frac{\partial}{\partial\eta}\left(\frac{v}{m}\right)\right]; \tag{255}$$

and the equation of continuity (126) thus becomes

$$\frac{\partial}{\partial t}\left(\frac{\pi}{mn}\right) + \frac{\partial}{\partial\xi}\left(\pi\frac{u}{n}\right) + \frac{\partial}{\partial\eta}\left(\pi\frac{v}{m}\right) + \frac{\partial}{\partial\sigma}\left(\frac{\pi\dot{\sigma}}{mn}\right) = 0. \tag{256}$$

2. *The Equation of Motion*

The equation of motion (136) may be written as

$$\frac{\partial\mathbf{v}}{\partial t} + \dot{\sigma}\frac{\partial\mathbf{v}}{\partial\sigma} + (f + \zeta)\mathbf{k} \times \mathbf{v} + \nabla(\tfrac{1}{2}\mathbf{v}^2 + \Phi) + \sigma\alpha\,\nabla\pi = \mathbf{F}, \tag{257}$$

where, by an argument similar to that for the divergence, the vorticity $\zeta = \mathbf{k} \cdot \nabla \times v$ can be expressed as

$$mn\left[\frac{\partial}{\partial\xi}\left(\frac{v}{n}\right) - \frac{\partial}{\partial\eta}\left(\frac{u}{m}\right)\right]. \tag{258}$$

The $\xi$ component of (257) is then

$$\frac{\partial u}{\partial t} + \dot{\sigma}\frac{\partial u}{\partial\sigma} - \left[f + mn\left(\frac{\partial}{\partial\xi}\left(\frac{v}{n}\right) - \frac{\partial}{\partial\eta}\left(\frac{u}{m}\right)\right)\right]v$$
$$+ m\frac{\partial}{\partial\xi}(\tfrac{1}{2}u^2 + \tfrac{1}{2}v^2 + \Phi) + m\sigma\alpha\frac{\partial\pi}{\partial\xi} = F_\xi. \tag{259}$$

Rearranging the terms,

$$\frac{\partial u}{\partial t} + mu\frac{\partial u}{\partial\xi} + nv\frac{\partial u}{\partial\eta} + \dot{\sigma}\frac{\partial u}{\partial\sigma} - \left[f + mn\left(v\frac{\partial}{\partial\xi}\frac{1}{n} - u\frac{\partial}{\partial\eta}\frac{1}{m}\right)\right]v$$
$$+ m\left[\frac{\partial\Phi}{\partial\xi} + \sigma\alpha\frac{\partial\pi}{\partial\xi}\right] = F_\xi. \tag{260}$$

Similarly,

$$\frac{\partial v}{\partial t} + mu\frac{\partial v}{\partial \xi} + nv\frac{\partial v}{\partial \eta} + \dot{\sigma}\frac{\partial v}{\partial \sigma} + \left[f + mn\left(v\frac{\partial}{\partial \xi}\frac{1}{n} - u\frac{\partial}{\partial \eta}\frac{1}{m}\right)\right]u + n\left[\frac{\partial \Phi}{\partial \eta} + \sigma\alpha\frac{\partial \pi}{\partial \eta}\right] = F_\eta. \tag{261}$$

Combining (256) and (260) gives the flux form for the $u$ momentum equation

$$\frac{\partial}{\partial t}\left(\frac{\pi}{mn}u\right) + \frac{\partial}{\partial \xi}\left(\frac{\pi u}{n}u\right) + \frac{\partial}{\partial \eta}\left(\frac{\pi v}{m}u\right) + \frac{\partial}{\partial \sigma}\left(\frac{\pi\dot{\sigma}}{mn}u\right) - \left[\frac{f}{mn} + \left(v\frac{\partial}{\partial \xi}\frac{1}{n} - u\frac{\partial}{\partial \eta}\frac{1}{m}\right)\right]\pi v + \frac{\pi}{n}\left[\frac{\partial \Phi}{\partial \xi} + \sigma\alpha\frac{\partial \pi}{\partial \xi}\right] = \frac{\pi}{mn}F_\xi; \tag{262}$$

and the flux form for $v$ is similarly obtained,

$$\frac{\partial}{\partial t}\left(\frac{\pi}{mn}v\right) + \frac{\partial}{\partial \xi}\left(\frac{\pi u}{n}v\right) + \frac{\partial}{\partial \eta}\left(\frac{\pi v}{m}v\right) + \frac{\partial}{\partial \sigma}\left(\frac{\pi\dot{\sigma}}{mn}v\right) + \left[\frac{f}{mn} + \left(v\frac{\partial}{\partial \xi}\frac{1}{n} - u\frac{\partial}{\partial \eta}\frac{1}{m}\right)\right]\pi u + \frac{\pi}{m}\left[\frac{\partial \Phi}{\partial \eta} + \sigma\alpha\frac{\partial \pi}{\partial \eta}\right] = \frac{\pi}{mn}F_\eta. \tag{263}$$

The general circulation model uses the spherical coordinates $\xi = \lambda$ (longitude) and $\eta = \varphi$ (latitude) where $1/m = a\cos\varphi$ and $1/n = a$. Thus, from this point on, consideration will be restricted to those coordinate systems such as spherical (or cylindrical) in which $m$ and $n$ do not depend on $\xi$.

From (262) the (relative) angular momentum equation can be obtained,

$$\frac{\partial}{\partial t}\left(\frac{\pi}{mn}\frac{u}{m}\right) + \frac{\partial}{\partial \xi}\left(\frac{\pi u}{n}\frac{u}{m}\right) + \frac{\partial}{\partial \eta}\left(\frac{\pi v}{m}\frac{u}{m}\right) + \frac{\partial}{\partial \sigma}\left(\frac{\pi\dot{\sigma}}{mn}\frac{u}{m}\right) - \left[\frac{f}{mn}\frac{\pi v}{m}\right] + \pi\left[\frac{\partial}{\partial \xi}\frac{\Phi}{mn} + \sigma\alpha\frac{\partial}{\partial \xi}\frac{\pi}{mn}\right] = \frac{\pi}{mn}\frac{F_\xi}{m}. \tag{264}$$

3. *The Thermodynamic Energy Equation*

The thermodynamic energy equation (143) can be expressed in curvilinear coordinates as

$$\frac{\partial}{\partial t}\left(\frac{\pi}{mn} c_p T\right) + \frac{\partial}{\partial \xi}\left(\frac{\pi u}{n} c_p T\right) + \frac{\partial}{\partial \eta}\left(\frac{\pi v}{m} c_p T\right) + p^{\kappa} \frac{\partial}{\partial \sigma}\left(\frac{\pi \dot{\sigma}}{mn} c_p \theta\right)$$

$$= \pi\sigma\alpha\left(\frac{\partial}{\partial t}\left(\frac{\pi}{mn}\right) + \frac{u}{n}\frac{\partial \pi}{\partial \xi} + \frac{v}{m}\frac{\partial \pi}{\partial \eta}\right) + \frac{\pi}{mn} Q. \tag{265}$$

## B. Horizontal Differencing of the Governing Equations

Despite the introduction here of the use of the $\sigma$ coordinate and the presence of metric factors, the manner of derivation of a difference scheme for the continuity equation and for the advection and coriolis terms in the $u$ and $v$ momentum equations so closely parallels the methods presented in Section III, C that the representation chosen for these terms will be presented here without elaboration. The new considerations introduced by the requirement of total energy conservation in a three-dimensional domain will be explained more fully.

1. *The Continuity Equation*

For the continuity equation (256) multiplied by $\Delta\xi\Delta\eta$, the following form is chosen

$$\frac{\partial \prod_{i,j}}{\partial t} + (\delta_\xi F)^k_{i,j} + (\delta_\eta G)^k_{i,j} + \frac{1}{\Delta\sigma_k}(\delta_\sigma \dot{S})^k_{i,j} = 0, \tag{266}$$

where

$$\prod \equiv \pi(\Delta\xi\ \Delta\eta/mn), \qquad F \equiv \pi u(\Delta\eta/n),$$
$$G \equiv \pi v(\Delta\xi/m), \qquad \dot{S} \equiv \prod\dot{\sigma}. \tag{267}$$

The vertical index $k$ now appears as a superscript on all variables except $\pi$ and $\prod$. Although $\pi$ and $\prod$ do have different definitions for $k < k_{\mathrm{I}}$ and $k > k_{\mathrm{I}}$, the superscript is dropped for simplicity.

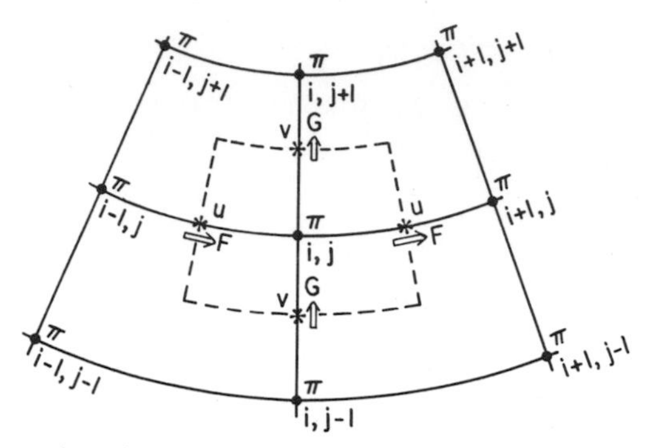


FIG. 24. A $\pi$-centered portion of the spherical grid showing points of definition of the fluxes introduced in Eq. (266).

For the mass fluxes $F$ and $G$, shown in Fig. 24, the following forms are used:

$$F^k_{i+1/2,\,j} = \overline{\left(u\,\frac{\Delta\eta}{n}\right)^k_{i+1/2,\,j}}\,\overline{\pi}^{\xi}_{i+1/2,j}, \tag{268}$$

where

$$[u(\Delta\eta/n)]^k_{i+1/2,\,j} \equiv u^k_{i+1/2,\,j}(\Delta\eta/n)_j \tag{269}$$

and

$$G^k_{i,\,j+1/2} = [v(\Delta\xi/m)]^k_{i,\,j+1/2}\overline{\pi}^{\eta}_{i,\,j+1/2}, \tag{270}$$

where

$$[v(\Delta\xi/m)]^k_{i,\,j+1/2} \equiv v^k_{i,\,j+1/2}(\Delta\xi/m)_{j+1/2}. \tag{271}$$

The superior bar operator, which is a linear smoothing operator in $\xi$, should be ignored for the present. The form and role of this operator will be described in the next subsection.

## 2. *The Momentum Advection Terms*

The form chosen for the terms

$$\frac{\partial}{\partial t}\left(\pi\,\frac{\Delta\xi\,\Delta\eta}{mn}\,u\right) + \Delta\xi\,\frac{\partial}{\partial\xi}\left(\pi u\,\frac{\Delta\eta}{\eta}\,u\right)$$
$$+ \Delta\eta\,\frac{\partial}{\partial\eta}\left(\pi v\,\frac{\Delta\xi}{m}\,u\right) + \frac{\partial}{\partial t}\left(\pi\dot{\sigma}\,\frac{\Delta\xi\,\Delta\eta}{mn}\,u\right)$$

from Eq. (262) multiplied by $\Delta\xi\ \Delta\eta$ is

$$\frac{\partial}{\partial t}(\prod{}^{(u)}u^k)_{i,j} + \Big[\delta_\xi(\mathscr{F}^{(u)}\overline{u}^{\xi}) + \delta_\eta(\mathscr{G}^{(u)}\overline{u}^{\eta}) + \delta_{\xi'}(\tilde{\mathscr{F}}^{(u)}\overline{u}^{\xi'}) + \delta_{\eta'}(\tilde{\mathscr{G}}^{(u)}\overline{u}^{\eta'}) + \frac{1}{\Delta\sigma_k}\,\delta_\sigma(\dot{S}^{(u)}\overline{u}^{\sigma})\Big]^k_{i,j}. \tag{272}$$

If the variables $\prod^{(u)}$ and $\dot{S}^{(u)}$ at $u$ points and $\mathscr{F}^{(u)}$, $\mathscr{G}^{(u)}$, $\tilde{\mathscr{F}}^{(u)}$ and $\tilde{\mathscr{G}}^{(u)}$ shown in Fig. 25 satisfy the constraint

$$\frac{\partial}{\partial t}\prod{}^{(u)}_{i,j} + \Big[\delta_\xi\mathscr{F}^{(u)} + \delta_\eta\mathscr{G}^{(u)} + \delta_{\xi'}\tilde{\mathscr{F}}^{(u)} + \delta_{\eta'}\tilde{\mathscr{G}}^{(u)} + \frac{1}{\Delta\sigma_k}\,\delta_\sigma\dot{S}^{(u)}\Big]^k_{i,j} = 0, \tag{273}$$

then conservation of kinetic energy under a pure advective process is maintained.

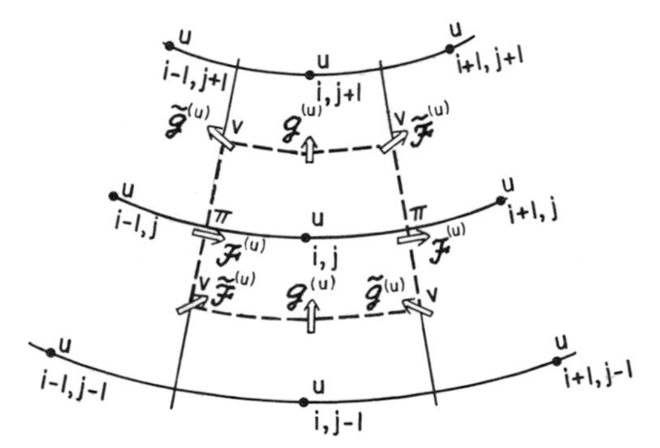


FIG. 25. A $u$-centered portion of the spherical grid showing points of definition of the flux terms introduced in Eq. (272).

With $F^*$ and $G^*$ defined by

$$F^*_{i,j} = (\overline{F}^{\xi})_{i,j}, \qquad G^*_{i,j} = (\overline{G}^{\eta})_{i,j}, \tag{274}$$

the following choice for the fluxes in Eq. (272) is guided by Eq. (95):

$$\begin{aligned}
\mathscr{F}^{(u)}_{i+1/2,\,j} &= \tfrac{2}{3}(\overline{F^*}^{\eta\eta})_{i+1/2,\,j} \\
\mathscr{G}^{(u)}_{i,\,j+1/2} &= \tfrac{2}{3}(\overline{G^*}^{\eta\xi})_{i,\,j+1/2} \\
\tilde{\mathscr{F}}^{(u)}_{i+1/2,\,j+1/2} &= \tfrac{1}{6}(\overline{G^*}^{\eta} + \overline{F^*}^{\eta})_{i+1/2,\,j+1/2} \\
\tilde{\mathscr{G}}^{(u)}_{i-1/2,\,j+1/2} &= \tfrac{1}{6}(\overline{G^*}^{\eta} - \overline{F^*}^{\eta})_{i-1/2,\,j+1/2}.
\end{aligned} \tag{275}$$

Then Eq. (273) is consistent with the continuity equation (266) if

$$\textstyle\prod_{i,j}^{(u)} = (\overline{\prod}^{\xi\eta\eta})_{i,j} \tag{276}$$

and

$$\dot{S}_{i,j}^{(u)} = (\bar{\dot{S}}^{\xi\xi\eta})_{i,j}. \tag{277}$$

For the $v$-momentum equation (263) multiplied by $\Delta\xi\,\Delta\eta$, a form identical to (272) is used, but with $u$ replaced by $v$. The definitions corresponding to (275)–(277) are then

$$\begin{aligned}
\mathscr{F}^{(v)}_{i+1/2,j} &= \tfrac{2}{3}(\overline{F^*}^{\xi\eta})_{i+1/2,j},\\
\mathscr{G}^{(v)}_{i,j+1/2} &= \tfrac{2}{3}(\overline{G^*}^{\xi\xi})_{i+1/2,j+1/2},\\
\tilde{\mathscr{F}}^{(v)}_{i+1/2,j+1/2} &= \tfrac{1}{6}(\overline{G^*}^{\xi} + \overline{F^*}^{\xi})_{i+1/2,j+1/2},\\
\tilde{\mathscr{G}}^{(v)}_{i-1/2,j+1/2} &= \tfrac{1}{6}(\overline{G^*}^{\xi} - \overline{F^*}^{\xi})_{i-1/2,j+1/2},
\end{aligned} \tag{278}$$

$$\textstyle\prod_{i,j}^{(v)} = (\overline{\prod}^{\xi\xi\eta})_{i,j}, \tag{279}$$

and

$$\dot{S}_{i,j}^{(v)} = (\bar{\dot{S}}^{\xi\eta\eta})_{i,j}. \tag{280}$$

3. *The Coriolis Terms*

From Eq. (262) the contribution from the coriolis force plus the metric term to $\partial(\prod u)/\partial t$ is

$$\left[f\,\frac{\Delta\xi\,\Delta\eta}{mn} - u\,\Delta\xi\,\Delta\eta\,\frac{\partial}{\partial\eta}\frac{1}{m}\right]\pi v \tag{281}$$

and the corresponding contribution to $\partial(\prod v)/\partial t$ is

$$-\left[f\,\frac{\Delta\xi\,\Delta\eta}{mn} - u\,\Delta\xi\,\Delta\eta\,\frac{\partial}{\partial\eta}\frac{1}{m}\right]\pi u. \tag{282}$$

A variable $C_{ij}^k$ is defined at $\pi$ points by

$$C_{i,j}^k = f_j\left(\frac{\Delta\xi\,\Delta\eta}{mn}\right) - (\bar{u}^{\xi})_{i,j}^k\,\delta_\eta\left(\frac{\Delta\xi}{m}\right)_j \tag{283}$$

Then, the following form is used for (281) at a $u$ point $(i, j)$:

$$(\overline{\pi C \overline{v}^{\eta}}^{\xi})_{i,j}^{k}; \tag{284}$$

and for (282) at a $v$ point $(i + 1/2, j + 1/2)$,

$$-(\overline{\pi C \overline{u}^{\xi}}^{\eta})_{i+1/2,j+1/2}^{k}. \tag{285}$$

This choice of differencing does not lead to any false generation of total kinetic energy.

4. *The Pressure Gradient Force*

As in Eq. (262), the pressure gradient force in the $\xi$ direction can be written as

$$-\frac{\pi}{n}\left[\frac{\partial \Phi}{\partial \xi} + \sigma\alpha \frac{\partial \pi}{\partial \xi}\right]. \tag{286}$$

The form chosen for the first term is

$$-\left(\frac{\pi}{n}\frac{\partial \Phi}{\partial \xi}\right)_{i+1/2,j}^{k} = -\frac{1}{\Delta\xi\,\Delta\eta}\frac{\Delta\eta}{n_j}(\overline{\pi}^{\xi}\,\delta_{\xi}\Phi^{k})_{i+1/2,j}. \tag{287}$$

Continue, for the time being, to ignore the bar operator.

Corresponding to the relation

$$-\frac{\pi}{n}\frac{\partial \Phi}{\partial \xi} = \frac{1}{n}\left[\frac{\partial}{\partial \xi}(\pi\Phi) - \Phi\frac{\partial \pi}{\partial \xi}\right],$$

Eq. (287) can be rewritten in the form

$$-\left(\frac{\pi}{n}\frac{\partial \Phi}{\partial \xi}\right)_{i+1/2,j}^{k} = -\frac{1}{\Delta\xi\,\Delta\eta}\frac{\Delta\eta}{n_j}\overline{[\delta_{\xi}(\pi\Phi) - \overline{\Phi}^{\xi}\,\delta_{\xi}\pi]}_{i+1/2,j}^{k}. \tag{288}$$

To be consistent with Eq. (288), the following form is chosen for the second term in Eq. (286),

$$-\left(\frac{\pi}{n}\sigma\alpha\frac{\partial \pi}{\partial \xi}\right)_{i+1/2,j}^{k} = -\frac{1}{\Delta\xi\,\Delta\eta}\frac{\Delta\eta}{n_j}\overline{[\overline{(\pi\sigma\alpha)}^{\xi}\,\delta_{\xi}\pi]}_{i+1/2,j}^{k} \tag{289}$$

where, through the application of Eq. (194) at each grid point,

$$(\pi\sigma\alpha)_{ij}{}^{k} = \pi_{ij}c_p \left[ \frac{1}{P^k} \frac{\partial P^k}{\partial \pi} T^k \right]_{i,j}, \tag{290}$$

where $P^k_{i,j}$ is defined by Eq. (251).

Adding (289) to (288) gives a form that corresponds to

$$-\frac{1}{n}\left[ \frac{\partial(\pi\Phi)}{\partial\xi} - \frac{\partial}{\partial\sigma}(\Phi\sigma)\frac{\partial\pi}{\partial\xi} \right],$$

from which it can readily be shown that the properties of the vertically integrated pressure gradient force discussed in Section V, A are maintained.

In summary, the pressure gradient force which contributes to $(\partial/\partial t) \times (\prod^{(u)}u)^k_{i+1/2,j}$ is

$$-\frac{\Delta\eta}{n_j}\overline{[\bar{\pi}^{\xi}\,\delta_\xi\Phi^k + \overline{(\pi\sigma\alpha)}^{\xi}\,\delta_\xi\pi]}^{k}_{i+1/2,j}. \tag{291}$$

Similarly, the pressure gradient force which contributes to $\partial/\partial t\,(\prod^{(v)}v)^k_{i,j+1/2}$ is

$$-\frac{\Delta\xi}{m_{j+1/2}}[\bar{\pi}^{\eta}\,\delta_\eta\Phi^k + \overline{(\pi\sigma\alpha)}^{\eta}\,\delta_\eta\pi]^{k}_{i,j+1/2}, \tag{292}$$

where $(\pi\sigma\alpha)^k_{i,j}$ is given by Eq. (290).

### 5. *Kinetic Energy Generation and the Thermodynamic Energy Equation*

The contribution of the pressure gradient force to the kinetic energy generation, $\partial/\partial t\,(\prod^{(u)}\frac{1}{2}u^2)^k_{i+1/2,j}$, is obtained by multiplying Eq. (291) by $u^k_{i+1/2,j}$. Then the kinetic energy generation is

$$-\left(u\frac{\Delta\eta}{n}\right)^k_{i+1/2,j}\overline{[\bar{\pi}^{\xi}\,\delta_\xi\Phi + \overline{(\pi\sigma\alpha)}^{\xi}\,\delta_\xi\pi]}^{k}_{i+1/2,j}. \tag{293}$$

From the form of the superior bar operator, it can be shown that Eq. (293) is equivalent to

$$\overline{\left(u\frac{\Delta\eta}{n}\right)}^k_{i+1/2,j}[\bar{\pi}^{\xi}\,\delta_\xi\Phi + \overline{(\pi\sigma\alpha)}^{\xi}\,\delta_\xi\pi]^{k}_{i+1/2,j}, \tag{294}$$

in the sense that the difference $R_{i+1/2}$ defined by

$$R_{i+1/2} \equiv (293) - (294)$$

vanishes when the summation over all $i$ is taken; i.e.,

$$\sum_i R_{i+1/2} = 0.$$

Using the definition of $F$ given by Eq. (268), Eq. (294) becomes

$$-[F\,\delta_\xi\Phi + u(\Delta\eta/n)\overline{(\pi\sigma\alpha)}^\xi\,\delta_\xi\pi]^k_{i+1/2,\,j}. \tag{295}$$

It can then be shown that

$$\sum_i (295) = \sum_i \left[\Phi\delta_\xi F - \overline{\overline{\left(u\frac{\Delta\eta}{n}\right)\overline{(\pi\sigma\alpha)}^\xi\,\delta_\xi\pi}}^\xi\right]^k_{i,\,j}. \tag{296}$$

Similarly, the contribution of the gradients of $\Phi$ and $\pi$ to $\partial/\partial t\,(\prod^{(v)}\frac{1}{2}v^2)^k_{i,\,j}$ is given by

$$\sum_j \left[\Phi\delta_\eta G - \overline{\left(v\frac{\Delta\xi}{m}\right)\overline{(\pi\sigma\alpha)}^\eta\,\delta_\eta\pi}^\eta\right]^k_{ij}. \tag{297}$$

Now $\sum_j (296) + \sum_i (297)$ give the discrete form of the total kinetic generation by the pressure gradient force, $-\pi\mathbf{v}\cdot[\nabla\sigma\Phi + \sigma\alpha\,\nabla\pi]$. It is useful to note that $\sum_i$ [first term of (296)] $+ \sum_i$ [first term of (297)], which gives the contribution of ${\Phi_{ij}}^k$ to the kinetic energy generation, can be written using the equation of continuity (266) as

$$-\left[\frac{\partial\prod_{ij}}{\partial t} + \frac{1}{\Delta\sigma_k}(\dot{S}^{k+1}_{i,\,j} - \dot{S}^{k-1}_{i,\,j})\right]\Phi^k_{i,\,j}. \tag{298}$$

The derivation given in Eq. (185) leading to a finite-difference expression for $\omega\alpha$ can be followed exactly, but now with the horizontal differencing specified as well. Using such an expression in a manner completely analogous to the procedure in Section V, B, 8, allows us to write the thermodynamic energy equation given in Eq. (209), with the horizontal differencing incorporated, as

$$\frac{\partial}{\partial t}(\prod T)^k_{i,j} + [\delta_\xi(F\overline{T}^\xi) + \delta_\eta(G\overline{T}^\eta)]^k_{i,j} + \frac{1}{\Delta\sigma_k} P^k_{i,j}\,\delta_\sigma(\dot{S}\hat{\theta})^k_{i,j}$$

$$= \frac{1}{c_p}\left[(\pi\sigma\alpha)\frac{\partial\prod}{\partial t} + \overline{u\,\overline{\frac{\Delta\eta}{n}}\,\overline{(\pi\sigma\alpha)}^\xi\,\delta_\xi\pi}^\xi + \overline{v\,\overline{\frac{\Delta\xi}{m}}\,\overline{(\pi\sigma\alpha)}^\eta\,\sigma_\eta\pi}^\eta + \prod Q\right]^k_{i,j}, \quad (299)$$

where $(\pi\sigma\alpha)^k_{i,j}$ is defined by Eq. (290).

## C. Modification of the Horizontal Differencing Near the Poles

### 1. *Modification of the Difference Equations*

The poles are singular points of the spherical coordinates and the velocity components cannot be defined there; the poles are thus taken as $\pi$-points. The value of $\pi$ at the poles must change as a result of the meridional mass flux $G$, defined by Eq. (270), at all the points on the nearest latitude circle where the meridional velocity component $v$ is carried.

Consider the case of the North Pole, identified by $j = p$ in Fig. 26. To simplify the computation, the pole is treated as if it were a group of points. Each point has index $i$ and represents the shaded area shown. Defining $\prod_{i,p}$ and $\dot{S}_{i,p}$ based on that area, the equation of continuity (266) is applied to $j = p$, omitting all horizontal mass flux terms except $G^k_{i,p-1/2}$. After computing $\partial\prod/\partial t$ and $\dot{S}$ for all $i$, the average is taken.

At the grid point $(i, p - 1/2, k)$, the form chosen for the first line of (263) multiplied by $\Delta\xi\,\Delta\eta$ is

$$\frac{\partial}{\partial t}(\prod^{(v)}v)^k_{i,p-1/2} + \delta_\xi(\mathscr{F}^{*(v)}\overline{v}^\xi)^k_{i,p-1/2} - (\mathscr{G}^{(v)}\overline{v}^\eta)^k_{i,p-1}$$

$$- (\tilde{\mathscr{F}}^{(v)}\overline{v}^{\xi'})^k_{i-1/2,p-1} - (\tilde{\mathscr{G}}^{(v)}\overline{v}^{\eta'})^k_{i+1/2,p-1} + \delta_\sigma(\dot{S}\overline{v}^\sigma)^k_{i,p-1/2}. \quad (300)$$

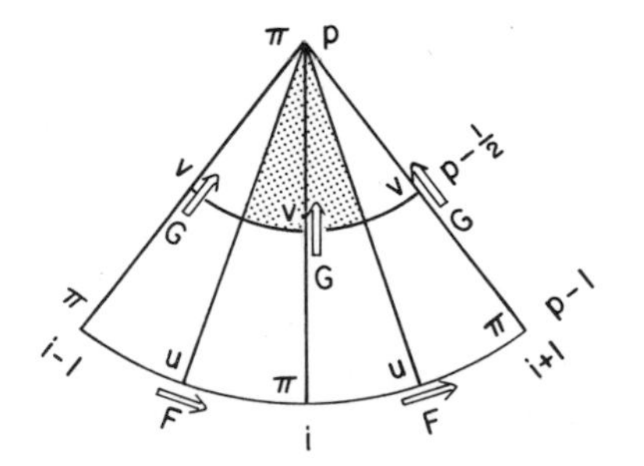


FIG. 26. A polar segment of the spherical grid showing mass fluxes.

Invoking the consistency requirement that the global sum of $\prod^{(v)}$ be equal to the global sum of $\prod$ gives

$$\begin{aligned}\prod\nolimits^{(v)}_{i,p-1/2} \equiv \tfrac{1}{8}\{&\prod\nolimits_{i-1,p} + \prod\nolimits_{i-1,p-1} + 2(\prod\nolimits_{i,p} + \prod\nolimits_{i,p-1}) + \prod\nolimits_{i+1,p}\\ &+ \prod\nolimits_{i+1,p-1}\} + \tfrac{1}{2} \times \tfrac{1}{4}(\prod\nolimits_{i-1,p} + 2\prod\nolimits_{i,p} + \prod\nolimits_{i+1,p}).\end{aligned} \tag{301}$$

The definition of $\dot{S}^{(v)}$ is readily obtained by replacing $\prod$ everywhere in Eq. (301) by $\dot{S}$. The requirement for kinetic energy conservation during advective processes alone given in Section III, C can be shown equivalent to the requirement that the new variable $\mathscr{F}^*$ be chosen such that (300) vanish when $v$ is replaced by a constant value, which can be taken as unity. The resulting equation can be shown consistent with the continuity equation (266) and $\partial/\partial t\,\overline{\prod}^i_{i,p} = \overline{G}^i_{i,p-1/2}$ if

$$\mathscr{F}^{*(v)}_{i-1/2,p-1/2} \equiv \tfrac{1}{3}\overline{F^*}^{\xi}_{i-1/2,p-1}, \tag{302}$$

where $F^*$ is defined by Eq. (274).

In a similar manner, at the grid point $(i, p-1, k)$, the form chosen for the first line of Eq. (262) multiplied by $\Delta\xi\,\Delta\eta$ is

$$\begin{aligned}&\frac{\partial}{\partial t}(\prod\nolimits^{(u)}u)^k_{i,p-1} + \delta_\xi(\mathscr{F}^{*(u)}\overline{u}^{\xi})^k_{i,p-1} - (\mathscr{G}^{(u)}\overline{u}^{\eta})^k_{i,p-3/2}\\ &\quad - (\tilde{\mathscr{F}}^{(u)}\overline{u}^{\xi'})^k_{i-1/2,p-3/2} - (\tilde{\mathscr{G}}^{(u)}\overline{u}^{\eta'})^k_{i+1/2,p-3/2} + \frac{1}{\Delta\sigma_k}(\dot{S}\overline{u}^{\sigma})^k_{i,p-1}.\end{aligned} \tag{303}$$

Again, it is required that the global sum of $\prod^{(u)}$ be equal to the global sum of $\prod$, which gives

$$\begin{aligned}\prod\nolimits^{(u)}_{i,p-1} \equiv \tfrac{1}{8}\{&\prod\nolimits_{i-1/2,p} + \prod\nolimits_{i+1/2,p} + 2(\prod\nolimits_{i-1/2,p-1} + \prod\nolimits_{i+1/2,p-1})\\ &+ \prod\nolimits_{i-1/2,p-2} + \prod\nolimits_{i+1/2,p-2}\} + \tfrac{1}{8}\{3(\prod\nolimits_{i-1/2,p} + \prod\nolimits_{i+1/2,p})\\ &+ \prod\nolimits_{i-1/2,p-1} + \prod\nolimits_{i+1/2,p-1}\}.\end{aligned} \tag{304}$$

The definition of $\dot{S}^{(u)}$ is obtained by replacing $\prod$ everywhere in Eq. (304) by $\dot{S}$. For kinetic energy conservation during advective processes alone, it is necessary that (303) vanish when $u$ is replaced by unity. The resulting equation is consistent with Eq. (266) and $(\partial/\partial t)\,\overline{\prod}^i_{i,p} = \overline{G}^i_{i,p-1/2}$ if

$$\mathscr{F}^*_{i-1/2,p-1} \equiv \tfrac{1}{6}(4F^*_{i-1/2,p-1} + F^*_{i-1/2,p-2}). \tag{305}$$

2. *Longitudinal Averaging of Selected Terms Near the Poles*

To avoid the use of the extremely short time interval necessary for computational stability due to the convergence of the meridians toward the poles, a longitudinal averaging is done of *selected terms* in the prognostic equations.

For the purpose of illustration, consider the simple system of linearized equations that governs a gravity wave on the spherical earth:

$$\frac{\partial u}{\partial t} + \frac{1}{a \cos \varphi} \frac{\partial \phi}{\partial \lambda} = 0, \tag{306}$$

$$\frac{\partial v}{\partial t} + \frac{1}{a} \frac{\partial \phi}{\partial \varphi} = 0, \tag{307}$$

$$\frac{\partial \phi}{\partial t} + \frac{gH}{a \cos \varphi} \left( \frac{\partial u}{\partial \lambda} + \frac{\partial (v \cos \varphi)}{\partial \varphi} \right) = 0, \tag{308}$$

where $H$ is the equivalent depth. Other symbols are as defined in Section V, C. Because our concern here is only with waves that have frequencies sufficiently higher than the earth's rotation rate, the coriolis force has been omitted for simplicity.

With the grid shown in Fig. 26 and a space finite differencing consistent with that of the model, the discrete analogs of Eqs. (306)–(308) can be written as follows:

$$\frac{\partial u_{i+1/2,\, j}}{\partial t} + \frac{1}{a \cos \varphi} \frac{1}{\Delta \lambda} (\delta_\lambda \phi)_{i+1/2,\, j} = 0 \tag{309}$$

$$\frac{\partial v_{i,\, j+1/2}}{\partial t} + \frac{1}{a} \frac{1}{\Delta \varphi_j} (\delta_\varphi \phi)_{i,\, j+1/2} = 0, \tag{310}$$

$$\frac{\partial \phi_{i,\, j}}{\partial t} + \frac{gH}{a \cos \varphi} \left[ \frac{1}{\Delta \lambda} (\delta_\lambda u) + \frac{1}{\Delta \varphi} \delta_\varphi (v \cos \varphi) \right]_{ij} = 0. \tag{311}$$

Let us consider a solution of the form

$$u_{i+1/2,\, j} = \mathrm{Re} \left\{ \hat{u}_j \exp \left[ \bar{i}(s(i + 1/2)\, \Delta \lambda + \sigma t) \right] \right\} \tag{312}$$

$$v_{i,\, j+1/2} = \mathrm{Re} \left\{ \hat{v}_{j+1/2} \exp \left[ \bar{i}(si\, \Delta \lambda + \sigma t) \right] \right\} \tag{313}$$

$$\phi_{i,\, j} = \mathrm{Re} \left\{ \hat{\phi}_j \exp \left[ \bar{i}(si\, \Delta \lambda + \sigma t) \right] \right\} \tag{314}$$

where $\bar{\imath} \equiv \sqrt{-1}$. Substituting Eqs. (312), (313), and (314) into Eqs. (309), (310), and (311), we obtain

$$\bar{\imath}\sigma\hat{u}_j + \frac{\bar{\imath}s}{a\cos\varphi_j}\left(\frac{\sin(s\,\Delta\lambda/2)}{s\,\Delta\lambda/2}\right)S_j(s)\hat{\phi}_j = 0, \tag{315}$$

$$\bar{\imath}\sigma\hat{v}_{j+1/2} + (1/a\,\Delta\varphi)(\hat{\phi}_{j+1} - \hat{\phi}_j) = 0, \tag{316}$$

$$\bar{\imath}\sigma\hat{\phi}_j + \frac{gH}{a\cos\varphi_j}\left[\bar{\imath}s\left(\frac{\sin(s\,\Delta\lambda/2)}{s\,\Delta\lambda/2}\right)S_j(s)\hat{u}_j + \frac{1}{\Delta\varphi}\{(\hat{v}\cos\varphi)_{j+1/2} - (\hat{v}\cos\varphi)_{j-1/2}\}\right] = 0, \tag{317}$$

where $S_j(s) \equiv 1$ for present purposes. Eliminating $\hat{u}$ and $\hat{v}$ from Eqs. (315)–(317) gives

$$C^2\left[\frac{s}{a\cos\varphi_j}\frac{\sin(s\,\Delta\lambda/2)}{(s\,\Delta\lambda/2)}S_j(s)\right]^2\hat{\phi}_j + \frac{C^2}{(a\,\Delta\varphi)^2}\left[(\hat{\phi}_j - \hat{\phi}_{j-1})\frac{\cos\varphi_{j-1/2}}{\cos\varphi_j} - (\hat{\phi}_{j+1} - \hat{\phi}_j)\frac{\cos\varphi_{j+1/2}}{\cos\varphi_j}\right] = \sigma^2\hat{\phi}_j, \tag{318}$$

which is the discrete analog of the meridional structure equation for $\hat{\phi}$. Here $C^2 \equiv gH$.

For a given $s$, with the boundary condition $\hat{\phi} = 0$ at the poles, possible values of $\sigma^2$ are obtained as eigenvalues from the matrix equation represented by Eq. (318) applied to all $j$ interior to the poles. When $s$ is large, the matrix is very close to diagonal for $j$'s near the poles and, therefore, the maximum eigenvalue can be only slightly larger than the maximum diagonal component, which is approximately the maximum value of the coefficient of the first term of Eq. (318). If $S_j(s) = 1$, this argument gives

$$|\sigma|_{\max} \doteqdot \frac{2|C|}{a\,\Delta\lambda}\frac{1}{(\cos\varphi_j)_{\min}}\sin\frac{s\,\Delta\lambda}{2}. \tag{319}$$

For most conditionally stable time difference schemes, the stability criterion is given by

$$|\sigma|\,\Delta t < \varepsilon, \tag{320}$$

where $\varepsilon$ is a constant ($\varepsilon = 1$ for the leapfrog scheme). From Eq. (319), this stability criterion is approximately equivalent to

$$\frac{|C|\,\Delta t}{a\,\Delta\lambda}\sin\frac{s\,\Delta\lambda}{2} < \frac{\varepsilon}{2}(\cos\varphi_j)_{\min}. \tag{321}$$

To make the scheme stable for all resolvable waves, it is necessary to require

$$\frac{|C|\,\Delta t}{a\,\Delta\lambda} < \frac{\varepsilon}{2}(\cos\varphi_j)_{\min}. \tag{322}$$

Thus, since $(\cos\varphi_j)_{\min} \ll 1$, an extremely small $\Delta t$ must be used to ensure stability.

The method devised to allow the use of a longer $\Delta t$ in the model is to smooth the longitudinal pressure gradient in the momentum equation and the longitudinal divergence in the continuity equation with a longitudinal averaging operator. If the amplitude of the longitudinal pressure gradient and divergence are modified by the factor

$$S_j(s) = (a\,\Delta\lambda/d^*)[\cos\varphi_j/\sin(s\,\Delta\lambda/2)], \tag{323}$$

where $d^*$ is a specified constant length, then Eq. (318) becomes

$$\frac{4C^2}{d^{*2}}\hat{\phi}_j + \frac{C^2}{(a\,\Delta\varphi)^2}\left[(\hat{\phi}_j - \hat{\phi}_{j-1})\frac{\cos\varphi_{j-1/2}}{\cos\varphi_j} - (\hat{\phi}_{j+1} - \hat{\phi}_j)\frac{\cos\varphi_{j+1/2}}{\cos\varphi_j}\right] = \sigma^2\hat{\phi}_{ij}, \tag{324}$$

for all $j$. Thus the first term now contributes to the eigenvalue $\sigma^2$ a constant amount $4C^2/d^{*2}$. The dependence of $|\sigma|_{\max}$ on $(\cos\varphi_j)_{\min}$ is eliminated, and a $\Delta t$ satisfying the stability criterion depends now on the constant length $d^*$. In the model, $d^*$ is taken as the latitudinal grid size, $a\,\Delta\varphi$.

In practice, it is sufficient to perform the smoothing only at higher latitudes. Then

$$S_j(s) = (\Delta\lambda/\Delta\varphi)[\cos\varphi_j/\sin(s\,\Delta\lambda/2)] \tag{325}$$

when the right-hand side is less than 1; $S_j(s) = 1$ otherwise.

To apply the operator, the zonal pressure gradient and the zonal mass flux are expanded into Fourier series and the amplitude of each wave component reduced by the factor $S_j(s)$. This is the bar operation shown in

Section VI, B. The form of the smoothing of the mass flux given by Eq. (268) is chosen to maintain the energy conservation.

It is important to note that this smoothing operation does not smooth or truncate the Fourier expansions of the fields of the variables. It is simply a generator of multiple point difference quotients in the space difference scheme. For the example given above, the solution of Eqs. (315), (316), and (317) is still a neutral oscillation.

## VII. Vertical and Horizontal Differencing of the Water Vapor and Ozone Continuity Equations

### A. Vertical Differencing

#### 1. *ln q-Conserving Scheme*

Let $q$ be the mixing ratio of water vapor or ozone. The corresponding continuity equation is given by Eq. (145). The vertical differencing given by

$$\frac{\partial}{\partial t}(\pi_k q_k) + \nabla \cdot (\pi_k \mathbf{v}_k q_k) + \frac{1}{\Delta\sigma_k}[(\pi\dot{\sigma})_{k+1}\hat{q}_{k+1} - (\pi\dot{\sigma})_{k-1}\hat{q}_{k-1}] = \pi_k S_k \quad (326)$$

guarantees the conservation of total water vapor or total ozone, when there are no sources and sinks, for any choice of $\hat{q}$.

The ozone mixing ratio varies in the vertical over a wide range of magnitudes and, as with potential temperature, shows a highly skewed "mass density function." Applying the considerations of Section V, B, 3 to the ozone mixing ratio, a $\hat{q}$ can be chosen that leads to conservation of a discrete analog of the global integral of ln $q$ with respect to mass. From Eq. (189), such a $\hat{q}$ must be of the form

$$\hat{q}_{k+1} = \frac{\ln q_k - \ln q_{k+2}}{(1/q_{k+2}) - (1/q_k)}. \quad (327)$$

Further discussion of this scheme has been presented by Schlesinger (1976).

The same form for $\hat{q}_{k+1}$ could also be used for the water vapor mixing ratio. Release of heat of condensation, however, makes the choice of $\hat{q}$ for water vapor more difficult, as discussed below.

#### 2. *Moist Adiabatic Process—Continuous Case*

Consider, first, a moist adiabatic process in the continuous atmosphere. Let the air be saturated and remain saturated, and let there be no heating other than the heat of condensation.

Let $q$ be the mixing ratio and $q^*(T, p)$ be the saturation mixing ratio of water vapor. Then the water vapor continuity equation, when condensation is occurring, is

$$dq/dt = dq^*/dt = -C, \tag{328}$$

where $C$ is the sink of water vapor per unit mass of dry air. This can also be written as

$$\left(\frac{\partial q^*}{\partial T}\right)_{\mathrm{p}} \frac{dT}{dt} + \left(\frac{\partial q^*}{\partial p}\right)_{\mathrm{T}} \omega = -C. \tag{329}$$

The thermodynamic energy equation is

$$(d/dt)c_{\mathrm{p}}T = \omega\alpha + LC, \tag{330}$$

where $L$ is the heat of condensation per unit mass. Then Eqs. (329) and (330) give

$$C = -\frac{\omega}{1 + [(L/c_{\mathrm{p}})(\partial q^*/\partial T)_{\mathrm{p}}]}\left[\left(\frac{\partial q^*}{\partial p}\right)_{\mathrm{T}} + \frac{\alpha}{c_{\mathrm{p}}}\left(\frac{\partial q^*}{\partial T}\right)_{\mathrm{p}}\right]. \tag{331}$$

Substituting Eq. (331) into Eq. (330) gives

$$dT/dt = \omega(\partial T/\partial p)_{\mathrm{m}}, \tag{332}$$

or

$$[(\partial/\partial t) + \mathbf{v} \cdot \nabla]_{\mathrm{p}} T = \omega[(\partial T/\partial p)_{\mathrm{m}} - (\partial T/\partial p)]. \tag{333}$$

where

$$\left(\frac{\partial T}{\partial p}\right)_{\mathrm{m}} \equiv \left[\frac{\alpha}{c_{\mathrm{p}}} - \frac{L}{c_{\mathrm{p}}}\left(\frac{\partial q^*}{\partial p}\right)_{\mathrm{T}}\right] \Big/ \left[1 + \frac{L}{c_{\mathrm{p}}}\left(\frac{\partial q^*}{\partial T}\right)_{\mathrm{p}}\right], \tag{334}$$

Here $\partial/\partial p$ without a subscript is the derivative under constant horizontal coordinates and constant time.

The corresponding equation with the $\sigma$ coordinate can be readily obtained by using the following relations:

$$\left(\frac{\partial}{\partial t} + \mathbf{v} \cdot \nabla\right)_{\mathrm{p}} = \left(\frac{\partial}{\partial t} + \mathbf{v} \cdot \nabla\right)_{\sigma} - \frac{\sigma}{\pi}\left(\frac{\partial}{\partial t} + \mathbf{v} \cdot \nabla\right) \pi \frac{\partial}{\partial \sigma},$$

and

$$\omega = \sigma\left(\frac{\partial}{\partial t} + \mathbf{v}\cdot\nabla\right)\pi + \pi\dot{\sigma}.$$

With these, Eq. (333) becomes

$$\left(\frac{\partial}{\partial t} + \mathbf{v}\cdot\nabla\right)_{\sigma} T = \left(\frac{\partial T}{\partial p}\right)_{\mathrm{m}} \sigma\left(\frac{\partial}{\partial t} + \mathbf{v}\cdot\nabla\right)\pi + \pi\dot{\sigma}\left[\left(\frac{\partial T}{\partial p}\right)_{\mathrm{m}} - \frac{\partial T}{\partial p}\right], \tag{335}$$

where

$$\frac{\partial}{\partial p} = \frac{1}{\pi}\frac{\partial}{\partial \sigma}.$$

Now making use of the relation

$$\frac{\partial q^*}{\partial p} = \left(\frac{\partial q^*}{\partial p}\right)_{\mathrm{T}} + \left(\frac{\partial q^*}{\partial T}\right)_{\mathrm{p}}\frac{\partial T}{\partial p}, \tag{336}$$

the last term in Eq. (335) can be written as

$$\begin{aligned}\left(\frac{\partial T}{\partial p}\right)_{\mathrm{m}} - \frac{\partial T}{\partial p} &= \frac{1}{1 + [(L/c_{\mathrm{p}})(\partial q^*/\partial T)_{\mathrm{p}}]}\left[\frac{\alpha}{c_{\mathrm{p}}} - \frac{\partial T}{\partial p} - \frac{L}{c_{\mathrm{p}}}\frac{\partial q^*}{\partial p}\right] \\ &= \frac{1}{1 + [(L/c_{\mathrm{p}})(\partial q^*/\partial T)_{\mathrm{p}}]}\left[-\left(\frac{p}{p_0}\right)^{\kappa}\frac{\partial \theta}{\partial p} - \frac{L}{c_{\mathrm{p}}}\frac{\partial q^*}{\partial p}\right].\end{aligned} \tag{337}$$

With this expression, Eq. (335) becomes

$$\begin{aligned}\left(\frac{\partial}{\partial t} + \mathbf{v}\cdot\nabla\right)_{\sigma} T &= \left(\frac{\partial T}{\partial p}\right)_{\mathrm{m}} \sigma\left(\frac{\partial}{\partial t} + \mathbf{v}\cdot\nabla\right)\pi \\ &\quad - \pi\dot{\sigma}\left[\left(\frac{p}{p_0}\right)^{\kappa}\frac{\partial \theta}{\partial p} + \frac{L}{c_p}\frac{\partial q^*}{\partial p}\right]\Bigg/\left[1 + \frac{L}{c_p}\left(\frac{\partial q^*}{\partial T}\right)_p\right].\end{aligned} \tag{338}$$

From the form of the hydrostatic equation given by Eq. (122) the following equation can be derived

$$\left(\frac{p}{p_0}\right)^{\kappa} c_{\mathrm{p}}\left(\frac{\partial \theta}{\partial p}\right) = \frac{\partial}{\partial p}(c_{\mathrm{p}}T + \Phi). \tag{339}$$

If the moist static energy $h$ and the saturation moist static energy $h^*$ are defined by

$$h \equiv c_p T + \Phi + Lq, \qquad h^* \equiv c_p T + \Phi + Lq^*, \tag{340}$$

Eq. (339) can be written as

$$\left(\frac{p}{p_0}\right)^\kappa \frac{\partial \theta}{\partial p} + \frac{L}{c_p}\frac{\partial q^*}{\partial p} = \frac{1}{c_p}\frac{\partial h^*}{\partial p}. \tag{341}$$

Using this expression, Eq. (338) can be put in the final form

$$\left(\frac{\partial}{\partial t} + \mathbf{v}\cdot\nabla\right)_\sigma T = \left(\frac{\partial T}{\partial p}\right)_m \sigma\left(\frac{\partial}{\partial t} + \mathbf{v}\cdot\nabla\right)_\sigma \pi - \frac{\partial h^*/\partial p}{c_p + L(\partial q^*/\partial T)_p}\pi\dot{\sigma}, \tag{342}$$

where $\partial h^*/\partial p = 0$ when the lapse rate is moist adiabatic.

### 3. *Moist Adiabatic Process—Discrete Case*

The derivation of the vertically differenced form of the water vapor continuity equation is completely analogous to that for the continuous case presented in the previous subsection.

Let $q_k^* \equiv q^*(T_k, p_k)$. When level $k$ is saturated and remains saturated, Eq. (326) may be rewritten as

$$\left(\frac{\partial}{\partial t} + \mathbf{v}_k\cdot\nabla\right) q_k^* + \frac{1}{(\pi\,\Delta\sigma)_k}[(\pi\dot{\sigma})_{k+1}(\hat{q}_{k+1} - q_k^*) + (\pi\dot{\sigma})_{k-1}(q_k^* - \hat{q}_{k-1})] = -C_k, \tag{343}$$

and then as

$$\left(\frac{\partial q^*}{\partial T}\right)_{pk}\left(\frac{\partial}{\partial t} + \mathbf{v}_k\cdot\nabla\right) T_k + \left(\frac{\partial q^*}{\partial p}\right)_{Tk}\sigma_k\left(\frac{\partial}{\partial t} + \mathbf{v}_k\cdot\nabla\right)\pi_k$$
$$+ \frac{1}{(\pi\,\Delta\sigma)_k}[(\pi\dot{\sigma})_{k+1}(\hat{q}_{k+1} - q_k^*) + (\pi\dot{\sigma})_{k-1}(q_k^* - \hat{q}_{k-1})] = -C_k, \tag{344}$$

where

$$\left(\frac{\partial q^*}{\partial T}\right)_{pk} \equiv \left(\frac{\partial q_k^*}{\partial T_k}\right)_{pk}, \qquad \left(\frac{\partial q^*}{\partial p}\right)_{Tk} \equiv \left(\frac{\partial q_k^*}{\partial p_k}\right)_{Tk}.$$

The thermodynamic energy equation is, from Eq. (209),

$$\left(\frac{\partial}{\partial t} + \mathbf{v}_k \cdot \nabla\right) T_k + \frac{1}{(\pi\,\Delta\sigma)_k}\left[(\pi\dot{\sigma})_{k+1}(P_k\hat{\theta}_{k+1} - T_k) + (\pi\dot{\sigma})_{k-1}(T_k - P_k\hat{\theta}_{k-1})\right]$$
$$= \frac{1}{c_p}\alpha_k\sigma_k\left(\frac{\partial}{\partial t} + \mathbf{v}_k \cdot \nabla\right)\pi_k + \frac{L}{c_p}C_k, \tag{345}$$

where $\alpha_k \equiv c_p\theta_k(\partial P_k/\partial\pi_k)/\sigma_k$ and $\sigma_k \equiv (p_k - p_I)/\pi_k$. Eqs. (344) and (345) give

$$C_k = -\frac{1}{1 + [(L/c_p)(\partial q^*/\partial T)_{pk}}\left[\left\{\left(\frac{\partial q^*}{\partial p}\right)_{Tk} + \frac{\alpha_k}{c_p}\left(\frac{\partial q^*}{\partial T}\right)_{pk}\right\}\sigma_k\left(\frac{\partial}{\partial t} + \mathbf{v}_k \cdot \nabla\right)\pi_k\right.$$
$$- \left(\frac{\partial q^*}{\partial T}\right)_{pk}\frac{P_k}{(\pi\,\Delta\sigma)_k}\{(\pi\dot{\sigma})_{k+1}(\hat{\theta}_{k+1} - \theta_k) + (\pi\dot{\sigma})_{k-1}(\theta_k - \hat{\theta}_{k-1})\}$$
$$\left.+ \frac{1}{(\pi\,\Delta\sigma)_k}\{(\pi\dot{\sigma})_{k+1}(\hat{q}_{k+1} - q_k^*) + (\pi\dot{\sigma})_{k-1}(q_k^* - \hat{q}_{k-1})\}\right]. \tag{346}$$

Substituting Eq. (346) into Eq. (345) gives

$$\left(\frac{\partial}{\partial t} + \mathbf{v}_k \cdot \nabla\right) T_k = \left(\frac{\partial T}{\partial p}\right)_{mk}\sigma_k\left(\frac{\partial}{\partial t} + \mathbf{v}_k \cdot \nabla\right)\pi_k$$
$$- \frac{1}{1 + [(L/c_p)(\partial q^*/\partial T)_{pk}]}\frac{1}{(\pi\,\Delta\sigma)_k}\left[(\pi\dot{\sigma})_{k+1}\left(P_k\hat{\theta}_{k+1}\right.\right.$$
$$\left.+ \frac{L}{c_p}\hat{q}_{k+1} - P_k\theta_k - \frac{L}{c_p}q_k^*\right) + (\pi\dot{\sigma})_{k-1}\left(P_k\theta_k + \frac{L}{c_p}q_k^*\right.$$
$$\left.\left.- P_k\hat{\theta}_{k-1} - \frac{L}{c_p}\hat{q}_{k-1}\right)\right], \tag{347}$$

where

$$\left(\frac{\partial T}{\partial p}\right)_{mk} \equiv \left[\frac{\alpha_k}{c_p} - \frac{L}{c_p}\left(\frac{\partial q^*}{\partial p}\right)_{Tk}\right]\Bigg/\left[1 + \frac{L}{c_p}\left(\frac{\partial q^*}{\partial T}\right)_{pk}\right]. \tag{348}$$

Equation (347) is an analog of Eq. (338).

The coefficient of $(\pi\dot{\sigma})_{k+1}$ in Eq. (347) is

$$P_k(\hat{\theta}_{k+1} - \hat{\theta}_k) + \frac{L}{c_p}(\hat{q}_{k+1} - q_k^*) = \frac{1}{c_p}[(c_p\hat{T}_{k+1} + \hat{\Phi}_{k+1} + L\hat{q}_{k+1}) - (c_p T_k + \Phi_k + Lq_k^*)] \equiv \frac{1}{c_p}(\hat{h}_{k+1} - h_k^*),$$

where Eq. (199) has been used. Similarly, the coefficient of $(\pi\dot{\sigma})_{k-1}$ in Eq. (347) is

$$(1/c_p)(h_k^* - \hat{h}_{k-1}).$$

Thus Eq. (347) can be written as

$$\left(\frac{\partial}{\partial t} + \mathbf{v}_k \cdot \nabla\right) T_k = \left(\frac{\partial T}{\partial p}\right)_{mk} \sigma_k \left(\frac{\partial}{\partial t} + \mathbf{v}_k \cdot \nabla\right) \pi_k - \frac{1}{c_p + L(\partial q^*/\partial T)_{pk}} \frac{1}{(\pi\,\Delta\sigma)_k} [(\pi\dot{\sigma})_{k+1}(\hat{h}_{k+1} - h_k^*) + (\pi\dot{\sigma})_{k-1}(h_k^* - \hat{h}_{k-1})]. \tag{349}$$

Equation (349) is an analog of Eq. (342), but the choice of $\hat{q}$ (or equivalently, $\hat{h}$) at the even levels remains to be specified.

4. *Choice of $\hat{q}$ for Water Vapor*

From Eq. (349) it is clear that a negative $(\pi\dot{\sigma})$ has a warming effect for $\hat{h}_{k+1} > h_k^*$. This may occur even when $h_{k+2}^* < h_k^*$, that is, when no conditional instability exists between the odd levels $k + 2$ and $k$, which carry the temperatures. (The same effect can similarly occur for a negative $(\pi\dot{\sigma})_{k-1}$ when $h_k^* > \hat{h}_{k-1}$, even when $h_k^* < h_{k-2}^*$.) Any moist convective instability produced by such a warming effect is the result merely of a poor choice of $\hat{q}_{k+1}$ and should be regarded as a kind of computational instability, which may be termed "conditional instability of a computational kind" (CICK).

The CICK phenomenon can be avoided if the choice of $\hat{q}_{k+1}$ and thus $\hat{h}_{k+1}$ satisfies the following requirements when $h_{k+2}^* < h_k^*$:

$$\hat{h}_{k+1} < h_k^* \qquad \text{when} \quad r_k = 1$$

and

$$h_{k+2}^* < \hat{h}_{k+1} \qquad \text{when} \quad r_{k+2} = 1, \tag{350}$$

where $r_k$ is the relative humidity of the level $k$, given by

$$r_k = q_k/q_k^*. \tag{351}$$

One definition of $\hat{h}_{k+1}$ that satisfies the above requirements is

$$\hat{h}_{k+1} = \hat{r}_{k+1}(\hat{h}_{k+1}^* - \hat{s}_{k+1}) + \hat{s}_{k+1}, \tag{352}$$

where

$$\hat{s}_{k+1} \equiv c_p \hat{T}_{k+1} + \hat{\Phi}_{k+1}, \tag{353}$$

$$\hat{r}_{k+1} = \frac{r_k + r_{k+2} - 2r_k r_{k+2}}{2 - r_k - r_{k+2}}; \tag{354}$$

and $\hat{h}_{k+1}^*$ is an interpolation of $h^*$ from the levels $k$ and $k + 2$ to the level $k + 1$ which guarantees $h_{k+2}^* < \hat{h}_{k+1}^* < h_k^*$ if $h_{k+2}^* < h_k^*$. Equation (354) gives $\hat{r}_{k+1} = 1$ (and, therefore, $\hat{h}_{k+1} = \hat{h}_{k+1}^*$) when either $r_k = 1$ or $r_{k+2} = 1$; then $h_{k+2}^* < \hat{h}_{k+1} < h_k^*$ is guaranteed regardless of the form of $\hat{h}^*$.

The form of the interpolation used to obtain $\hat{h}^*$ is important, however, in relation to that chosen for $\hat{s}_{k+1}$. Since

$$h_k^* = Lq_k^* + s_k,$$

an interpolation for $\hat{h}_{k+1}^*$ independent of that for $\hat{s}_{k+1}$ could in theory allow the implicit generation of a negative $\hat{q}_{k+1}^*$. To avoid this, the interpolation for $\hat{h}_{k+1}^*$ is chosen proportional to that for $\hat{s}_{k+1}$:

$$\begin{aligned} \hat{h}_{k+1}^* - h_k^* &= A(\hat{s}_{k+1} - s_k), \\ h_{k+2}^* - \hat{h}_{k+1}^* &= A(s_{k+2} - \hat{s}_{k+1}), \end{aligned} \tag{355}$$

and

$$A \equiv \frac{h_{k+2}^* - h_k^*}{s_{k+2} - s_k}. \tag{356}$$

Recall that Eqs. (198) and (199) give

$$\begin{aligned} \hat{s}_{k+1} - s_k &= P_k c_p(\hat{\theta}_{k+1} - \theta_k), \\ s_{k+2} - \hat{s}_{k+1} &= P_{k+2} c_p(\theta_{k+2} - \hat{\theta}_{k+1}), \end{aligned}$$

and thus

$$s_{k+2} - s_k = P_k c_p(\hat{\theta}_{k+1} - \theta_k) + P_{k+2} c_p(\theta_{k+2} - \hat{\theta}_{k+1}),$$

where

$$\hat{\theta} \equiv \frac{\ln \theta_k - \ln \theta_{k+2}}{(1/\theta_{k+2}) - (1/\theta_k)}.$$

Equation (352) gives

$$\hat{q}_{k+1} = (1/L)\hat{r}_{k+1}(\hat{h}^*_{k+1} - \hat{s}_{k+1}). \tag{357}$$

There is no reason to choose this $\hat{q}$, however, if the air is not near saturation. Presumably, the application of Eq. (327) to water vapor mixing ratio is a better choice for the relatively dry case. The final form chosen for use in the model is a weighted mean of Eqs. (357) and (327), given by

$$\hat{q}_{k+1} = \hat{r}_{k+1}\left[\frac{1}{L}(\hat{h}^*_{k+1} - \hat{s}_{k+1})\right] + (1 - \hat{r}_{k+1})\left[\frac{\ln q_k - \ln q_{k+2}}{(1/q_{k+2}) - (1/q_k)}\right]. \tag{358}$$

The CICK is still prevented because Eq. (358) becomes identical to Eq. (357) when $\hat{r}_{k+1} = 1$.

The use of Eq. (358) for $\hat{q}_{k+1}$, however, does not guarantee that $q$ at odd levels remains positive or zero. For example, if $q_k = 0$, $\hat{q}_{k+1} > 0$ and $(\pi\dot{\sigma})_{k+1} > 0$, then the downward current removes a positive amount from zero. To avoid generation of a negative mixing ratio, $\hat{q}_{k+1}$ is replaced by zero when $(\pi\dot{\sigma})_{k+1} > 0$ and $q_k \leqslant 0$ [or when $(\pi\dot{\sigma})_{k+1} < 0$ and $q_{k+2} \leqslant 0$].

## B. Horizontal Transport of Water Vapor and Ozone

The finite-difference scheme for the divergence of the horizontal transport of water vapor and ozone is similar to the corresponding scheme for temperature, given at the point $(i, j)$ by the second term of Eq. (299), except that $\overline{q}^{\xi}$ and $\overline{q}^{\eta}$ are replaced by the harmonic mean when the corresponding mass flux, $F$ or $G$, is outwardly directed from the grid point $(i, j)$ under consideration. This guarantees zero transport out of the grid points where the mixing ratio is zero.

## C. Large-Scale Condensation and Precipitation

In the model there is water vapor condensation and release of latent heat not only by the parameterized cumulus convection, which does not require that the air be saturated on the scale of the grid, but also when the air becomes super-saturated on the scale of the grid; the latter phenomenon is called "large-scale condensation." The excess water removed from an atmospheric layer in this way precipitates into the layer immediately below. The falling precipitation either evaporates completely in that layer or brings the layer to saturation and then passes to the next layer below, where the process is repeated. When the lowermost layer is saturated, the condensed water precipitates onto the ground as rain or snow.

Large-scale condensation occurs when $q_{ij}{}^{k}$ is greater than $q_{ij}^{*k}$, where $q_{ij}{}^{k}$ is the provisional value of the water vapor mixing ratio predicted by the advective process only, and $q_{ij}^{*k}$ is the saturation mixing ratio at the temperature $T_{ij}{}^{k}$ and the pressure $p_{ij}{}^{k}$.

Let $C\,\Delta t$ denote the amount of condensation at level $k$ per unit mass of dry air when $q_{ij}{}^{k} > q_{ij}^{*k}$. Then

$$(q_{ij}{}^{k})' = q_{ij}{}^{k} - C\,\Delta t, \tag{359}$$

$$(T_{ij}{}^{k})' = T_{ij}{}^{k} + \frac{L}{c_{p}}\,C\,\Delta t, \tag{360}$$

$$(q_{ij}{}^{k})' = q^{*}[(T_{ij}{}^{k})', p_{ij}{}^{k}], \tag{361}$$

where the primes denote values modified by condensation. Equation (361) describes the saturation condition for the modified moisture and temperature. From these three equations an equation for the modified temperature can be obtained

$$q_{ij}{}^{k} - \frac{c_{p}}{L}[(T_{ij}{}^{k})' - T_{ij}{}^{k}] = q^{*}[(T_{ij}{}^{k})', p_{ij}{}^{k}]. \tag{362}$$

With $q_{ij}{}^{k}$, $T_{ij}{}^{k}$, $p_{ij}{}^{k}$ and the functional form of $q^{*}(T, p)$ given, the transcendental equation (362) can be solved iteratively for $(T_{ij}{}^{k})'$. After $(T_{ij}{}^{k})'$ is obtained, $C\,\Delta t$ and $(q_{ij}{}^{k})'$ may be computed from Eqs. (360) and (359).

Choosing

$$T_{0} = T_{ij}{}^{k} \qquad \text{and} \qquad q_{0} = q_{ij}{}^{k}, \tag{363}$$

$(T_{\nu+1}, q_{\nu+1})$ are determined recursively by

$$\begin{aligned} T_{\nu+1} &= T_\nu + (L/c_p)C_{\nu+1}\,\Delta t \\ q_{\nu+1} &= q_\nu - C_{\nu+1}\,\Delta t, \end{aligned} \tag{364}$$

where $\nu = 0, 1, 2, \ldots,$ and

$$C_{\nu+1}\,\Delta t \equiv \frac{q_\nu - q^*(T_\nu)}{1 + (L/c_p)(\partial q^*/\partial T)_{T=T_\nu}}. \tag{365}$$

In summary,

$$(T_{ij}{}^k)' = T_{ij}{}^k + \frac{L}{c_p}\sum_{\nu=1}^{\nu_{max}} C_\nu\,\Delta t, \tag{366}$$

$$(q_{ij}{}^k)' = q_{ij}{}^k - \sum_{\nu=1}^{\nu_{max}} C_\nu\,\Delta t, \tag{367}$$

where $\nu_{max}$ is the maximum number of iterations in the layer. A value of $\nu_{max} = 3$ seems to give sufficient accuracy for present purposes.

The effect of evaporation of the falling precipitation on the layer immediately below is incorporated in the following expressions:

$$(T_{ij}^{k+2})' = T_{ij}^{k+2} - \frac{L}{c_p}\sum_{\nu=1}^{\nu_{max}} C_\nu\,\Delta t \pi_k\,\Delta\sigma_k/(\pi_{k+2}\,\Delta\sigma_{k+2}), \tag{368}$$

$$(q_{ij}^{k+2})' = q_{ij}^{k+2} + \sum_{\nu=1}^{\nu_{max}} C_\nu\,\Delta t \pi_k\,\Delta\sigma_k/(\pi_{k+2}\,\Delta\sigma_{k+2}). \tag{369}$$

If the layer becomes supersaturated due to the evaporation, the entire process is repeated for that layer.

## VIII. Time Differencing

To explain the procedure, the equations can be written symbolically in the following form:

$$\frac{\partial}{\partial t}\pi = f(\pi, \mathbf{A}), \tag{370}$$

$$\frac{\partial}{\partial t}(\pi\mathbf{A}) = g(\pi, \mathbf{A}). \tag{371}$$

Equation (370) represents the continuity equation and Eq. (371) represents the prognostic equations for the other variables described in the previous sections.

The leapfrog scheme (L) is given by

$$\frac{\pi^{n+1} - \pi^{n-1}}{2\Delta t} = f(\pi^n, \mathbf{A}^n),$$

$$\frac{\pi^{n+1}\mathbf{A}^{n+1} - \pi^{n-1}\mathbf{A}^{n-1}}{2\Delta t} = g(\pi^n, \mathbf{A}^n),$$

where the superscript denotes a time level. The Matsuno scheme (M), which is sometimes called the Euler-backward scheme, is given by

$$\frac{\pi^{(n+1)*} - \pi^n}{\Delta t} = f(\pi^n, \mathbf{A}^n)$$

$$\frac{\pi^{(n+1)*}\mathbf{A}^{(n+1)*} - \pi^{(n)*}\mathbf{A}^{(n)*}}{\Delta t} = g(\pi^n, \mathbf{A}^n),$$

$$\frac{\pi^{n+1} - \pi^n}{\Delta t} = f(\pi^{(n+1)*}, \mathbf{A}^{(n+1)*})$$

$$\frac{\pi^{n+1}\mathbf{A}^{n+1} - \pi^n\mathbf{A}^n}{\Delta t} = g(\pi^{(n+1)*}, \mathbf{A}^{(n+1)*}).$$

The time differencing used in the model for the basic dynamical terms is essentially the leapfrog scheme, but with a periodic insertion of the Matsuno scheme, as shown in Fig. 27.

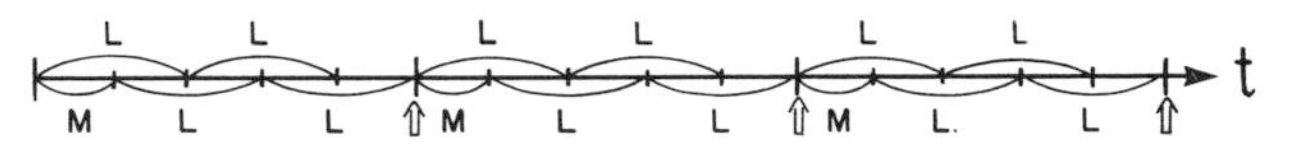


FIG. 27. Schematic representation of the time differencing of the model showing sequence of use of leapfrog (L) and Matsuno (M) schemes. Arrows indicate calculation of the heating and friction terms.

At present, the source and sink terms described in the Introduction and the vertical flux convergence term of the moisture equation are calculated every five time steps, as shown by the arrows in the figure. Those calculations are followed by a single step of the Matsuno scheme.

## IX. Summary and Conclusions

In this chapter, only the computational design of the basic dynamical processes of the current UCLA general circulation model has been described. To determine the heating and friction, the model includes many important physical processes, such as those associated with radiation, photochemistry, the boundary layer, the thermodynamics and hydrology of the ground, as well as processes associated with grid- and subgrid-scale clouds. These physical processes could not be adequately described in a single chapter and, therefore, with the exception of the advective processes for water vapor and ozone and the large-scale condensation processes, were not included here.

Section I gives a brief outline of the model, whose 12 layers represent both troposphere and stratosphere. The prognostic variables of the model are the surface pressure, horizontal velocity, temperature, water vapor and ozone of each layer; the planetary boundary layer (PBL) depth and magnitudes of the temperature, moisture and momentum discontinuities at the PBL top; the ground temperature and water storage; and the mass of snow on the ground. It should be noted that the degree of freedom added by the PBL makes the model effectively equivalent to a 13-layer model.

Section II describes the principles of mathematical modeling that were followed in the computational design of the basic dynamical processes of the model. The basic principle employed in selecting a space finite-difference scheme from the many that share the same order of accuracy was a requirement that the scheme maintain discrete analogs of a number of physically important integral constraints of the continuous atmosphere. Energy propagation properties in physical space, as well as in spectral space, were also considered in the selection of a scheme.

Section III describes space finite-difference schemes for homogeneous incompressible flow, with and without a free surface. Section III, A shows that the dispersion properties of inertia-gravity waves are highly scheme-dependent and that from the point of view of geostrophic adjustment there is only one satisfactory distribution (staggering) of the dependent variables into grid points.

Section III, B discusses finite-difference schemes for nonlinear two-dimensional nondivergent flow and replaces Part II of the paper by Arakawa (1966), which was originally planned as a separate publication. A drastic difference in the energy cascade exists between solutions obtained by schemes that conserve enstrophy and by those that do not. Due to the relatively small amount of energy in the high wave number range with enstrophy-conserving schemes, the overall error is expected to be small. This subsection also derives, for the cartesian grid, the momentum advection scheme consistent with the energy and enstrophy conserving vorticity advection scheme for two-dimensional nondivergent flow. The total momentum is also conserved with this scheme.

Section III, C generalizes the momentum advection scheme for nondivergent flow to a scheme that maintains conservation of total energy and momentum for divergent flow. It should be pointed out, however, that this generalization is not unique and is not necessarily the best choice from the standpoint of potential vorticity advection when the lower boundary has relatively steep topography. In general circulation models, horizontal discretization errors should be small for planetary-scale waves *after they are generated* because their horizontal scales are sufficiently large compared to the usual horizontal grid size. However, horizontal discretization errors can be very serious for the *generation* of planetary-scale waves by longitudinally narrow (but meridionally wide) mountain ridges. A search for a generalization to divergent flow that is better from this point of view is now in progress.

Section IV describes the vertical coordinate of the model. It is a version of the $\sigma$ coordinate below 100 mb and the pressure coordinate above 100 mb. The basic governing equations in terms of that vertical coordinate are presented.

Section V describes the vertical difference scheme. Various integral properties are presented in Section V, A; Section V, B then discusses the logical procedure for deriving a scheme that maintains discrete analogs of these properties. Section V, C presents the final determination of the vertical difference scheme based on considerations of accuracy in both the vertical propagation of wave energy and the hydrostatic equation.

Section VI presents the horizontal difference scheme of the model. The scheme for three-dimensional motion on a sphere is a generalization, although not unique, of the scheme developed in Section III, C. With the current scheme, however, enstrophy is not conserved for two-dimensional incompressible flow on a sphere, and solutions from the model show some computational quasi-stationary noise near the poles that would seem to correspond to a false production of enstrophy. The new generalization now being sought should be better from this point of view also. Section VI, C, 2 describes the method devised to avoid the use of the extremely short time interval required for computational stability due to the convergence of meridians toward the poles. The method employs an operator to smooth, in a longitudinal sense, selected terms of the prognostic equations that involve longitudinal differences. The result is equivalent to the use of multipoint finite-difference quotients and the space finite-difference scheme remains energy conserving.

Section VII gives the space finite-difference schemes for the advection of water vapor and ozone. Special advection schemes are necessary both in that the mixing ratios of these atmospheric constituents vary in space over a wide range of orders of magnitude, and also in that the release of latent heat through condensation of water vapor can cause a false moist convective instability. Our method for the calculation of the large-scale condensation is also described in this section.

In Section VIII is described the time differencing of the model. The heating and friction terms are calculated every fifth time step. For the basic dynamical processes, at the steps which immediately follow the calculations of heating and friction, the Matsuno scheme is inserted; for all other time steps, the leapfrog scheme is used.

Descriptions of physical and computational aspects of the model related to those physical processes that determine the heating and friction will be published separately elsewhere. The most complete documentation currently available for the radiation and photochemical processes is given in Schlesinger (1976) and for the boundary layer and stratus cloud processes in Randall (1976). The parameterization of cumulus convection is based on the theory proposed by Arakawa and Schubert (1974). Some computational problems associated with the application of the theory were discussed by Schubert (1973). A more complete description of this aspect of the model, including the more recent revisions, is now being prepared for publication.

## Acknowledgments

The development of the UCLA atmospheric general circulation models has been carried out, over a number of years, with Professor Yale Mintz as the principal investigator. The authors also wish to acknowledge the considerable contributions of Drs. Akira Katayama, Jeong-Woo Kim, David Randall, Wayne Schubert, Michael Schlesinger, and Tatsushi Tokioka, and of Winston Chao and Stephen Lord to the development of the general circulation model.

We express our appreciation to Donna Hollingworth and Christopher Kurasch for their substantial assistance in programming the model. Appreciation is also due to Mrs. Grace McMurray for the typing and to Mrs. Beverly Gladstone for the drafting of the figures.

The research reported here was supported by the National Science Foundation under Grant GA-34306; the Department of Transportation Climatic Impact Assessment Program under Grant GA-34306X; and the National Aeronautics and Space Administration, Institute for Space Studies, Goddard Space Flight Center under Grant NGR 05-007-328.

Computing assistance was obtained from the UCLA Campus Computing Network.

## References

Arakawa, A. (1966). *J. Comput. Phys.* **1**, 119–143.

Arakawa, A. (1970). *In* "SIAM-AMS Proceedings of the Symposium in Applied Mathematics" (G. Birkhoff and S. Varga, eds.), Vol. 2, pp. 24–40. American Mathematical Society, Providence, Rhode Island.

Arakawa, A. (1972). "Numerical Simulation of Weather and Climate," Tech. Rep. No. 7. Dept. Meteorol., University of California, Los Angeles.

Arakawa, A., and Schubert, W. H. (1974). *J. Atmos. Sci.* **31**, 674–701.

Cahn, A. (1945). *J. Meteorol.* **2**, 113–119.

Fjørtoft, R. (1953). *Tellus* **5**, 225–230.

Lorenz, E. N. (1955). *Tellus* **7**, 157–167.

Lorenz, E. N. (1960). *Tellus* **12**, 364–373.

Mintz, Y. (1965). *W.M.O. Tech. Notes* **66**, 141–167.

Mintz, Y. (1968). *Am. Meteorol. Soc., Monogr.* **8**, 20–36.

Phillips, N. A. (1957). *J. Meteorol.* **14**, 184–185.

Phillips, N. A. (1959). *In* "The Atmosphere and the Sea in Motion," (Bert Bolin, ed.) pp. 501–504. Rockefeller Inst. Press, New York.

Phillips, N. A. (1974). Natl. Meteorol. Cent. Off. Note 104. Natl. Weather Service, Washington, D.C.

Randall, D. A. (1976). "The Interaction of the Planetary Boundary Layer with Large-Scale Circulations." Ph.D. Thesis, Dept. Atmos. Sci., University of California, Los Angeles.

Schlesinger, M. E. (1976). "A Numerical Simulation of the General Circulation of Atmospheric Ozone." Ph.D. Thesis, Dept. Atmos. Sci., University of California, Los Angeles.

Schubert, W. H. (1973). "The Interaction of a Cumulus Cloud Ensemble with the Large-Scale Environment." Ph.D. Thesis, Dept. Meteorol., University of California, Los Angeles.

Winninghoff, F. J. (1968). "On the Adjustment toward a Geostrophic Balance in a Simple Primitive Equation Model with Application to the Problems of Initialization and Objective Analysis." Ph.D. Thesis, Dept. of Meteorol., University of California, Los Angeles.

# Global Modeling of Atmospheric Flow by Spectral Methods

WILLIAM BOURKE, BRYANT MCAVANEY, KAMAL PURI, AND ROBERT THURLING

AUSTRALIAN NUMERICAL METEOROLOGY RESEARCH CENTRE, MELBOURNE, AUSTRALIA

I. Introduction . . . 268
A. Preview . . . 268
B. Development of Spectral Models . . . 268
C. Relative Merits of Spectral and Finite Difference Models . . . 270
D. Survey of Current Status of Spectral Models . . . 270
II. Spectral Algebra . . . 271
A. Barotropic Nondivergent Model . . . 271
B. Silberman's Method . . . 272
C. Integral Constraints . . . 275
D. Transform Method . . . 276
E. Experiments with Barotropic Spectral Transform Models . . . 279
III. Multilevel Spectral Model . . . 285
A. Model Formulation . . . 285
B. Equations of Motion in Spectral Form . . . 289
C. Semi-implicit Time Integration . . . 295
D. Inclusion of Topography . . . 297
E. Model Computer Coding . . . 299
F. Application of Time-Splitting Technique . . . 301
IV. Numerical Weather Prediction via a Spectral Model . . . 302
A. Operational Model Configuration . . . 303
B. Spectral Data Analysis and Processing . . . 303
C. Model Initialization . . . 304
D. Typical Synoptic Results . . . 304
E. General Comments . . . 310
V. General Circulation via a Spectral Model . . . 311
A. Model Configuration . . . 311
B. Radiative Transfer Calculation . . . 312
C. Lower Boundary Conditions . . . 312
D. Mean January Simulation . . . 313
E. General Comments . . . 319
VI. Conclusion . . . 319
Appendix . . . 320
A. Matrix $G$ . . . 320
B. Matrices $\mathbf{V}$ and $\mathbf{V}'$ . . . 321
References . . . 323

## I. Introduction

### A. Preview

Numerical modeling of the large-scale atmospheric flow has progressed rapidly since the pioneering experiments in numerical weather prediction by Charney *et al.* (1950). Subsequent evolution in computational and numerical procedures coupled with enhanced understanding of physical processes governing the atmospheric flow has now enabled realistic simulations of, for example, the seasonal variation of the global atmospheric circulation (Manabe and Holloway, 1975) and the operational application of numerical weather prediction in many centers throughout the world. The initiation of the Global Atmospheric Research Programme (GARP) in 1967 has further accelerated the study of large scale atmospheric fluctuations, such as those that control changes in the weather and the study of the general circulation of the atmosphere with a view to understanding the earth's climate.

A central aspect of the GARP is the further development of those numerical and computational methods that will enhance both the simulation and prediction capability of large-scale models of the atmosphere.

The numerical schemes for atmospheric modeling commonly employ one of two methods. The more usual approach represents the dynamic meteorological variables in space and time on a finite difference grid. As is apparent from the other contributions to this volume the grid-point method has proved to be very successful both for numerical simulation and prediction. An alternative method represents the variance of the dynamic fields, in part, by truncated expansions in terms of analytic spectral functions. A spectral representation of the horizontal variance of dynamic fields in terms of orthogonal surface spherical harmonics, in particular, offers a number of significant advantages in global and hemispheric scale simulation and prediction. This alternative, commonly referred to in the literature as the spectral method, has now become a very competitive approach and the present contribution aims to illustrate (i) the basic spectral modeling approach as embodied in two-dimensional flow; (ii) the extension of the spectral method to the three-dimensional domain by incorporation of a discrete vertical representation; and (iii) the application of the method in operational numerical weather prediction and large-scale general circulation studies.

### B. Development of Spectral Models

The development of the spectral method of numerical integration of the equations of motion of the atmosphere can be traced to Silberman (1954); he considered the integration of the barotropic vorticity equation in spherical geometry. Subsequent study of the spectral method was performed by Lorenz (1960), Platzman (1960), Kubota *et al.* (1961), Baer and Platzman (1961), and Elsaesser (1966). The studies of Lorenz demonstrated that the truncated spectral equations for nondivergent barotropic flow have the remarkable property that the mean square kinetic energy and mean square vorticity are invariants, just as in the exact differential equation of motion. Platzman pointed out that this property automatically eliminated nonlinear instability, which at that time was a substantial difficulty in grid-point models.

The spectral method may be broadly described as (i) representing the variance of dynamic fields as a superposition of a finite sum of component waves; and (ii) reducing the equations of motion to prognostics for the amplitude and phase of each of the component waves.

The essential nontrivial aspect of the calculation becomes then the evaluation of the nonlinear contributions to spectral component tendencies. A characteristic of spectral models based on Silberman's procedure was the explicit treatment of nonlinear interaction, i.e., explicit term by term multiplication when forming the product of two spectral series.

Silberman expressed the stream function of the flow field in terms of orthogonal surface spherical harmonics and, in explicitly treating nonlinearity, required the use of interaction coefficients. The interaction coefficient method survived unchallenged until Robert (1966) suggested the use of low-order nonorthogonal spectral functions based on elements of the spherical harmonics. This approach eliminated the complexity of interaction coefficients at the expense of an orthogonalization procedure at each time step; however, the basic strategy here was still that of spectral or nonlocal multiplication as in Silberman's method.

In 1970 Orszag and Eliasen *et al.*, working independently, suggested the approach commonly referred to as the transform method of spectral modeling. In this scheme the evaluation of nonlinear products is readily simplified by (i) a transform of appropriate fields to a spatial grid; (ii) operation on these grid point fields to evaluate requisite nonlinear terms at each grid point; and (iii) an inverse transform on these required terms back into the spectral domain.

The advent of the transform method has now greatly enhanced the efficiency and enlarged the capability of the spectral method. The inherent advantages of the spectral approach had always been attractive, but the method was generally considered to be at a disadvantage with respect to the computational efficiency available in finite difference models. In addition, the traditional spectral method only permitted a consideration of nonlinearities that were at the most quadratic. This limitation and lack of computational efficiency greatly dampened progress with the spectral method in atmospheric modeling.

## C. Relative Merits of Spectral and Finite Difference Models

The spectral method, in the present context of discussing global or hemispheric simulation of the atmosphere, will be understood to imply the use of surface spherical harmonics to specify the horizontal variance of dynamic fields.

Foremost of the advantages of the spectral method relative to finite difference methods are (i) the intrinsic accuracy of evaluation of horizontal advection; (ii) the elimination of aliasing arising from quadratic nonlinearity; (iii) the ease of modeling flow over the entire globe; and (iv) the ease of incorporation of semi-implicit time integration. These advantages are now well known and have been discussed in detail by Platzman (1960), Elsaesser (1966), Robert (1969), and Orszag (1970, 1974).

The inclusion of nonlinearities, which are more complicated than quadratic, is perhaps the least explored and most difficult aspect of spectral modeling. Such problems are common to both the spectral and finite difference approach, although they seem to appear somewhat more explicitly in the spectral method.

The above-mentioned characteristics of the spectral method afford in practice (i) highly accurate and stable numerics and (ii) efficient and simple computer coding.

## D. Survey of Current Status of Spectral Models

The current status of spectral modeling and an indication of the recent rejuvenation of interest in the method can be gauged from the advent of the GARP-sponsored International Symposium on Spectral Methods in Numerical Weather Prediction held in Copenhagen in August 1974 (GARP, 1974).

Orszag (1970) formulated the transform method in terms of a full two-dimensional transform, i.e., incorporating Legendre and Fourier transforms, and proposed a scheme suitable for integration of the nondivergent barotropic model. Eliasen *et al.* (1970) independently demonstrated the applicability of the Legendre transform approach for integration of a divergent barotropic model (i.e., a single-level primitive equation model). Subsequently, Machenhauer and Rasmussen (1972) and Bourke (1972) demonstrated the full two-dimensional transform for the divergent barotropic model incorporating Legendre transforms as proposed by Eliasen *et al.* and the fast Fourier transform as proposed by Orszag. The approach by Machenhauer employed prognostics for momentum, weighted by cosine of latitude, while that of Bourke employed prognostics for vorticity and divergence. In extending the transform method to three dimensions, Bourke (1974) employed a discrete representation in the vertical, while Machenhauer and Daley (1972) have employed a spectral representation in the vertical as well as in the horizontal. Daley *et al.* (1974), Hoskins and Simmons (1975), and Gordon and Stern (1974) have followed the approach of Bourke and employ prognostics for vorticity and divergence and a discrete finite difference representation in the vertical. Most of these authors are well represented in the report of the Copenhagen Symposium.

The three-dimensional spectral models cited above are being employed at the various institutions for studies of the general circulation, numerical weather prediction, and the dynamics of the atmospheric fluid. The application of the spectral method at the Australian Numerical Meteorology Research Centre (ANMRC) in the areas of general circulation and numerical weather prediction forms the basis of discussion in Sections IV and V.

## II. Spectral Algebra

Many of the mathematical properties of the equations of motion describing the global atmosphere are embodied in simpler subsets of equations such as those describing barotropic flow. It is with reference to nondivergent barotropic flow that the basic spectral algebra will be illustrated.

### A. Barotropic Nondivergent Model

The flow of a nondivergent barotropic atmosphere may be described in terms of conservation of absolute vorticity as

$$\partial\zeta/\partial t = -\mathbf{V}\cdot\nabla(\zeta + f) \tag{1}$$

where $\mathbf{V}$ is the horizontal wind vector with eastward and northward components $u$ and $v$, respectively, $f$ is the Coriolis parameter, $\nabla$ is the horizontal gradient operator, and $\zeta$ is the vertical component of relative vorticity $\mathbf{k}\cdot\nabla\times\mathbf{V}$; $\mathbf{k}$ denotes the vertical unit vector.

Robert (1966) pointed out that the components $u$ and $v$ of the wind field constitute pseudo-scalar fields on the globe, and as such are not well suited to representation in terms of scalar spectral expansions; he suggested that the variables

$$U = u\cos\phi \qquad \text{and} \qquad V = v\cos\phi, \tag{2}$$

where $\phi$ denotes latitude, would be more appropriate for global spectral representation.

The nondivergent flow field may be represented in terms of a scalar stream function $\psi$ as

$$\mathbf{V} = \mathbf{k} \times \nabla\psi. \tag{3}$$

Accordingly $\zeta$ is seen to be expressible as

$$\zeta = \mathbf{k} \cdot \nabla \times \mathbf{V} = \nabla^2\psi \tag{4}$$

and the quantities $U$ and $V$ as

$$U = -\frac{\cos\phi}{a}\frac{\partial\psi}{\partial\phi} \quad \text{and} \quad V = \frac{1}{a}\frac{\partial\psi}{\partial\lambda}. \tag{5}$$

Equation (1) may now be rewritten, with substitution of Eqs. (2) and (4), and an expansion into spherical polar coordinates as

$$\frac{\partial}{\partial t}\nabla^2\psi = -\frac{1}{a\cos^2\phi}\left[\frac{\partial}{\partial\lambda}(U\nabla^2\psi) + \cos\phi\frac{\partial}{\partial\phi}(V\nabla^2\psi)\right] - 2\Omega\frac{V}{a}, \tag{6}$$

where $\phi$, $\lambda$, $a$, and $\Omega$ denote, respectively, latitude, longitude, and the earth's radius and rotation rate. The linear equations (5) provide specification of the diagnostic quantities $U$ and $V$ in terms of the prognostic $\psi$.

The prognostic equation (6) constitutes a form especially appropriate for the transform method. The more usual form of the barotropic vorticity equation is seen on substitution of Eqs. (5) into Eq. (6) to be

$$\frac{\partial}{\partial t}(\nabla^2\psi) = \frac{1}{a^2\cos\phi}\left[\frac{\partial\nabla^2\psi}{\partial\lambda}\frac{\partial\psi}{\partial\phi} - \frac{\partial\psi}{\partial\lambda}\frac{\partial\nabla^2\psi}{\partial\phi}\right] - \frac{2\Omega}{a^2}\frac{\partial\psi}{\partial\lambda} \tag{7}$$

### B. Silberman's Method

Equation (7) is as used by Silberman in his pioneering application of the spectral method. Some discussion of the algebra arising in Silberman's method is warranted since it illustrates important properties of the spectral method and will facilitate subsequent comparison with the transform method.

A spectral form of Eq. (7) may be obtained via representation of the variable $\psi$ in terms of orthogonal surface spherical harmonics. These harmonics are the eigenfunctions of the two-dimensional Laplacian ($\nabla^2$) operator in spherical coordinates and are thus very appropriate in the prognostic for quantities such as $\nabla^2\psi$.

A truncated expansion for approximating the stream function is as follows

$$\psi(\lambda, \phi) = a^2 \sum_{m=-J}^{+J} \sum_{l=|m|}^{|m|+J} \psi_l^m(t) Y_l^m(\lambda, \phi) \tag{8}$$

where
1. The term $\psi_l^m$ denotes time-dependent, generally complex, expansion coefficients; the reality of the field requires $(\psi_l^m)^* = (-)^m \psi_l^{-m}$.
2. $Y_l^m(\lambda, \phi) = P_l^m(\sin\phi)e^{im\lambda}$; $P_l^m(\sin\phi)$ is an associated Legendre polynomial of the first kind normalized to unity; that is,

$$\int_{-\pi/2}^{+\pi/2} P_l^m(\sin\phi) P_l^m(\sin\phi) \cos\phi \, d\phi = 1.$$

3. The truncation parameter $J$ denotes rhomboidal wave number truncation (Elsaesser, 1966); $m$ denotes a planetary wave number, and $l - m$ denotes a meridional wave number.

(In the following the arguments $\lambda$, $\phi$, and $t$ are omitted for clarity of presentation.)

If now a spectral representation of the nonlinear advection term is introduced, as follows,

$$C = \frac{1}{a^2 \cos\phi}\left[\frac{\partial \nabla^2 \psi}{\partial \lambda}\frac{\partial \psi}{\partial \phi} - \frac{\partial \nabla^2 \psi}{\partial \phi}\frac{\partial \psi}{\partial \lambda}\right]$$

$$= \sum_{m=-J}^{+J} \sum_{l=|m|}^{|m|+J} C_l^m Y_l^m \tag{9}$$

then the vorticity prognostic may be reduced, on application of the orthogonality property of the harmonic functions, to the form

$$-l(l+1)(\partial \psi_l^m / \partial t) = C_l^m - 2i\Omega m \psi_l^m \tag{10}$$

On substituting in Eq. (9) for the spectral expansion of $\psi$ it follows that

$$C = -\frac{1}{a^2 \cos\phi}\left\{\frac{\partial}{\partial \lambda}\sum_{m_1}\sum_{l_1} l_1(l_1+1)\psi_{l_1}^{m_1} Y_{l_1}^{m_1} \times \frac{\partial}{\partial \phi}\sum_{m_2}\sum_{l_2} a^2 \psi_{l_2}^{m_2} Y_{l_2}^{m_2}\right.$$

$$\left. -\frac{\partial}{\partial \phi}\sum_{m_1}\sum_{l_1} l_1(l_1+1)\psi_{l_1}^{m_1} Y_{l_1}^{m_1} \times \frac{\partial}{\partial \lambda}\sum_{m_2}\sum_{l_2} a^2 \psi_{l_2}^{m_2} Y_{l_2}^{m_2}\right\} \tag{11}$$

Further manipulation of Eq. (11) yields

$$C = \frac{i}{2} \sum_{m_1} \sum_{l_1} \sum_{m_2} \sum_{l_2} \{l_2(l_2 + 1) - l_1(l_1 + 1)\} \psi_{l_1}{}^{m_1} \psi_{l_2}{}^{m_2}$$
$$\times \frac{1}{\cos\phi} \left\{ m_1 P_{l_1}{}^{m_1} \frac{dP_{l_2}{}^{m_2}}{d\phi} - m_2 P_{l_2}{}^{m_2} \frac{dP_{l_1}{}^{m_1}}{d\phi} \right\} \exp\left[i(m_1 + m_2)\lambda\right] \quad (12)$$

On multiplying both sides of Eq. (12) by the complex conjugate $Y_l^{m*}$ and integrating over the sphere, the orthogonality condition finally yields the expression for $C_l^m$ in the form

$$C_l^m = \frac{i}{2} \sum_{m_1} \sum_{l_1} \sum_{m_2} \sum_{l_2} \{l_2(l_2 + 1) - l_1(l_1 + 1)\} \psi_{l_1}{}^{m_1} \psi_{l_2}{}^{m_2} L(l_1 l_2 l; m_1 m_2) \quad (13)$$

Here

$$L(l_1 l_2 l; m_1 m_2) = \int_{-\pi/2}^{+\pi/2} P_l^m \left\{ m_1 P_{l_1}{}^{m_1} \frac{dP_{l_2}{}^{m_2}}{d\phi} - m_2 P_{l_2}{}^{m_2} \frac{dP_{l_1}{}^{m_1}}{d\phi} \right\} d\phi \quad (14)$$

is an interaction coefficient defining the coupling via the scalar triple product (i.e., the Jacobian interaction) between two components $\psi_{l_1}{}^{m_1}$ and $\psi_{l_2}{}^{m_2}$ interacting to contribute to the tendency of $\psi_l{}^m$.

As shown by Silberman, this Jacobian interaction coefficient $L(l_1 l_2 l; m_1 m_2)$ may be specified analytically in terms of integrals of triple products of the associated Legendre polynomials. Such triple integrals are well known and occur in the product theorem for spherical harmonics and give rise to the selection rules to be expected from the vector addition of angular momentum $l_1$ and $l_2$ and the algebraic addition of $m_1 + m_2$. These scalar interaction selection rules then define selection rules for the Jacobian interaction coefficient $L(l_1 l_2 l; m_1 m_2)$ such that $|l_1 - l_2| < l < l_1 + l_2$ and $m = m_1 + m_2$. A detailed discussion of these rules and their implications is given by Elsaesser (1966); such rules allow identification of principal interactions and a specification of null interactions. For example, the interaction of a single wave component $\psi_l{}^m$ with the component $\psi_1{}^0$ yields only solid body advection of the $\psi_l{}^m$ wave and is in fact an example of a Rossby wave, i.e., the solution of the linearized vorticity equation in spherical geometry (Haurwitz, 1940).

Equation (13) conveys readily the complexity of this interaction coefficient approach. In evaluating each of the $\sim J^2$ tendencies it is necessary to evaluate a triple summation; the quadruple summation has an implied redundancy as $m = m_1 + m_2$. Thus with each summation containing $\sim J$ terms there is a requirement for the evaluation of $0(J^5)$ products to evaluate all spectral tendencies (Orszag, 1970). Orszag notes that the increase in the number of operations due to the complex algebra is compensated for by the decrease in the number of actual interactions upon taking into account the selection rules. In addition to this evaluation it is necessary to precompute the interaction coefficients and thereafter sequentially access them at successive time steps of numerical integration.

### C. Integral Constraints

As alluded to in Section I, B, Lorenz (1960) demonstrated that a truncated set of spectral equations such as defined by Eqs. (10) and (13) rather remarkably retain the quadratic invariants of the exact differential equation for nondivergent barotropic flow. As an example, the conservation of kinetic energy is considered in the following.

The domain mean kinetic energy is given by the area integral over the sphere of

$$K = \mathbf{V} \cdot \mathbf{V}/2 = \nabla\psi \cdot \nabla\psi/2$$

which reduces to

$$\bar{K} = \frac{a^2}{4} \sum_l \sum_m l(l+1) \psi_l^m \psi_l^{m*}.$$

The tendency then for the total kinetic energy is

$$\frac{\partial \bar{K}}{\partial t} = \frac{a^2}{4} \sum_l \sum_m l(l+1) \left( \psi_l^m \frac{\partial \psi_l^{m*}}{\partial t} + \psi_l^{m*} \frac{\partial \psi_l^m}{\partial t} \right). \tag{15}$$

As described by Lorenz, on substituting for the time derivatives $\partial\psi_l^{m*}/\partial t$ and $\partial\psi_l^m/\partial t$, as given by Eqs. (10) and (13), it may be shown that the tendency of the mean kinetic energy vanishes. Merilees (1968) explicitly illustrates this algebra and the result that the truncated set of spectral equations retains as invariant the domain integral of kinetic energy via conservative redistribution of energy within each individual interaction. It may be similarly shown that other invariants such as mean squared vorticity are also invariant in the truncated system. Lorenz (1960) utilized these properties to study highly simplified nonlinear systems that were still characterized by the same integral constraints as the full differential equations. Platzman (1960) noted, as mentioned earlier, that these quadratic invariants guarantee the absence of nonlinear instability.

The quadratic invariants of the truncated system may be considered in less rigorous terms than described above. Having nominated some finite truncation of the flow field, e.g., wave number $J$, it is seen that although in Eq. (15) there may be nonzero values for $\partial\psi_l^m/\partial t$ outside this truncation due to nonlinearity, these do not contribute to $\partial\bar{K}/\partial t$ since they are multiplied by $\psi_l^{m*}$, which is zero; outside the truncation $J$ the amplitudes $\psi_l^m$ are defined to be zero. Now the spectral tendencies within the truncation $J$ are evaluated exactly, i.e., all interactions contributing to these tendencies are explicitly considered; thus the only tendencies contributing to the change of kinetic energy are considered exactly and accordingly retain the invariant property of the exact equation.

### D. Transform Method

The vorticity equation in the concise form of Eq. (6) is more appropriate for application of the transform method than Eq. (7). In the simple example of the nondivergent barotropic model, the two forms could be considered with approximately equal efficiency; however, the fully differentiated vorticity equation is considerably less appropriate in a divergent barotropic model or a full three-dimensional model. The advantage of the concise form is that differentiation may be performed after evaluation of the nonlinear product.

In addition to the spectral expansion for $\psi$, the spectral form of Eq. (6) requires a spectral representation of the diagnostic variables $U$ and $V$ as follows

$$U = a \sum_{m=-J}^{+J} \sum_{l=|m|}^{|m|+J+1} U_l^m Y_l^m$$

$$V = a \sum_{m=-J}^{+J} \sum_{l=|m|}^{|m|+J} V_l^m Y_l^m \tag{16}$$

In expressing $U$ and $V$ spectrally, it is essential that these derived wind fields be identical to those implied by the truncated expansion for $\psi$. Substitution of expansions (8) and (16) into Eq. (5) yields, on application of a standard recurrence relation and the orthogonality property of the spherical harmonics, the relationships

$$U_l^m = (l-1)\varepsilon_l^m \psi_{l-1}^m - (l+2)\varepsilon_{l+1}^m \psi_{l+1}^m$$

and

$$V_l^m = im\psi_l^m \tag{17}$$

where

$$\varepsilon_l^m = [(l^2 - m^2)/(4l^2 - 1)]^{1/2}.$$

It becomes apparent from Eqs. (17) that the expansion for $U$, as implied in Eq. (16), must extend to one degree above that defined for $\psi$ in Eq. (8); for example nonzero values of $U^{|m|}_{|m|+J+1}$ are implied by variance of $\psi^{|m|}_{|m|+J}$.

In Eq. (6) the nonlinear products are $U\ \nabla^2\psi$ and $V\ \nabla^2\psi$. The spectral representation of these terms may be accomplished by (i) a transform of spectral fields $U$, $V$, and $\nabla^2\psi$ to a two-dimensional latitude–longitude grid on the sphere, (ii) evaluation of the requisite products at each grid point, (iii) an inverse transform.

Having obtained the grid-point values of $U\nabla^2\psi$ and $V\nabla^2\psi$ it is appropriate to represent these values in terms of truncated Fourier series at each latitude circle as

$$\begin{aligned} U\ \nabla^2\psi &= a \sum_{m=-J}^{+J} A_m e^{im\lambda} \\ V\ \nabla^2\psi &= a \sum_{m=-J}^{+J} B_m e^{im\lambda} \end{aligned} \tag{18}$$

The Fourier amplitudes $A_m$ and $B_m$ are obtained by an inverse transform. After substituting Eqs. (18) into Eq. (6) and again using the orthogonality property of the spherical harmonics, the spectral prognostic for vorticity is given by Eq. (10) but with $C_l^m$ now given by the integral

$$C_l^m = \int_{-\pi/2}^{+\pi/2} \frac{1}{\cos^2\phi}\left(imA_m + \cos\phi\,\frac{\partial B_m}{\partial\phi}\right) P_l^m(\sin\phi)\cos\phi\, d\phi$$

The second term in this integrand is seen to imply a differential with respect to $\phi$ of the quantities $B_m$, which are known only at discrete latitudinal intervals. Upon integrating by parts, following Eliasen *et al.* (1970), and invoking the boundary condition $B_m(\pm\pi/2) = 0$ since $V(\pm\pi/2) = 0$, the integral may be simplified to

$$C_l^m = \int_{-\pi/2}^{+\pi/2} \frac{1}{\cos^2\phi}\left[imA_m P_l^m(\sin\phi) - B_m\cos\phi\,\frac{\partial P_l^m}{\partial\phi}(\sin\phi)\right]\cos\phi\, d\phi \tag{19}$$

where the differential with respect to $\phi$ may now be treated analytically.

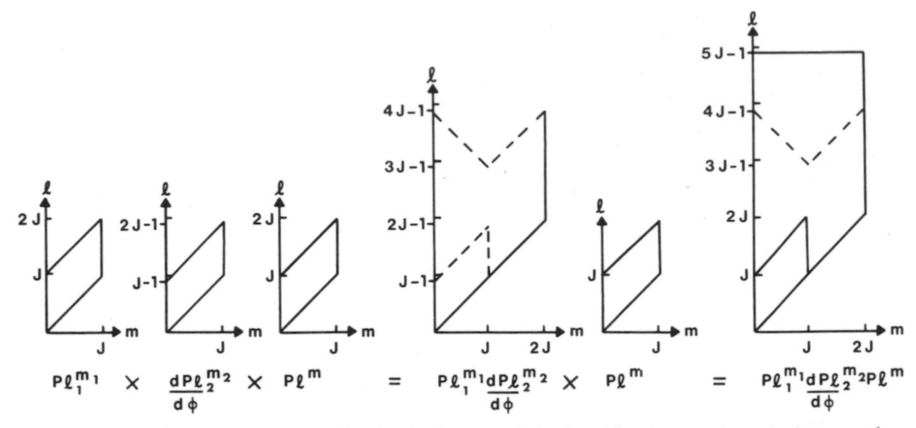


FIG. 1. A schematic representation in $(l, m)$ space of the Jacobian interaction; the integrands of Eqs. (14) and (19) can be entirely defined in terms of associated Legendre polynomials from within the $(l, m)$ domain at the extreme right. The selection rules $|l_1 - l_2| < l < |l_1 + l_2|$ and $m = m_1 + m_2$ define the domain of influence.

If the transform procedure for evaluating $C_l^m$ is to be equivalent to the nonlocal evaluation employing interaction coefficients, the grid-point multiplications and transforms must resolve all possible interactions, i.e., be free of aliasing. Orszag (1970) noted that $4J$ equispaced grid points around each latitude circle would allow alias-free evaluation of the Fourier transforms, such as required in Eq. (18), and has subsequently pointed out (Orszag, 1971), as have Machenhauer and Rasmussen (1972), that $(3J + 1)$ equispaced points are in fact sufficient to ensure alias-free evaluation of the requisite Fourier transforms. Eliasen *et al.* (1970) discussed in detail the requirements of resolution in the meridional direction in performing numerical integration of integrals such as in (19); they suggested the use of Gaussian quadrature which provides when applied over $\kappa$ angles, exact integration for integrands that are polynomials of degree less than $2\kappa - 1$. The degree of polynomial in the integrand of (19) can be determined upon inspection and with rhomboidal truncation at wave number $J$ is seen to be $\leqslant 5J - 1$. The domain in $(l, m)$ space influenced by the nonlinearity of the prognostic is illustrated in Fig. 1 and is defined by the selection rules $|l_1 - l_2| < l < l_1 + l_2$ and $m_1 + m_2 = m$ governing the interaction coefficient $L(l_1 l_2 l; m_1 m_2)$ given by Eq. (14). Thus the requisite number of latitude circles necessary for exact Gaussian quadrature is seen to be $\geqslant 5J/2$.

The number of arithmetic operations inherent in the transform method is seen to be the sum of

i. $0(2J^2)$ operations to derive Fourier amplitudes from spherical harmonic coefficients.

ii. $0[((3J + 1)/2)^2]$, or if using a fast Fourier transform $0[((3J + 1)/2) \log_2 ((3J + 1)/2)]$ operations to generate grid point values.

for each latitude circle and for each field that is subject to transform and subsequent inverse transform. There are $\sim 5J/2$ latitudes and the equivalent of five fields to be transformed. That is,

$$0[5 \times 5J/2 \times (2J^2 + 1.5J \log_2 1.5J)]$$

products which is

$$0(25J^3 + 20J^2 \log_2 1.5J).$$

The interaction coefficient method scales as $0(J^5)$ as seen earlier and the transform method is thus seen to be a considerable improvement at large $J$, say $J \sim 25$, where there would be a requirement of $0(J^4)$ operations. Such estimates are only a guide to the relative computer requirements, since it is difficult to account for all factors as, for example, the overhead in accessing interaction coefficients. It should be remarked that the interaction coefficient method becomes more and more laborious with increased complexity in the prognostic equations. For example, the interaction coefficient method would require an order of magnitude more operations if the model were divergent barotropic; in contrast the corresponding increase with the transform method would be only approximately 2-fold.

## E. Experiments with Barotropic Spectral Transform Models

The initial pioneering study of Eliasen *et al.* (1970) and subsequent experiments by Machenhauer and Rasmussen (1972) and Bourke (1972) demonstrated the suitability of the transform method for integration of the equations describing a divergent barotropic spectral model (DBSM). The extension of the discussion of the preceding section (II, D) on nondivergent barotropic spectral models (NDBSM) to that appropriate to divergent models is straightforward and will not be detailed here since full details are available in the literature cited above. Eliasen *et al.* formulated their model in terms of prognostics for the quantities $U$ and $V$ as defined in Eq. (2) following the approach of Robert (1966). Bourke (1972) (subsequently denoted as I) formulated his model in terms of prognostics for vorticity and divergence. These two approaches differ principally in the method for truncation of the spectral tendencies; the truncation for $U$ and $V$ requires a correction procedure, while the truncation of vorticity and divergence prognostics is quite straightforward since they are scalars. As the $U$, $V$ models in fact simulate the truncation that would pertain for prognostics for vorticity and divergence, they are in fact mathematically equivalent methods of integration.

1. *Numerical Stability*

The DBSM integrations were undertaken without a guarantee of long-term integral constraint on quadratics such as applies in the NDBSM integration. This arises since the kinetic energy, and mean square vorticity, now contain cubic terms due to the correlation of the mass with the wind field. However, in practice it has been found that these DBSM integrations conserve the requisite quadratics to a high degree. In I a global DBSM, truncated at $J = 15$, was integrated to 116 days and the available energy, angular momentum, and square potential vorticity varied by $+2$, $-0.01$, and $-4\%$, respectively. In the DBSM the equivalent expression to that of Eq. (15) incorporates both available potential and kinetic energy and may be written as

$$\frac{\partial}{\partial t}(K + P) = \frac{\partial}{\partial t}\frac{a^2}{4\bar{\Phi}}\sum_l \sum_m (E_l^m \Phi_l^{m*} + \Phi_l^m \Phi_l^{m*})$$

$$= \frac{a^2}{4\bar{\Phi}}\sum_l \sum_m \left(E_l^m \frac{\partial \Phi_l^{m*}}{\partial t} + \Phi_l^{m*}\frac{\partial E_l^m}{\partial t} + \Phi_l^{m*}\frac{\partial \Phi_l^m}{\partial t} + \Phi_l^m \frac{\partial \Phi_l^{m*}}{\partial t}\right) \tag{20}$$

where $E_l^m$ is defined by

$$\sum_l \sum_m E_l^m Y_l^m = \frac{U^2 + V^2}{a^2}$$

and $\Phi$, the geopotential height of the free surface of the barotropic fluid, is expressed as

$$\Phi = a^2 \sum_l \sum_m \Phi_l^m Y_l^m \qquad \text{and} \qquad \bar{\Phi} = \Phi_0^0/\sqrt{2}$$

On inspection then of Eq. (20) it is seen that the first product within the summation, i.e., $E_l^m(\partial \Phi_l^{m*}/\partial t)$ is finite outside the truncation $J$ since both terms are in general finite through nonlinear interaction. Thus it is to be expected that there is no constraint on the energy in the truncated system.

Weigle (1972) has made a detailed study of the conservation properties of DBSM for both spherical and Cartesian geometry. He demonstrated that the truncated set of spectral equations did not offer any quadratic energy constraint by detailed expansion of expressions of the form of Eq. (20).

2. *Numerical Accuracy*

The accuracy of a numerical integration of nonlinear equations can be established by examining the convergence of the numerical solution as a function of increasing resolution in both the space and time domain.

Spectral models of the type discussed hitherto are not subject to linear spatial truncation error or aliasing error. Of course, nonlinear truncation error occurs and the questions arise as to how accurate are spectral models and what is the magnitude of truncation error in the space and time domain.

A useful method for testing DBSM and its finite difference counterpart is the initialization, suggested by Phillips (1959), where a particular solution of the NDBSM is employed. The DBSM solution as expected parallels closely the analytic nondivergent solution with some retardation of phase speed due to the presence of divergence. Puri and Bourke (1974) demonstrated convergence of numerical solution for a Phillips wave initialization of planetary wave number $m = 4$; the contribution of time truncation was shown to be negligible. In these DBSM calculations, the approach to a convergent solution was seen to be discontinuous as model resolution was increased to include dominant first-order interactions and even subsequent second-order interactions.

Doron *et al.* (1974) have compared the spectral and finite difference methods in integrations of the divergent barotropic model. These integrations suggest that, for example, a $J = 16$ rhomboidal truncation is more accurate than second-order finite differencing on a $3° \times 5°$ latitude–longitude grid. The initialization used in these calculations was that of Phillips and the model resolution $J = 16$ just resolves the dominant wave–wave interaction of the planetary wave number $m = 8$ initial field; it is difficult to infer from these results the resolution required to give equivalent results with spectral and finite difference methods. An equivalent comparison with initialization via eigenfunctions of the linearized finite difference equations following Dickinson and Williamson (1972) would be of interest in this context.

Merilees (1972) has examined the magnitude and evolution of nonlinear truncation error in a NDBSM initialized with real data for the Northern Hemisphere. This study showed that for time scales of the order of 5 days, horizontal truncation error varied as $m^{-\alpha}$, where $\alpha = 1.4$ and $m$ is the planetary wave number. Merilees described these conclusions as tentative in view of the limited resolution of the model $m$ ($\leqslant 21$) and initial data ($m \leqslant 9$). Puri and Bourke (1974) considered truncation error in global integrations of a NDBSM with real data specified at $J = 15$ and the maximum model resolution ranging up to $J = 36$. In integrations to 4 days the truncation error decreased exponentially as the number of degrees of freedom ($\sim J^2$) was increased. Here again, time truncation error was negligible. Rasmussen (1974) has also evaluated truncation error in a DBSM initialized with hemispheric data at $J = 17$ with model resolution ranging to $J = 33$; he did not offer any estimate of the fall off of truncation error with increasing resolution, although clearly demonstrating that the magnitude of error is very dependent on the initial variance of the starting fields.

From the above it may be concluded that the exact sensitivity of barotropic spectral models to increasing horizontal resolution is not especially clear at this time; it remains a rather difficult task to estimate theoretically the magnitude and sensitivity to resolution of nonlinear truncation error. Accordingly, as will be apparent from subsequent discussion of experiments with multilevel models, a rather pragmatic view is taken, and model resolution is largely determined as with finite difference models by available computer power and a compromise between requirements of horizontal and vertical resolution. Orszag (1974) comments that finite difference methods, such as the commonly used second-order schemes, require at least a 2-fold increase in resolution in each spatial direction in comparison to spectral methods to approach a comparable level of simulation accuracy.

3. *Computational Efficiency*

Of primary concern in the development of transform spectral models was the question of computer efficiency relative to the more traditional interaction coefficient approach. As proposed by Orszag (1970) in estimating the number of requisite arithmetic operations and such as illustrated in Section II, D, the transform approach was expected to provide substantial gains in efficiency.

In I the computation time per time step of integration was considered as a function of wave number truncation $J$ for both an interaction coefficient and transform DBSM. Table I presents this timing information of the transform model for the range of $J$ considered and the corresponding transform

TABLE I

TRANSFORM MODEL TIMING

| Wave number truncation $J$ (rhomboidal) | Transform grid: Number of Gaussian latitudes | Transform grid: Number of points per latitude circle | Computation time/time step (sec) |
|---|---|---|---|
| 7 | 20 | 24 | 2.3 |
| 12 | 32 | 48 | 7.8 |
| 15 | 40 | 48 | 12.2 |
| 24 | 64 | 96 | 39 |
| 30 | 80 | 96 | 67 |

TABLE II

INTERACTION COEFFICIENT TIMING

| Wave number truncation $J$ (rhomboidal) | Total number of interaction coefficients[a] | | Computation time/time step (sec) |
|---|---|---|---|
| | Jacobian | Scalar | |
| 5 | 1625 | 2032 | 1.1 |
| 7 | 7471 | 8779 | 4.2 |
| 10 | 39222 | 43947 | 18.6 |
| 13 | 135702 | 147875 | 59 |
| 15 | 268802 | 289474 | 114 |

[a] Comprising interaction coefficients arising from scalar triple products and from scalar products.

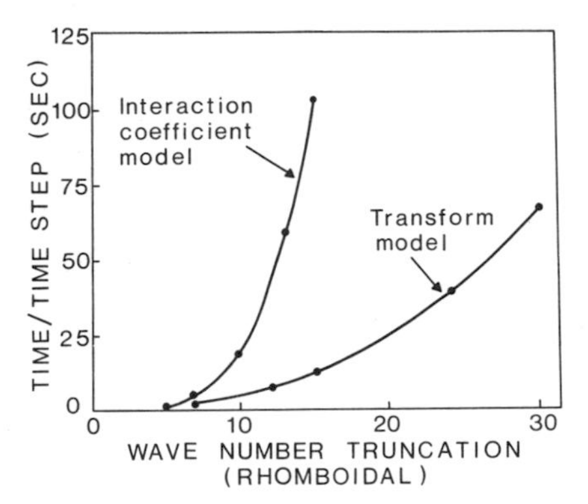


FIG. 2. Computation time per time step(s) as a function of spectral resolution. Integrations of a global spectral model employing a transform method and employing the interaction coefficient method are compared (Bourke, 1972).

grid resolution; Table II presents the timing information for the interaction coefficient model and the corresponding number of requisite interaction coefficients. Figure 2 displays this timing information as a function of $J$; the rapid loss of efficiency with the traditional method is quite apparent and even at the moderate resolution of $J = 15$ the transform approach is an order of magnitude more efficient.

The impact of the transform model with respect to efficiency is clearly substantial. At the same time the diagnostic evaluation of highly nonlinear quantities, such as the square potential vorticity as performed in I, is indicative of the much enhanced capability in spectral modeling as is discussed in the context of the multilevel model in Section III, B.

4. *Pseudo-Spectral Approach*

There is a variant of the transform method discussed hitherto which is referred to as the pseudo-spectral method by Orszag (1971). It is equivalent, in the present context, to permitting aliasing in the evaluation of nonlocal spectral products; this is achieved by employing a resolution in the latitude–longitude grid sufficient to resolve only the variance of the interacting fields themselves. Orszag found that this approximation does not generate serious error with the spectral method applied in rectangular geometry. The applicability of the pseudo-spectral approach to the spherical geometry of the multilevel model is discussed in Section III, B, 4.

An extension of the pseudo-spectral approach by Merilees (1973) incorporates the use of Fourier series in both the zonal and meridional directions. The advantage envisaged here is that the fast Fourier transform may also be employed in the meridional direction thereby eliminating the relatively inefficient Legendre transforms. A detailed assessment of the double Fourier series approach is given by Orszag (1974); he draws attention to deficiencies in the approach arising through lack of NDBSM quadratic constraints, polar singularities, and inhomogeneous time-step requirements. The major objection to using spherical harmonics is the lack of efficiency associated with Legendre transforms.

While the Legendre transform can be considered inefficient relative to the fast Fourier transform, the application of Gaussian quadrature greatly facilitates computer coding of spherical harmonic spectral models, since the nonlinear contribution to spectral tendencies may be accumulated successively at each Gaussian latitude. Notionally the fast Fourier transform for the meridional direction is appealing, but this would necessitate simultaneous grid-point representation of the full two-dimensional grid or alternatively substantial peripheral device usage. This may be feasible with barotropic models but seems less than trivial for multilevel models.

5. *Spectral Truncation*

Orszag (1974), in discussing the double Fourier series approach in spherical geometry, provided a very elegant discussion of the properties of the surface spherical harmonic representation. He pointed out that the rhomboidal wave number truncation is not invariant under rotation and that this can lead to phase errors in those components for which $l > J$. The property of rotational invariance, as in, for example, triangular truncation, has the appeal of mathematical elegance although it should be commented that (i) predominant zonal advection of the atmosphere may be evaluated, using rhomboidal truncation, with minimal phase error of the type noted by Orszag; (ii) the use of rhomboidal truncation provides, relative to triangular truncation with the same number of degrees of freedom, enhanced resolution of the atmosphere's inhomogeneous meridional variance and enhanced resolution of meridional gradients which define the primary atmospheric forcing; and (iii) errors occur in all truncations through nonlinear truncation error and even in the simple evaluation of the Coriolis contribution to tendencies [as evident in Eqs. (21) and (23) of I].

The question of optimum distribution of the available degrees of freedom in $(l, m)$ space is currently somewhat unresolved; the predominant usage is rhomboidal and all models at ANMRC employ this approach. Simmons and Hoskins (1975) have compared rhomboidal and triangular truncation for integrations of a five-level spectral model initialized analytically. They describe their results as inconclusive with respect to the relative merits of rhomboidal and triangular truncations.

## III. Multilevel Spectral Model

The application of the spectral transform method to three-dimensional representation of atmospheric flow has, as mentioned in the introductory discussion of Section I, D, been approached by use of (i) a discrete representation in the vertical; and (ii) a spectral representation in the vertical. The latter is discussed in detail by Eliasen and Machenhauer (1974); the former is the basis for the ensuing discussion.

Bourke (1974) has already described the application of the transform method in a multilevel global spectral model which considered the atmosphere as dry and adiabatic. Subsequent developments of that model have introduced the influence of (i) moisture, (ii) topography, (iii) thermal and dynamic contrast as defined by the distribution of land and sea, (iv) additional parameterizations of subtruncation scale processes; and (v) a radiative transfer code. A description of these newer aspects of the model is given in the following; where detail appears to be omitted, reference should be made to Bourke (1974) subsequently denoted by II.

### A. Model Formulation

The model is formulated in terms of the derived primitive equations with prognostics in terms of the vertical component of relative vorticity and horizontal divergence, extending the approach detailed in I and in the present contribution. Additional prognostics for temperature, moisture mixing ratio and surface pressure, and application of the hydrostatic relationship complete the model framework.

The vertical coordinate system of Phillips (1957) $\sigma = p/p_*$ is employed; here $p_*$ defines the surface pressure and $p$ the pressure within the atmospheric fluid.

1. *Prognostic Equations*

The prognostics in the spherical polar coordinate system for vorticity and divergence are

$$\frac{\partial \zeta}{\partial t} = -\frac{1}{a \cos^2 \phi}\left[\frac{\partial A}{\partial \lambda} + \cos \phi \frac{\partial B}{\partial \phi}\right] - 2\Omega\left(\sin \phi\, D + \frac{V}{a}\right) + k \cdot \nabla \times ({}_h\mathbf{F} + {}_v\mathbf{F}) \tag{21}$$

$$\frac{\partial D}{\partial t} = \frac{1}{a \cos^2 \phi}\left[\frac{\partial B}{\partial \lambda} - \cos \phi \frac{\partial A}{\partial \phi}\right] + 2\Omega\left(\sin \phi\, \zeta - \frac{U}{a}\right) - \nabla^2(E + \Phi' + RT_0 q) + \nabla \cdot ({}_h\mathbf{F} + {}_v\mathbf{F}) \tag{22}$$

where

$$A = \zeta U + \dot{\sigma}\frac{\partial V}{\partial \sigma} + \frac{RT'}{a}\cos \phi \frac{\partial q}{\partial \phi}$$

$$B = \zeta V - \dot{\sigma}\frac{\partial U}{\partial \sigma} - \frac{RT'}{a}\frac{\partial q}{\partial \lambda} \tag{23}$$

$$E = \frac{U^2 + V^2}{2 \cos^2 \phi}$$

Here $\zeta$ is the vertical component of relative vorticity, $D$ is the horizontal divergence, $U$ and $V$ are as in Eq. (2), $q$ is $\log p_*$, $T$ is the absolute temperature, $\Phi$ is the geopotential height, $\Omega$ is the earth's rotation rate, $R$ is the gas constant, $\nabla$ is the horizontal gradient operator, and $\dot{\sigma}$ is the total time derivative $d\sigma/dt$. A subscript zero denotes a horizontal mean value and the superscript prime the deviation from that mean.

The continuity equation appears as a prognostic for the logarithm of surface pressure in the form

$$\partial q/\partial t = -\mathbf{V} \cdot \nabla q - D - \partial \dot{\sigma}/\partial \sigma \tag{24}$$

and gives, via vertical integration and application of the boundary conditions $\dot{\sigma} = 0$ at $\sigma = 1$ and $\sigma = 0$,

(i) a prognostic for the surface pressure;

$$\partial q/\partial t = \bar{\mathbf{V}} \cdot \nabla q + \bar{D} \tag{25}$$

(ii) a diagnostic for the vertical velocity $\dot{\sigma}$

$$\dot{\sigma} = \{(1 - \sigma)\bar{D} - \bar{D}^{\sigma}\} + \{(1 - \sigma)\bar{V} - \bar{V}^{\sigma})\} \cdot \nabla q \tag{26}$$

where $\overline{(\ )}^{\sigma} = \int_{\sigma=1}^{\sigma} (\ )\, \partial\sigma$ and $\overline{(\ )}$ denotes the evaluation of this integral with the upper limit $\sigma = 0$. The hydrostatic relation is

$$\partial\Phi/\partial\sigma = -RT/\sigma. \tag{27}$$

[It should be noted that the present formulation omits consideration of the modification of the ideal gas law due to the presence of moisture; see, for example, Miyakoda (1973).]

The thermodynamic equation and the moisture mixing ratio prognostic are

$$\frac{\partial T}{\partial t} = -\frac{1}{a\cos^2\phi}\left[\frac{\partial}{\partial\lambda}(UT') + \cos\phi\frac{\partial}{\partial\phi}(VT')\right] + H + {}_hF_T + {}_vF_T, \tag{28}$$

where

$$H = T'D + \dot{\sigma}\gamma + RT/c_p[\bar{D} + (\mathbf{V} + \bar{\mathbf{V}})\cdot\nabla q] + H_c/c_p. \tag{29}$$

$$\frac{\partial M}{\partial t} = -\frac{1}{a\cos^2\phi}\left[\frac{\partial}{\partial\lambda}(UM') + \cos\phi\frac{\partial}{\partial\phi}(VM')\right] + I + {}_hF_M + {}_vF_M, \tag{30}$$

where

$$I = M'D - \dot{\sigma}(\partial M/\partial\sigma) + C. \tag{31}$$

Here $c_p$ is the specific heat at constant pressure, $\gamma = (RT/\sigma c_p) - (\partial T/\partial\sigma)$ is the static stability, $M$ is the moisture mixing ratio, $H_c$ is the rate of heating due to condensation and/or convection, and $C$ is the rate of change of $M$ due to condensation and convection.

2. *Horizontal Diffusion*

In Eqs. (21), (22), (28), and (30) ${}_h\mathbf{F}$, ${}_hF_T$, and ${}_hF_M$ denote parameterization of subtruncation scale (horizontal) processes via diffusion with

$$\begin{aligned} k\cdot\nabla\times{}_h\mathbf{F} &= K_h[\nabla^2\zeta + 2(\zeta/a^2)] \\ \nabla\cdot{}_h\mathbf{F} &= K_h[\nabla^2 D + 2(D/a^2)] \\ {}_hF_T &= K_h\,\nabla^2 T \\ {}_hF_M &= K_h\,\nabla^2 M \end{aligned} \tag{32}$$

The value of $K_h$ is taken to be $2.5 \times 10^5$ m$^2$ sec$^{-1}$. In general, from a numerical stability viewpoint, there is no requirement for these diffusive terms, but there is a requirement to inhibit spurious growth of amplitude at scales close to the point of truncation due to spectral blocking (Puri and Bourke, 1974). Accordingly, horizontal diffusion is applied only beyond a certain scale; somewhat arbitrarily the diffusive terms are permitted to contribute to tendencies for $l > J$, i.e., in the upper half of the rhomboid in $(l, m)$ space.

As is common in such formulations, the horizontal diffusion is applied in $\sigma$ coordinates, which introduces from a physical point of view spurious diffusion due solely to the topography of the $\sigma$ surface. The correction for this spurious effect is possible in principle, although not considered warranted in view of the arbitrariness of the parameterization. The horizontal diffusion in the divergence equation is omitted for the purposes of numerical weather prediction; as detailed in II, a term of the form $-KD$ is included in the divergence prognostic to control spurious inertia-gravity oscillations arising from initialization imbalances.

3. *Vertical Diffusion*

In Eqs. (21), (22), (28), and (30) the quantities ${}_{\mathrm{v}}\mathbf{F}$, ${}_{\mathrm{v}}F_T$, ${}_{\mathrm{v}}F_M$ denote diffusive parameterization of sub-grid scale vertical processes. The formulation is as follows

$$\{{}_{\mathrm{v}}\mathbf{F}, {}_{\mathrm{v}}F_T, {}_{\mathrm{v}}F_M\} = (g/p_*)(\partial/\partial\sigma)\{\boldsymbol{\tau}, \eta, \beta\} \tag{33}$$

where

$$\begin{aligned} \boldsymbol{\tau} &= \rho^2(g/p_*)K_{\mathrm{v}}(\partial\mathbf{V}/\partial\sigma) \\ \eta &= \delta\rho^2(g/p_*)K_{\mathrm{v}}(\partial\theta/\partial\sigma) \\ \beta &= \rho^2(g/p_*)K_{\mathrm{v}}(\partial M/\partial\sigma) \end{aligned} \tag{34}$$

Here $K_{\mathrm{v}}$ defines the vertical diffusion coefficient which is given in terms of a mixing length $\mu$ and the magnitude of the wind shear according to $K_{\mathrm{v}} = \rho(g/p_*)\mu^2|\partial\mathbf{V}/\partial\sigma|$. In the present usage $\mu$ is assumed to be 30 meters for $\sigma \leqslant 0.5$ and zero for $\sigma > 0.5$.

The lower boundary specification of the stress is given in terms of the bulk aerodynamic parameterization as follows

$$\begin{aligned} \boldsymbol{\tau}_* &= \rho_1 C_{\mathrm{d}}|\mathbf{V}_1|\mathbf{V}_1 \\ \eta_* &= \delta\rho_1 C_{\mathrm{d}}|\mathbf{V}_1|(\theta_* - \theta_1) \\ \beta_* &= \rho_1 C_{\mathrm{w}} C_{\mathrm{d}}|\mathbf{V}_1|(M_{\mathrm{S}}(T_*) - M_1) \end{aligned} \tag{35}$$

with the subscripts $*$ and 1 denoting the values at the surface and the lowest model level, respectively.

Here the potential temperature $\theta$ is defined as $\theta = T(p_0/p)R/c_p$, where the standard pressure $p_0$ is taken to be 1000 mb. The formulation for sensible heat flux is derived in terms of a contribution to a potential temperature tendency; accordingly the quantity $\delta = (p/p_0)R/c_p$ in Eq. (34) is introduced to approximately allow for the recasting of the thermodynamic equation in terms of a temperature tendency.

$\rho$ is the density, $C_{\mathrm{d}}$ is the drag coefficient, $C_{\mathrm{w}}$ is a wetness factor, and $M_{\mathrm{S}}$ is the saturation mixing ratio at $\sigma = 1$. In the present usage $C_d$ assumes values of 0.004 and 0.001 and $C_{\mathrm{w}}$ values of 0.25 and 1.0 over land and sea, respectively. Over the ocean the surface temperature is taken to be that of the current mean monthly sea surface temperature. Over land the surface temperature is diagnosed from either a heat balance equation in the case of the general circulation application of the model or from logarithmic extrapolation from within the fluid for short-term numerical weather prediction.

## B. Equations of Motion in Spectral Form

### 1. *Spectral Prognostics*

The prognostic variables are represented at discrete levels in the vertical as shown in Fig. 3. At each discrete full level truncated expansions in terms of the spherical harmonics $Y_l^m$ are required as follows:

$$\{\psi, \chi, T, M, q, \Phi\} = \sum_{m=-J}^{+J} \sum_{l=|m|}^{|m|+J} \{a^2\psi_l^m, a^2\chi_l^m, T_l^m, M_l^m, q_l^m, a^2\Phi_l^m\} Y_l^m$$

and

$$\{U, V\} = a \sum_{m=-J}^{+J} \sum_{l=|m|}^{|m|+J+1} \{U_l^m, V_l^m\} Y_l^m \tag{36}$$

The wind vector $\mathbf{V}$, in comparison with Eq. (3), is now defined in terms of a stream function $\psi$ and a velocity potential $\chi$

$$\mathbf{V} = \mathbf{k} \times \nabla\psi + \nabla\chi \tag{37}$$

Accordingly the quantities $\zeta$ and $D$ are seen to be expressible as

$$\xi = \mathbf{k} \cdot \nabla \times \mathbf{V} = \nabla^2\psi \qquad D = \nabla \cdot \mathbf{V} = \nabla^2\chi \tag{38}$$

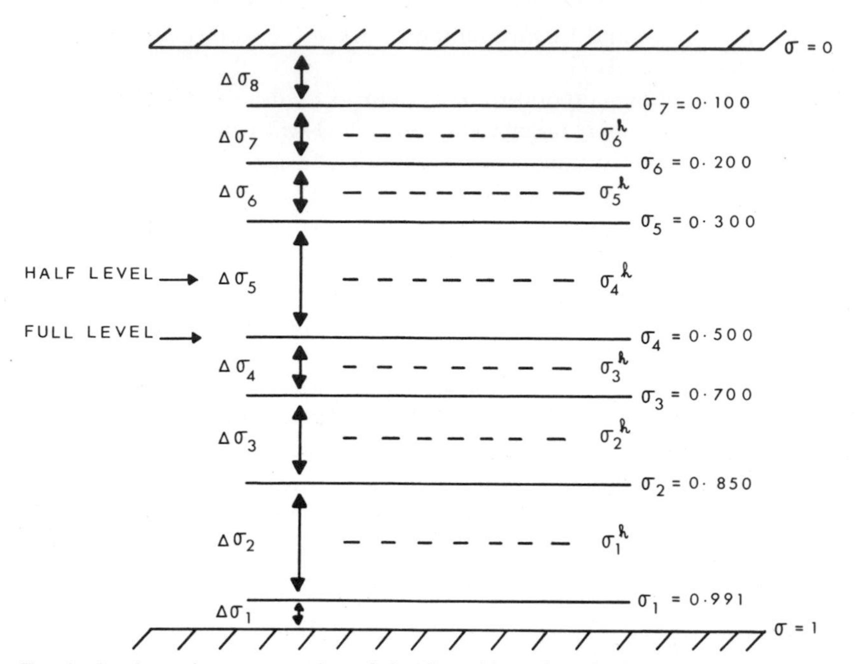


FIG. 3. A schematic representation of the disposition of model levels with respect to the vertical coordinate $\sigma$. The half levels denoted by dashed lines are midway between full levels and are identified by a superscript $h$. The full-level spacings identified are as used in the Appendix.

With the additional irrotational component of the wind, the diagnostic relationships for $U_l^m$ and $V_l^m$ become, in comparison with Eqs. (17),

$$\begin{aligned} U_l^m &= (l-1)\varepsilon_l^m \psi_{l-1}^m - (l+2)\varepsilon_{l+1}^m \psi_{l+1}^m + im\chi_l^m \\ V_l^m &= -(l-1)\varepsilon_l^m \chi_{l-1}^m + (l+2)\varepsilon_{l+1}^m \chi_{l+1}^m + im\psi_l^m \end{aligned} \tag{39}$$

It will be noted that now the expansions for both $U$ and $V$ are truncated at $l = |m| + J + 1$, thereby yielding the equivalence of representation between

$$\{\psi, \chi\} \qquad \text{and} \qquad \{U, V\}.$$

The hydrostatic relationship being linear is simply expressed in spectral form as

$$\partial \Phi_l^m / \partial \sigma = -(R/a^2)(T_l^m/\sigma). \tag{40}$$

The spectral representation of nonlinear products occurring in the prognostic equations is achieved, of course, via the transform method.

The requisite nonlinear products defined in Eqs. (21), (22), (25), (28), and (30) may be represented at each discrete model level, and at each latitude circle, in terms of truncated Fourier series as follows

$$\{A, B, E, UT', VT', UM', VM', H, I, \bar{\mathbf{V}} \cdot \nabla q\}$$

$$= \sum_{m=-J}^{+J} \{aA_m, aB_m, a^2E_m, aF_m, aG_m, aP_m, aQ_m, H_m, I_m, z_m\}e^{im\lambda} \quad (41)$$

The diagnostic evaluation of vertical velocity $\dot{\sigma}$ via Eq. (26) is only required at grid points and is not subsequently expressed spectrally.

The Legendre transform defined by

$$(\ )_l{}^m = \int_{-\pi/2}^{+\pi/2} (\ )_m P_l{}^m(\sin\phi)\cos\phi\, d\phi$$

is now applied to the quantities $E_m$, $H_m$, $I_m$, and $z_m$, thereby yielding $E_l{}^m$, $H_l{}^m$, $I_l{}^m$, and $z_l{}^m$. An additional operator, paralleling Eq. (19), is now introduced for brevity of notation and defined as follows:

$$L_l{}^m(x, y) = \int_{-\pi/2}^{\pi/2} \frac{1}{\cos^2\phi} \Big\{ imx_m P_l{}^m(\sin\phi)$$

$$- y_m \cos\phi \frac{\partial P_l{}^m}{\partial\phi}(\sin\phi)\Big\} \cos\phi\, d\phi \quad (42)$$

Thus the spectral prognostics may be expressed as

$$-l(l+1)(\partial\psi_l{}^m/\partial t) = -L_l{}^m(A, B) + 2\Omega\{l(l-1)\varepsilon_l{}^m\chi_{l-1}^m$$
$$+ (l+1)(l+2)\varepsilon_{l+1}^m\chi_{l+1}^m - V_l{}^m\} + {}_\psi S_l{}^m \quad (43)$$

$$-l(l+1)(\partial\chi_l{}^m/\partial t) = L_l{}^m(B, -A) - 2\Omega\{l(l-1)\varepsilon_l{}^m\psi_{l-1}^m$$
$$+ (l+1)(l+2)\varepsilon_{l+1}^m\psi_{l+1}^m + U_l{}^m\}$$
$$+ l(l+1)[E_l{}^m + \Phi_l{}^m + (RT_0/a^2)q_l{}^m] + {}_\chi S_l{}^m \quad (44)$$

$$\partial T_l{}^m/\partial t = -L_l{}^m(F, G) + H_l{}^m + {}_T S_l{}^m \quad (45)$$

$$\partial M_l{}^m/\partial t = -L_l{}^m(P, Q) + I_l{}^m + {}_M S_l{}^m \quad (46)$$

$$\partial q_l{}^m/\partial t = z_l{}^m - l(l+1)\overline{\chi_l{}^m} \quad (47)$$

Here ${}_{\psi}S_l^m$, etc., denote the spectral representations of the subtruncation scale diffusive terms; the vertical diffusion introduces coupling between the spectral tendencies at each level as described in the Appendix.

### 2. *Grid-Point Representation*

The nonlinear products implied in (41) are evaluated at grid points after representation at requisite Gaussian latitudes and longitude points of $\nabla^2\psi$, $\nabla^2\chi$, $T$, $M$, $U$, $V$, $q$, $\partial q/\partial\lambda$, and $\cos\phi(\partial q/\partial\phi)$. In addition, while in grid-point space, vertical integrals and differentials are evaluated. Vertical finite differences at full levels are obtained as the weighted average with respect to $\sigma$ of adjacent half-level values; the half levels are midway between the full levels to provide the first-order finite difference as representatively as possible. Vertical integrals are evaluated with one-sided approximations both below the first full level and above the topmost full level; intermediate contributions to integrals assume the half-level value in a centered approximation to the integrand between full levels. Where appropriate, the products, e.g., $\dot{\sigma}(\partial U/\partial\sigma)$ are evaluated at half levels prior to weighted averaging to full levels to facilitate incorporation of the upper and lower boundary conditions for $\dot{\sigma}$. Logarithmic variation of temperature with $\sigma$ between the first two full levels is assumed for the purposes of defining $\partial T/\partial\sigma$ at the first full and half levels.

The spectral amplitudes of the geopotential are derived from vertical integration of Eq. (40); the topography expressed spectrally ($\Phi_{*l}^m$) defines the lower limit of the hydrostatic integration. The first two full levels define a logarithmic dependence of temperature with respect to $\sigma$ for the purposes of hydrostatic integration. (See Appendix A of II.)

### 3. *Modeling of Physical Processes in Grid-Point Space*

The physical processes considered in grid-point space are

i. Evaluation of fluxes within the boundary layer via the formulation Eqs. (35) with an associated specification of parameters as function of a fixed distribution of land, sea, ice, or snow.

ii. Parameterization of convection and condensation via the convective and nonconvective adjustment scheme of Manabe *et al.* (1965) and the accumulation of precipitation on the latitude–longitude grid.

iii. Evaluation of temperature tendencies due to radiative processes and the evaluation of a heat balance calculation to specify the surface temperature distribution over land and ice-covered ocean.

The internal vertical diffusion of Eq. (33) is treated spectrally by linearization as discussed in Section III, C; the horizontal diffusion is linear and may be treated spectrally also.

4. *Transform Grid Resolution*

As discussed in Section II the transform method applied to quadratics may be considered equivalent to the analytic spectral multiplication provided the grid resolution is sufficient. The Jacobian term of Eq. (19) was seen to require a resolution of $\geqslant 5J/2$ Gaussian latitudes and $\geqslant 3J + 1$ equi-spaced longitudinal points. With the occurrence of scalar products in addition to the Jacobian in the primitive equations the Gaussian grid requires $\geqslant (5J + 1)/2$ latitudes to provide alias-free evaluation of all quadratics.

However, the model dynamics embody not only quadratics, but also triple products, as for example in the vertical advection $\dot{\sigma}(\partial U/\partial\sigma)$ where $\dot{\sigma}$ itself has a quadratic dependence on the interacting fields. In addition the parameterization of physical processes, as described in Section III, B, 3 above, introduces a degree of nonlinearity into the model that could be expected to be considerably more complicated than quadratic. Accordingly the grid resolution necessary to provide alias-free spectral representation in the multilevel model may be considerably higher than that suitable for the quadratic nonlinearity of the barotropic models. In the ANMRC models, however, as reported in II, it has been found satisfactory to define the grid resolution as that sufficient to resolve the quadratic interactions.

A quantitative assessment of the impact of aliasing introduced by the triple products and physical parameterizations has been conducted. These studies are described in the following in terms of overspecification and underspecification of resolution relative to that required to resolve quadratic interaction.

The influence of varying the grid resolution in the longitudinal direction was examined. A five-level hemispheric moist model truncated at $J = 15$ was integrated to 4 days with the grid resolution defined by (i) 20 Gaussian latitudes and (ii) a range of longitudinal resolutions varying from 32 to 64 points. A comparison of solutions has been made in terms of (i), the evolution in time of the quantity $\omega$ defined as $\omega = \overline{p_*(\int \dot{\sigma}^2\, \partial\sigma)^{1/2}}$ (here the overbars denote hemispheric mean values and $\omega$ is a measure of the mean vertical velocity in the model and provides a sensitive indication to model stability), and (ii), the root mean square deviation of the mean sea-level (MSL) pressure between the various integrations.

The time evolution of $\omega$ for resolutions employing 32, 36, 40, 48, and 64 points, respectively, in the longitudinal direction are displayed in Fig. 4a; from this figure it is evident that the lowest resolutions of 32 and 36 points are in fact unstable. With increasing resolution the solutions are seen to converge in terms of $\omega$. The time evolution of the root mean square deviation of the MSL pressure for integrations with 36, 40, and 48 points from the highest resolution calculation with 64 points is shown in Fig. 4b. The rms convergence of solution with increasing resolution is evident; the rms

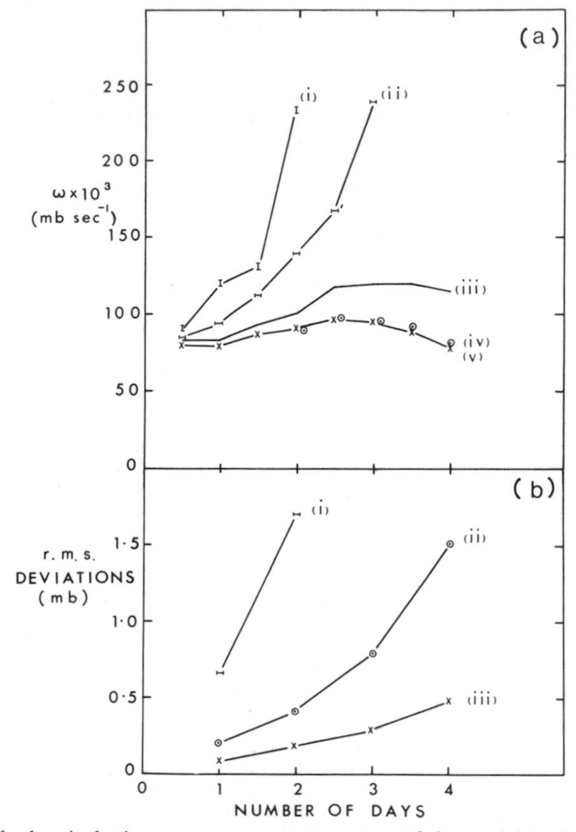


FIG. 4. (a) The hemispheric root mean square measure of the model vertical velocity as a function of the period of integration for zonal resolutions of the transform grid of (i) 32 points; (ii) 36 points; (iii) 40 points; (iv) 48 points; and (v) 64 points. (b) The hemispheric root mean square deviation of the MSL pressure as a function of the period of integration. The integrations that are compared to the solution with zonal resolution of 64 points are those with (i) 36 points; (ii) 40 points; and (iii) 48 points.

difference between integrations with 48 and 64 points is only 0.5 mb at 4 days. These results indicate that the underspecification of longitudinal resolution is unsatisfactory; the overspecification introduces a change that is considered negligible.

With respect to overspecification of the meridional grid resolution, a comparison has been made between integrations employing (i) 48 longitudes and 20 Gaussian latitudes and (ii) 64 longitudes and 28 Gaussian latitudes. Here the rms deviation in the MSL solution at day 4 is 0.5 mb. This result indicates that the combined overspecification of latitudinal and longitudinal resolution also introduces a negligible change.

Accordingly, the two-dimensional grid resolution is specified in terms of that necessary to resolve quadratics; in practice in the ANMRC models, truncated at wave number $J = 15$, there is slight overspecification relative to the quadratic requirement with the use of 48 points longitudinally and 20 Gaussian latitudes per hemisphere. The value of underspecification at least in the longitudinal direction is doubtful, however, and suggests that care should be taken in the pseudo-spectral approach.

### C. Semi-implicit Time Integration

The semi-implicit time integration algorithm, proposed by Robert (1969), provides a procedure for integrating primitive equations models with time steps limited by the highest frequency meteorological mode rather than the somewhat higher frequency gravity modes. Robert demonstrated the algorithm initially with a DBSM model and subsequently extended the scheme to multilevel spectral and grid models (Robert, 1970; Robert *et al.*, 1972). As described in II a modified semi-implicit algorithm has been designed that is more compatible with the vertical disposition of variables in ANMRC models. This algorithm closely parallels the DBSM scheme but with temperature replacing geopotential height. The ensuing discussion constitutes an extension of the scheme given in II thereby allowing implicit treatment of internal vertical stresses.

The prognostic equations (43)–(47) may be rewritten, omitting spectral subscripts, after some manipulation as

$$\partial \boldsymbol{\zeta}/\partial t = \mathbf{r} + \mathbf{V}\boldsymbol{\zeta} \tag{48}$$

$$\partial \mathbf{M}/\partial t = \mathbf{s} + \mathbf{VM} \tag{49}$$

$$\partial \mathbf{D}/\partial t = \mathbf{x} + l(l+1)(\boldsymbol{\Phi} + R'\mathbf{T}_0 q) - K\mathbf{D} + \mathbf{VD} \tag{50}$$

$$\partial \mathbf{T}/\partial t = \mathbf{y} + \mathbf{GD} + \mathbf{V}'\mathbf{T} \tag{51}$$

$$\partial q/\partial t = z + \mathbf{ID} \tag{52}$$

$$\boldsymbol{\Phi} = R'\mathbf{AT}, \tag{53}$$

$\boldsymbol{\zeta}$, $\mathbf{M}$, $\mathbf{D}$, $\mathbf{T}$, and $\boldsymbol{\Phi}$ are $N$-dimensional column vectors, $N$ being the number of model levels. $z$ and column vectors $\mathbf{r}$, $\mathbf{s}$, $\mathbf{x}$, $\mathbf{y}$ denote spectral representations of the nonlinear terms and other terms not explicit here. $\mathbf{V}$ and $\mathbf{V}'$, representing linearized internal vertical diffusion, and $\mathbf{G}$ are $N \times N$ matrices and are defined in the Appendix; the surface stress terms, which are nonlinear are treated explicitly and are included in $\mathbf{r}$, $\mathbf{s}$, $\mathbf{x}$, and $\mathbf{y}$. $\mathbf{A}$ is also an $N \times N$ matrix whose elements are derived in II, and $\mathbf{I}$ is an $N$-dimensional unit row vector (here $R' = R/a^2$).

As mentioned in Section III, A, 2 the divergence dissipation term in the divergence prognostic is only included in the weather prediction model.

The reduction of the thermodynamic equation into this form requires extraction of the linear components from the term $\dot{\sigma}\gamma + (RT/c_p)\bar{D}$ which contributes nonlinearly to the tendency. It should be noted, however, that the extent to which the thermodynamic equation requires linearization is not immediately obvious; indeed an alternative and simpler scheme in which $\dot{\sigma}\gamma + (RT/c_p)\bar{D}$ is rewritten as $-\dot{\sigma}(\partial T/\partial\sigma) + (RT/c_p)(\dot{\sigma}/\sigma + \bar{D})$ and only linear contributions from $(RT/c_p)(\dot{\sigma}/\sigma + \bar{D})$ are considered has been found to be well behaved in a limited series of integrations. A possible advantage of this approach is that the term $\dot{\sigma}(\partial T/\partial\sigma)$ may be evaluated in a fashion that is consistent with all other vertical advection terms.

The finite differencing in time of the vorticity equation may be written as

$$\zeta^{\tau+1} = \zeta^{\tau-1} + 2\Delta t\,\mathbf{r} + 2\Delta t\,\mathbf{V}\zeta^{\tau+1}$$

$\mathbf{r}$ contains terms which are both centered (time $\tau$; e.g., the advection terms) and lagged (time $\tau - 1$; e.g., surface stress and horizontal diffusion terms) in time. The equation can be solved for $\zeta^{\tau+1}$

$$\zeta^{\tau+1} = (\mathbf{1} - 2\Delta t\,\mathbf{V})^{-1}(\zeta^{\tau-1} + 2\Delta t\,\mathbf{r}) \tag{54}$$

The same procedure is applied to the mixing ratio equation to solve for $\mathbf{M}^{\tau+1}$.

Equations (50)–(52) are more complicated being coupled, and the finite differencing in time of these equations may be written as

$$\begin{aligned}\mathbf{D}^{\tau+1} &= \mathbf{D}^{\tau-1} + 2\Delta t\,\mathbf{x} + \Delta t\, l(l+1)[\mathbf{\Phi}^{\tau+1} + \mathbf{\Phi}^{\tau-1} + R'\mathbf{T}_0(q^{\tau+1} + q^{\tau-1})] \\ &\quad - 2\Delta t\,(K\mathbf{D}^{\tau+1} - \mathbf{V}\mathbf{D}^{\tau+1}) \end{aligned}\tag{55}$$

$$\mathbf{T}^{\tau+1} = \mathbf{T}^{\tau-1} + 2\Delta t\,\mathbf{y} + \Delta t\,\mathbf{G}(\mathbf{D}^{\tau+1} + \mathbf{D}^{\tau-1}) + 2\Delta t\,\mathbf{V}'\mathbf{T}^{\tau+1} \tag{56}$$

$$q^{\tau+1} = q^{\tau-1} + 2\Delta t\,z + \Delta t\,\mathbf{I}(\mathbf{D}^{\tau+1} + \mathbf{D}^{\tau-1}) \tag{57}$$

Applying matrix $R'\mathbf{A}$ of Eq. (53) to Eq. (56) gives

$$\mathbf{\Phi}^{\tau+1} = R'\mathbf{A}[\mathbf{1} - 2\Delta t\,\mathbf{V}']^{-1}[\mathbf{T}^{\tau-1} + 2\Delta t\,\mathbf{y} + \Delta t\,\mathbf{G}(\mathbf{D}^{\tau+1} + \mathbf{D}^{\tau-1})]$$

Substituting for $(q^{\tau+1} + q^{\tau-1})$ and $(\mathbf{\Phi}^{\tau+1} + \mathbf{\Phi}^{\tau-1})$ into Eq. (55) yields

$$\begin{aligned}&[(1 + 2\Delta t\,K)\mathbf{1} - 2\Delta t\,\mathbf{V} - l(l+1)\,\Delta t^2\,R'\{\mathbf{A}(\mathbf{1} - 2\Delta t\,\mathbf{V}')^{-1}\mathbf{G} + \mathbf{T}_0\mathbf{I}\}]\mathbf{D}^{\tau+1} \\ &\quad = [\mathbf{1} + l(l+1)\,\Delta t^2 R'\{\mathbf{A}(\mathbf{1} - 2\Delta t\,\mathbf{V}')^{-1} + \mathbf{T}_0\mathbf{I}\}]\mathbf{D}^{\tau-1} + \mathbf{x}'\end{aligned}\tag{58}$$

where **1** is a unit matrix and

$$\mathbf{x}' = 2\Delta t\, \mathbf{x} + l(l+1)\,\Delta t\,[R'\mathbf{A}\{\mathbf{T}^{\tau-1} + (\mathbf{1} - 2\Delta t\, \mathbf{V}')^{-1}[\mathbf{T}^{\tau-1} + 2\Delta t\, \mathbf{y}]\}] + 2l(l+1)\,\Delta t\, R'\mathbf{T}_0[q^{\tau-1} + \Delta t\, z].$$

Equation (58) can now be solved for $\mathbf{D}^{\tau+1}$ which in turn can be used in Eq. (57) to solve for $q^{\tau+1}$. Finally $\mathbf{T}^{\tau+1}$ is obtained by solving Eq. (56) written in the form

$$(\mathbf{1} - 2\Delta t\, \mathbf{V}')\mathbf{T}^{\tau+1} = \mathbf{T}^{\tau-1} + 2\Delta t\, \mathbf{y} + \Delta t\, \mathbf{G}(\mathbf{D}^{\tau+1} + \mathbf{D}^{\tau-1}). \tag{59}$$

The prognostic sequence requires application of the above algorithm for each spectral degree of freedom. In practice the solution for $\zeta^{\tau+1}$, $\mathbf{M}^{\tau+1}$, $\mathbf{D}^{\tau+1}$, and $\mathbf{T}^{\tau+1}$ is obtained by means of Gauss elimination and complete privoting. [Although not employed in the above formulation the inverse of Eq. (53), i.e., $\mathbf{T} = (1/R')\mathbf{A}^{-1}\mathbf{\Phi}$, should be treated cautiously; the inverse of **A** implies division by $\ln \sigma_1$, and for $\sigma_1 \sim 1$ this may introduce ill-conditioning.] The semi-implicit time step algorithm in comparison with a fully explicit or centered scheme (i) allows an increase in time step from $\sim$600 sec to 3600 sec (for $J = 15$); (ii) introduces time truncation error which is negligible; and (iii) introduces a computational overhead of $\sim$3%.

## D. Inclusion of Topography

The influence of topography in the model formulation with the Phillips $\sigma$ coordinate $\sigma = p/p_*$ enters explicitly only as lower boundary condition in the integration of the hydrostatic equation. Implicitly, of course, the mathematical surfaces of the model follow the variation of $p_*$ which varies as function of $\Phi_*$. $\Phi$ only appears explicitly in the divergence equation $\partial D/\partial t = \cdots - \nabla^2\Phi'$ which may be written as

$$\partial D/\partial t = \cdots - \nabla^2\Phi'_f - \nabla^2\Phi'_*$$

where $\Phi' = \Phi'_f + \Phi'_*$ defines the contribution to $\Phi'$ from the boundary $\Phi'_*$ and the fluid $\Phi'_f$ i.e., the topography appears explicitly as a linear term $-\nabla^2\Phi'_*$. It is thus seen that a spectral representation of $\Phi_*$ may be readily incorporated in the spectral model.

The spectral representation of the earth's topography is not straightforward, however. Thc variance of the scalar field defined by the height of the continents above sea level is greatly in excess of that resolvable by the finite number of scalar functions available at $J = 15$. Accordingly, prior to spherical harmonic analysis the topography is smoothed. A two-dimensional nine-point filter (Shapiro, 1970) was applied to the basic specification of the earth's topography, given in terms of the mean elevation over $5° \times 5°$ squares as obtained from Lee and Kaula (1967). The criterion for the number of applications of the filter was chosen such that spurious oceanic "ripples" should not exceed 60 meters. In addition, the power spectrum of this representation was inspected and visually assessed in terms of minimizing the discontinuity in the spectra at the point of truncation ($J = 15$). These two rather arbitrary criteria resulted in a need for 5 passes of the above-mentioned filter.

This topographical filtering produces distinct spreading of the dominant land masses; the mountain heights and continental boundaries are shown on a latitude–longitude grid in Fig. 5.

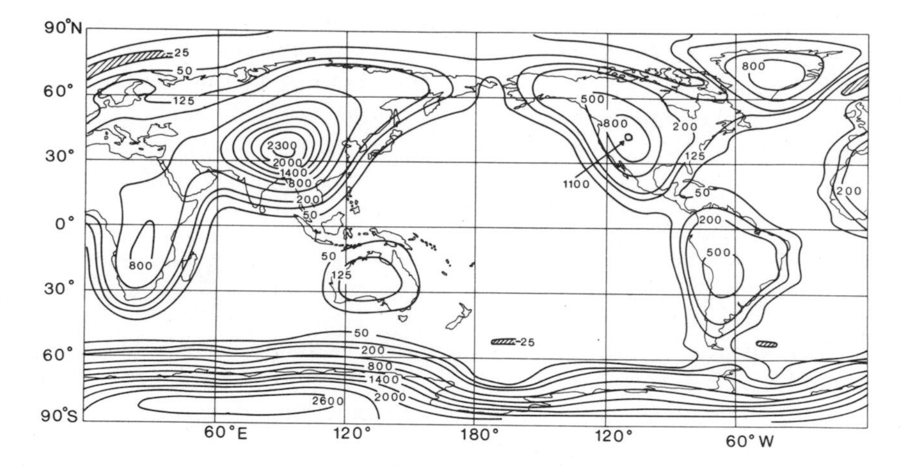


FIG. 5. A latitude–longitude grid display of the contour heights of the ($J = 15$) spectral representation of the earth's topography. The contour interval is 300 meters for elevations above 500 meters and 75 meters for elevations less than 500 meters. Areas with elevation below −25 meters, i.e., spurious ripples, are shaded; features such as the Rockies, Andes, and Himalayas are seen to be smoothed and broadened.

The linear nature of the topographical forcing suggests that the error introduced by this truncated representation of the topography will be orthogonal to those components and scales being represented in the model. This comment is, however, a simplification since (i) the equations are, of course, nonlinear and the finite truncation will generate nonlinear error and (ii) it is not clear that smoothed topography is the optimum parameterization since, for example, it introduces contributions from $\nabla^2\Phi_*$ in areas where this quantity should be in fact zero. However, for a given topographical representation, highly smoothed as it is, the influence of that "topography" may be treated exactly in a linear sense and with a minimum of aliasing.

In the present formulation, with the smoothed topography as shown in Fig. 5, there have been no difficulties in model integrations that are attributable to the incorporation of topography. A more detailed study of the impact of the spurious topographic ripples is currently in progress.

The specification of the land–sea boundaries within the model is not based on the spectral representation of topography but is determined from the correct geographical distribution to within the resolution of the transform latitude–longitude grid. This specification seems preferable in terms of defining the thermal contrast at the lower model boundary.

## E. MODEL COMPUTER CODING

The code developed to integrate numerically the spectral tendency equations (43) to (47) is represented in a flow diagram in Fig. 6. The major

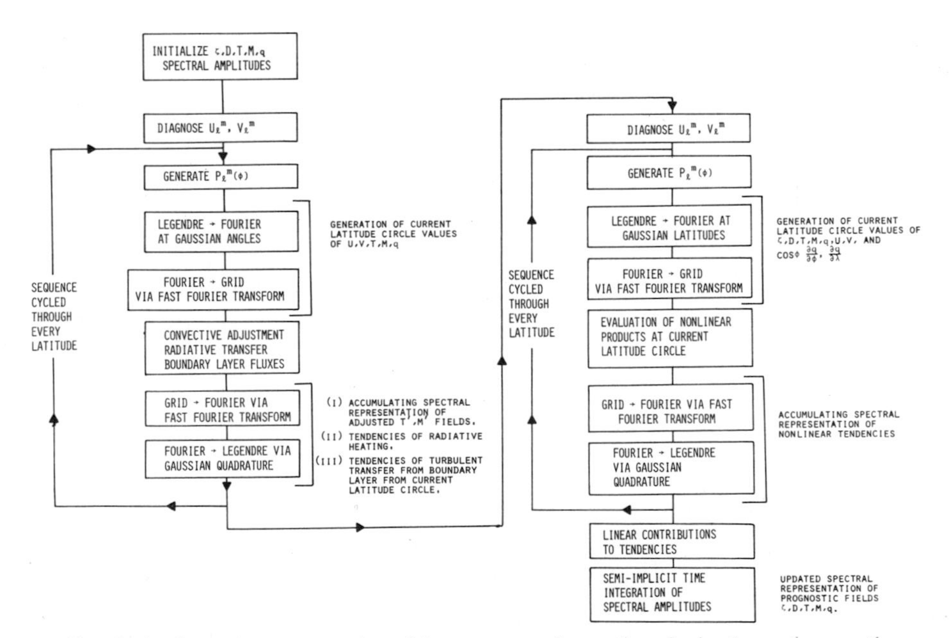


FIG. 6. A schematic representation of the sequence of operations for implementing one time step of the spectral model.

coding strategy has been to subdivide the model into two transform loops. The first loop handles physical processes and generates spectral representations of (i) convectively adjusted moisture and temperature fields; (ii) radiative heating tendencies; and (iii) tendencies at the lowest model level due to boundary layer turbulent stresses. The second loop handles the model dynamics and generates a spectral representation of the nonlinear contributions to the tendencies of the spectral components.

By using two transform loops it has been possible to limit the total core requirement of the model. For example, spectral representation of the adjusted and unadjusted moisture and temperature fields as well as their tendencies would be required simultaneously in core if the calculation were attempted in one loop. However, with two loops, only the adjusted and unadjusted form of the two fields are required in the first loop and the adjusted fields and their tendencies in the second loop.

With this strategy it has been possible to code the ANMRC models such that the use of peripheral devices is kept at a minimum. Quantities that are temporarily stored are the previous time-step dynamic fields, the boundary layer turbulent fluxes, radiative heating tendencies, weighted surface emissivities, net radiation at the surface, the surface temperature, and the accumulated precipitation.

A more general feature of this strategy has been the facility of representing nonlinear physical processes in spectral space prior to coupling their influences to the prognostics. In principle, the resolution of the latitude–longitude grid can be different with higher resolution available as required in either loop to enable a minimization of aliasing. In ANMRC codes it has not seemed warranted to employ this option, and the resolution in both transform loops has been such as to resolve quadratics.

An additional feature of this coding is the option of incorporating time-splitting algorithms in treating nonlinear physical processes; the strategy of two transform loops is ideally suited for this purpose. The incorporation of nonlinear vertical stresses via time splitting has been considered at some length in the ANMRC codes and is discussed in the next section.

The obvious deficiency of the present strategy is the overhead in, for example, twice transforming moisture and temperature fields from spectral to grid-point space. However, the computer coding of spectral models is considerably simplified if it is possible to retain the bulk of the spectral representation in core memory throughout the calculation. To achieve substantially higher model resolution it will be necessary to consider much more use of peripheral storage as is common to all grid-point models.

The hemispheric versions of the model assume reflective symmetry at the equator via the use of symmetric representation of all model fields except $\psi$

and $V$, which are antisymmetric. Accordingly, the computing task is halved both with respect to time taken and core memory required.

### F. Application of Time-Splitting Technique

The possibility of using a time-splitting technique, following Marchuk (1974), in the spectral model has been mentioned in the previous section. This will now be considered for the case of internal vertical diffusion which is currently linearized in the ANMRC spectral models; nonlinear terms have to be evaluated in grid-point space prior to being spectrally analyzed with the requirement thereby for additional core or peripheral usage. Linearized vertical diffusion can be readily made implicit, which ensures computational stability; by comparison implicit treatment of nonlinear vertical diffusion is not so straightforward. The time-splitting technique, however, provides a simple means for implicit treatment of nonlinear vertical diffusion without a requirement for additional core or peripheral usage.

Consider the prognostic equations written in the form

$$\partial \mathbf{S}/\partial t + \mathbf{\Lambda}_1 \mathbf{S} + \mathbf{\Lambda}_2 \mathbf{S} = 0 \tag{60}$$

where $\mathbf{S}$ can be $\zeta$, $\mathbf{D}$, $\mathbf{T}$, or $\mathbf{M}$; $\mathbf{\Lambda}_2$ is the operator representing internal vertical diffusion, and operator $\mathbf{\Lambda}_1$ represents the remaining terms. The splitting technique solves these equations in two steps. In the first step

$$\partial \mathbf{S}/\partial t + \mathbf{\Lambda}_1 \mathbf{S} = 0 \tag{61}$$

is solved by the standard method described in Section III, C. These solutions are then used as initial conditions in the second step to solve

$$\partial \mathbf{S}/\partial t + \mathbf{\Lambda}_2 \mathbf{S} = 0. \tag{62}$$

The nonlinear equation (62) is solved in grid-point space, which allows in the spectral model the term $\mathbf{S}$ to be made implicit, while retaining nonlinearity in $\mathbf{\Lambda}_2 \mathbf{S}$. The coding of the model into the physical processes loop and dynamics loop greatly facilitates this procedure. The equations are solved in the physical processes loop where, except for the winds above the lowest model level, all other variables are available. Hence the only overhead with this scheme is computational, i.e., a transform of the winds at the remaining levels to grid-point space and a reanalysis of the spectral amplitudes of the updated vorticity and divergence fields.

Equation (62) is solved using the following algorithm:

$$\mathbf{S}^{\tau+1} - \mathbf{S}' = -2\Delta t\, \mathbf{\Lambda}_2' \mathbf{S}^{\tau+1}$$

or (63)

$$\mathbf{S}^{\tau+1} = (\mathbf{1} + 2\Delta t\, \mathbf{\Lambda}_2')^{-1}\mathbf{S}'$$

where $\mathbf{S}'$ are the solutions of Eq. (61) and $\mathbf{\Lambda}_2'$ is the finite difference form of operator $\mathbf{\Lambda}_2$ based on the fields $\mathbf{S}'$. $\mathbf{\Lambda}_2'$ is a tridiagonal matrix and Eq. (63) is solved by using the Thomas algorithm.

The finite difference representation in time of Eq. (60) is of first-order accuracy; the splitting technique does not guarantee that the first-order accuracy is preserved if operators $\mathbf{\Lambda}_1$ and $\mathbf{\Lambda}_2$ do not commute. However, this does not appear to cause any problems, at least for model integrations up to 4 days. For example, the time truncation error, as measured by the root-mean-square difference between two integrations using time steps of 10 min. and 60 min, is negligible after 4 days. Also, the solution obtained with the splitting technique and linearized operator $\mathbf{\Lambda}_2$ shows negligible difference after 4 days from the solution obtained via the standard technique described in Section III, C.

The splitting technique was also considered as a means of treating boundary layer fluxes implicitly. However, it was found that although stable, the loss of first-order accuracy introduced nonnegligible time truncation error. This is due to the greater significance of boundary layer fluxes in comparison to internal vertical diffusion in defining model tendencies.

## IV. Numerical Weather Prediction via a Spectral Model

The spectral model, as formulated in Section III, was adopted for operational hemispheric use by the Melbourne World Meteorological Centre (WMC) in January 1976. This has followed (i) experimental evaluation of the performance of the model initialized with data of the GARP special observing period of November 1969 (see II), (ii) parallel real-time trials in mid-1974 as reported at the Copenhagen spectral model symposium (Bourke *et al.*, 1974; subsequently denoted as III), and (iii) recent parallel operational trials in October and November 1975; some of these results are presented in the following.

## A. Operational Model Configuration

The operational application of the spectral model is of necessity required to be somewhat economical in its use of computer core and time. The Melbourne WMC operational computer currently provides ~40K words of memory for numerical models; a 48 hr prognosis is required to take ~1 hr of central processor time. The optimum configuration satisfying these constraints has been chosen to be a model employing seven levels in the vertical with wave number truncation $J = 15$. In the interests of time economies a radiative transfer calculation is omitted in the operational model at present and replaced by the simple parameterization of Newtonian cooling.

Specific details of the operational model are as follows.

(i) Vertical levels are, as in Fig. 3, at

$$\sigma \equiv \{0.991, 0.850, 0.700, 0.500, 0.300, 0.200, 0.100\}$$

(ii) $C_d = 0.004$ and 0.001 over land and sea, respectively.

(iii) $D_w = 0.5$ and 1.0 over land and sea, respectively.

(iv) Nonconvective and convective condensation is assumed to occur as relative humidity exceeds 80%.

(v) Horizontal diffusion of $\zeta$, $T$, and $M$ with $K_h = 2.5 \times 10^5$ m$^2$ sec$^{-1}$ applied for $l > J$.

(vi) Inclusion of divergence dissipation.

(vii) Linearized vertical diffusion of $\zeta$, $D$, $T$, and $M$ with $\mu = 30$ meters for $\sigma \leqslant 0.5$ and zero for $\sigma > 0.5$.

(viii) Specification of the sea surface temperature in terms of mean monthly values.

In the general circulation simulations described in Section V condensation is determined to occur as the relative humidity exceeds 100%; in addition, the land wetness $D_w$ is set to 0.25. In the prediction model the decrease in the relative humidity threshold for condensation and the increased $D_w$ are attempts to enhance the precipitation and release of latent heat over continental areas.

## B. Spectral Data Analysis and Processing

The spectral analysis of meteorological data is currently generated via a spectral fitting of the WMC numerical analyses generated on a 47 × 47 polar stereographic grid of the Southern Hemisphere. The analysis scheme is similar to that described by Gauntlett *et al.* (1972). The numerical and spectral analyses are discussed in Appendix C of II. It is sufficient to note here that a moisture mixing ratio analysis is now generated. The $\sigma$ surfaces are defined in terms of the surface pressure as determined from a log-linear interpolation of pressure as a function of geopotential height; the geopotential height of the earth's surface corresponds to that regenerated from a spherical harmonic analysis of hemispheric topography smoothed in the manner discussed in Section III, D with respect to the global topography representation.

The post-prognosis processing and display of the model integration require (i) a spectral to grid transform to the latitude–longitude points of the usual polar stereographic grid and (ii) interpolation from the model $\sigma$ coordinates to pressure coordinates. In the presence of topography the MSL field is obtained via extrapolation from the surface temperature and pressure assuming a constant lapse rate of 6.5 °K/km and the hydrostatic relationship.

### C. Model Initialization

A very simple initialization scheme has been developed for the operational application of the model. This scheme is a modification of the scheme of R. Sadourny (unpublished manuscript, 1973) of forcing geostrophic adjustment by enhancing divergence diffusion. In the spectral model it has in fact been found, as detailed in II and III, that dissipation rather than diffusion of divergence is the more effective as regards controlling of spurious inertia-gravity oscillations. The coefficient of divergence dissipation decreases exponentially in the first 24 hr as $K = K_0 \exp(-10^{-4}t/2)$ and is thereafter held constant as $K = K_0 \exp(-4.32)$. Here $K_0 = 10^{10}/a^2 \text{ sec}^{-1}$ and $t$ is in seconds. The addition of divergence dissipation slightly reduces, as detailed in III, the mean square vertical velocity and the total precipitation. However, the preferred magnitude of these quantities is difficult to verify and the justification for divergence dissipation remains its efficient and successful suppression of oscillations which are nonmeteorological and accordingly are considered deleterious.

### D. Typical Synoptic Results

The performance of the spectral model in operational numerical prediction for the Southern Hemisphere has been assessed both qualitatively and quantitatively and has been compared to the hemispheric operational system hitherto employed by the Melbourne WMC. The Melbourne WMC has for several years employed a filtered baroclinic model which has six levels and is integrated on a polar stereographic projection with 30 points between Equator and pole; this model is an extension to the hemisphere of the operational regional model described by Maine (1972).

The most recent operational trials were conducted during October and November of 1975; these trials employed the model configuration described in Section IV, A. The model was integrated to 48 hr via 1-hr time steps; this prognosis requires 60 min CPU time on the IBM 360/65 computer.

TABLE III

TIME AVERAGED ROOT MEAN SQUARE ERRORS AND $S_1$ SKILL SCORES[a]

| | | 24 Hr | | | 48 Hr | | |
|---|---|---|---|---|---|---|---|
| | | Sp. | Pers. | F | Sp. | Pers. | F |
| Rms error of MSL (mb) | Aust. region | 4.8 | 5.2 | 5.3 | 5.2 | 7.8 | 6.7 |
| | Hemisphere | 5.8 | 6.8 | 6.9 | 7.2 | 8.5 | 8.8 |
| $S_1$ skill of MSL | Aust. region | 0.57 | 0.69 | 0.59 | 0.71 | 0.95 | 0.77 |
| | Hemisphere | 0.54 | 0.65 | 0.56 | 0.65 | 0.76 | 0.70 |
| Rms error of 500 mb geopotential (meters) | Aust. region | 40 | 61 | 60 | 54 | 84 | 84 |
| | Hemisphere | 66 | 89 | 76 | 81 | 107 | 97 |
| $S_1$ skill of 500 mb geopotential | Aust. region | 0.37 | 0.54 | 0.42 | 0.48 | 0.67 | 0.58 |
| | Hemisphere | 0.42 | 0.56 | 0.43 | 0.49 | 0.62 | 0.53 |

[a] At 24 and 48 hr for 24 prognoses from the operational trials of October and November 1975. Sp, Pers., and F denote, respectively, spectral model, persistence, and filtered model.

### 1. *Quantitative Verification*

A summary of the quantitative performance of the operational model during the above-mentioned trials, preceding the operational acceptance by the Melbourne WMC, is presented in Table III. The model is assessed here in terms of (i) root mean square error and $S_1$ skill score of the MSL pressure field; and (ii) the root mean square error and $S_1$ skill score of the 500 mb geopotential height. The $S_1$ skill score, as defined by Teweles and

Wobus (1954), provides an estimate of the normalized rms vector error of pressure gradient. Shuman (1972) states that in practice skill scores, over North America, of 0.30 at sea level and 0.20 at 500 mb are nearly perfect, while scores of 0.8 at sea level and 0.70 at 500 mb are worthless. The figures of Table III represent the results averaged over 24 days during October and November 1975. At both time intervals of 24 and 48 hr and in both the Australian region and the hemispheric region (defined by an annulus from 20°S to 60°S), the spectral model is seen to be consistently superior to the finite difference filtered model and to provide considerable gains over persistence (i.e., the forecast obtained by invoking no change in the ensuing 24 and 48 hr). It is also noteworthy that the $S_1$ skill scores here are still within the limits of usefulness suggested by Shuman; the NWP problem is clearly more difficult in the Southern Hemisphere, and in fact the expectation in skill is accordingly somewhat lower than in the Northern Hemisphere.

2. *Qualitative Assessment*

The visual qualitative assessment of the spectral model during the October–November 1975 trials identified several characteristics:

i. the phase movement of midlatitude troughs and ridges are especially well captured at 24 hr and provide reasonable guidance at 48 hr.
ii. the model performance deteriorates Equatorward of 30°S throughout the 48 hr prognosis period.
iii. the intensity of the polar front jet is reasonably well maintained while the subtropical jet is generally underestimated.
iv. the model fails to cope with smaller scale features such as low pressure cutoff systems and low pressure systems in the subtropical easterlies.

These general characteristics are consistent with the expectations from the horizontal resolution of the model and the data difficulties inherent in the Southern Hemisphere.

A prognosis example is presented in Figs. 7a, b, and c. The MSL and 500 mb analyses for 00GMT October 10 and 11 are presented together with the 24 hr prognoses from the spectral model. (These analyses are the operational WMC analyses which at 00Z are subject to substantial manual intervention based on infrared and visual satellite imagery; thus the influence of prognosis analysis feedback of the then operational filtered baroclinic model was accordingly minimal in these analyses.)

In the prognosis example, the spectral model correctly predicts at 24 hr the movement of all major systems over the hemisphere. Some of the general characteristics of performance mentioned above are evident in this prognosis example. The rms and $S_1$ values for this example are given in Table IV for comparison with the averaged results of Table III.

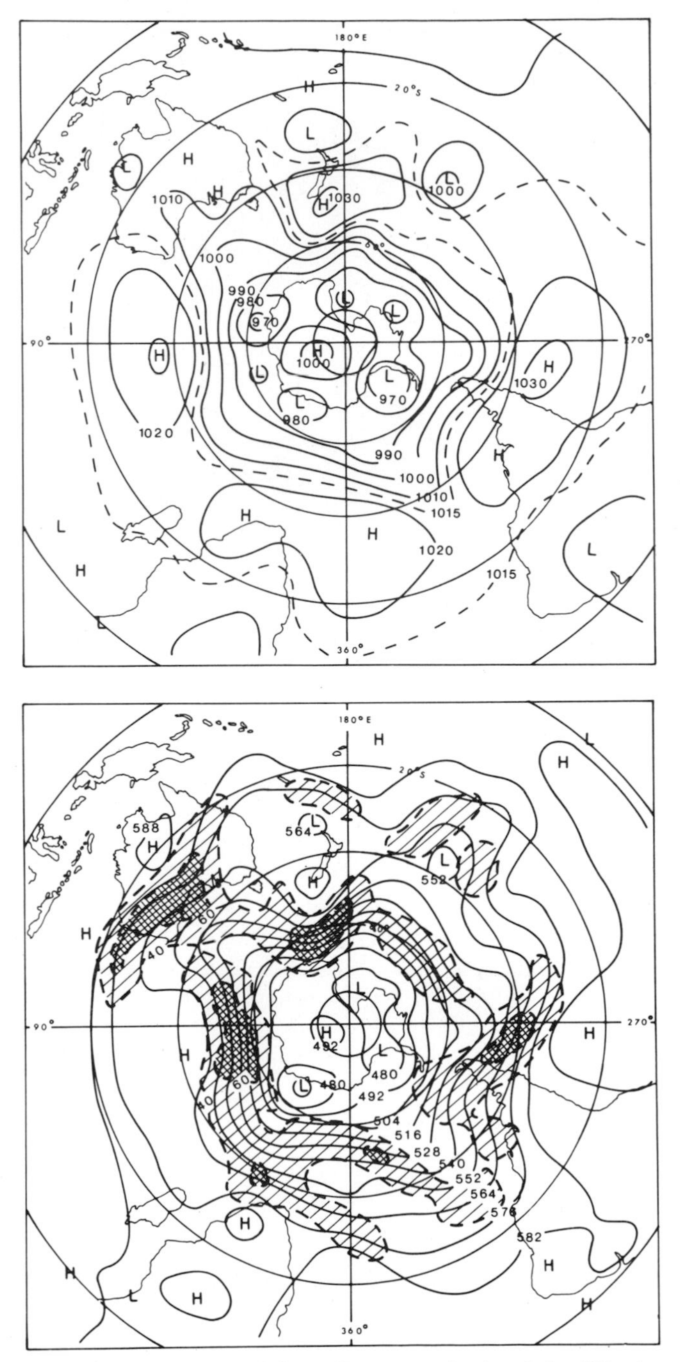


FIG. 7a. The initializing analysis of the MSL pressure (top) and the 500 mb geopotential heights and isotachs for 00GMT October 10, 1975. The MSL pressure is contoured at 10 mb intervals with the 1015 mb line dashed; the 500 mb heights are contoured at 12 dekameter intervals with the isotachs contoured at 20 knot intervals. Isotach contours above 40 knots are hatched.

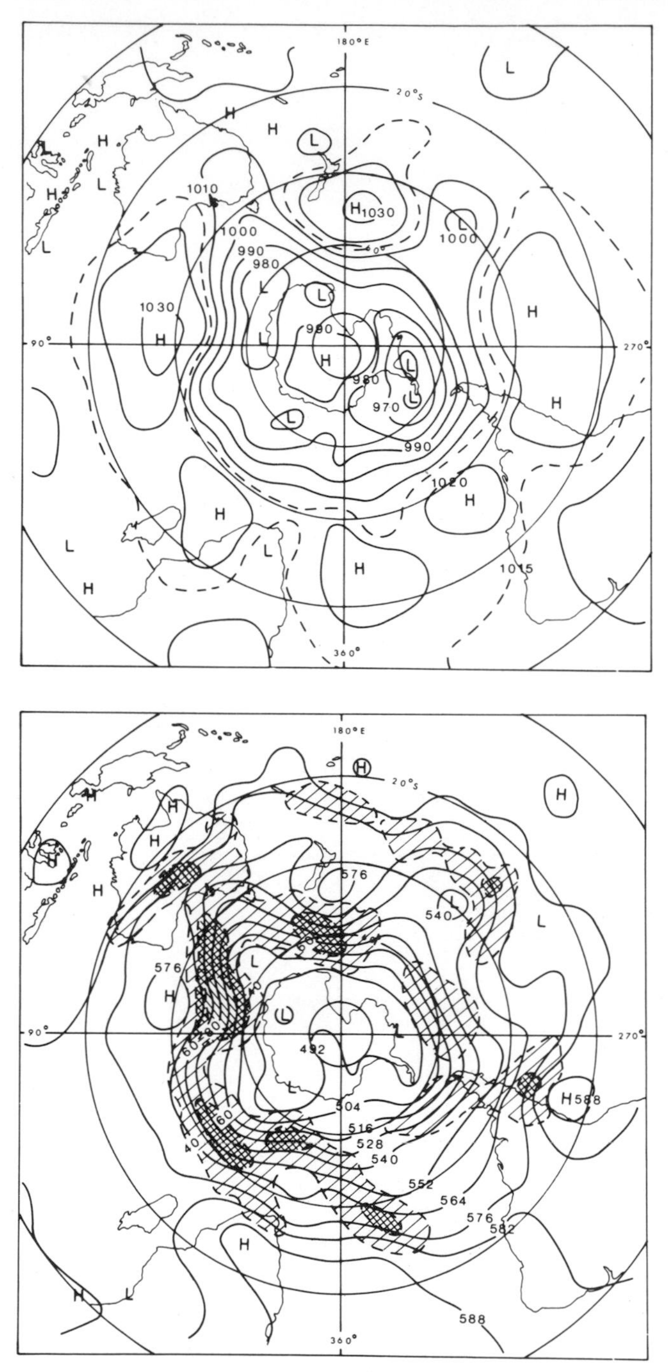


FIG. 7b. As in Fig. 7a but for 00GMT October 11, 1975.

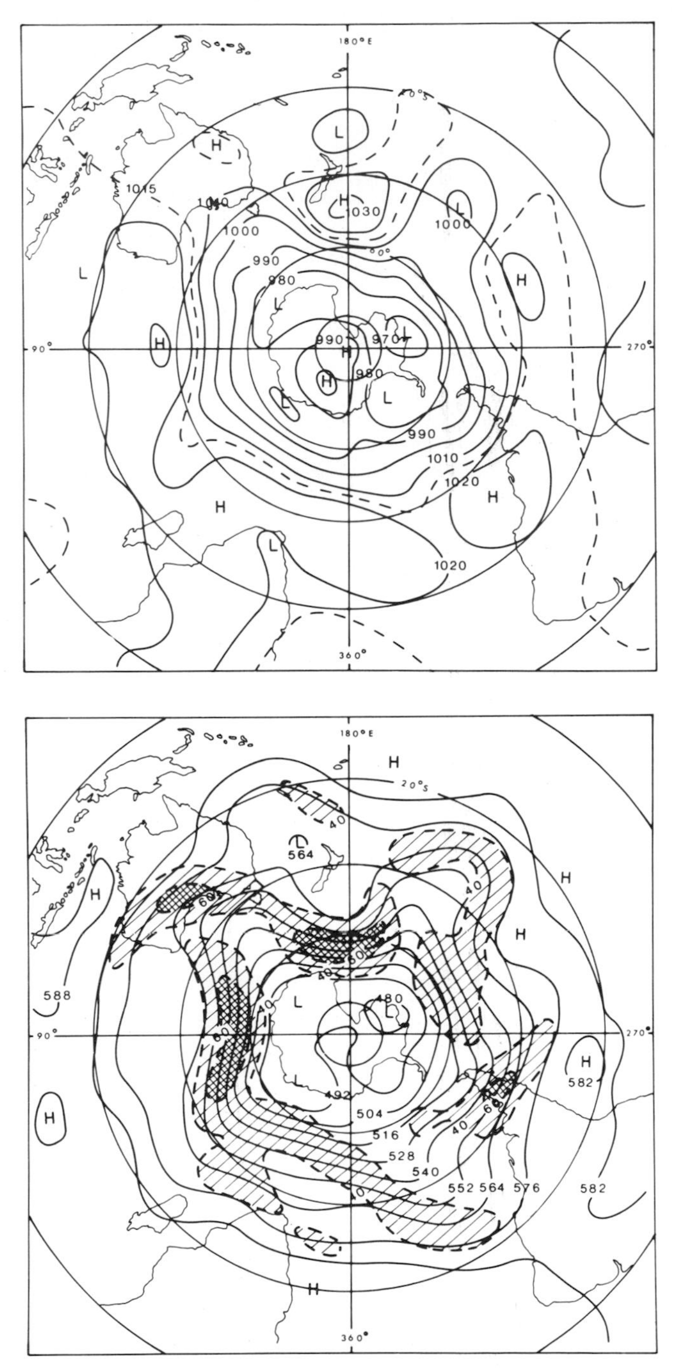




FIG. 7c. As in Fig. 7a but for the 24 hr spectral model prognosis initialized from 00GMT October 10, 1975.

TABLE IV

ROOT MEAN SQUARE ERRORS AND $S_1$ SKILL SCORES[a]

| | | 24 hr | |
|---|---|---|---|
| | | Sp. | Pers. |
| Rms error of MSL (mb) | Aust. region | 3.3 | 5.7 |
| | Hemisphere | 5.1 | 6.8 |
| $S_1$ skill of MSL | Aust. region | 0.51 | 0.60 |
| | Hemisphere | 0.48 | 0.62 |
| Rms error of 500 mb geopotential (meters) | Aust. region | 33 | 88 |
| | Hemisphere | 54 | 94 |
| $S_1$ skill of 500 mb geopotential | Aust. region | 0.35 | 0.53 |
| | Hemisphere | 0.37 | 0.52 |

[a] For the 24 hr prognosis example of Fig. 7. Sp. and Pers. denote, respectively, spectral model and persistence.

## E. GENERAL COMMENTS

It is perhaps surprising that truncation at wave number 15 provides a predictive capability that is superior to a filtered finite difference model with 30 points from Equator to pole on the stereographic projection. Here the spectral model has employed $\sim 256$ degrees of freedom relative to $\sim 3000$ degrees of freedom in the finite difference model. This result illustrates the potential of the spectral method as regards achieving a given level of accuracy with a minimum number of degrees of freedom. It is considered, however, that $J = 15$ is near to the minimum resolution that could be realistically used for hemispheric scale prediction.

The operational prediction model development is viewed as a prelude to employing the spectral model with higher resolution in extended prediction to $\sim 4$–5 days. Truncation experiments such as those of Puri and Bourke (1974) suggest that $J = 15$ is far from satisfactory as far as minimizing numerical error in integration to 4 days. Indeed it would seem that accurate numerical solution at that time scale will require truncation at $J \sim 30$, i.e., a quadrupling of the number of degrees of freedom hitherto considered in the ANMRC models.

The consideration of extended prediction presupposes especially in the Southern Hemisphere an improved capability in specifying the initial atmospheric state. The advent of the GARP experiment and the opportunity to consider an enhanced observing network via remotely sensed temperatures, and the measurements from ocean buoys and floating balloons will provide an opportunity for a much expanded study of the predictive capability of numerical models. The extent to which improved prediction emerges will hinge very much on the success in utilizing the data from these new observing systems.

## V. General Circulation via a Spectral Model

The model formulated in Section III has been employed for a simulation of the global circulation. This study has been undertaken to establish the climatology of the spectral model and to identify and improve on systematic deficiencies in physical parameterizations. The initial simulation has considered radiative forcing appropriate for mean January conditions. The model configuration has employed nine vertical levels and truncation at wave number $J = 15$. This global simulation has been integrated on the CSIRO Cyber 7600 computer and has been initialized from real meteorological data.

### A. Model Configuration

(i) The disposition of the vertical levels has followed that devised for the Geophysical Fluid Dynamics Laboratory (GFDL) nine-level models; see Manabe *et al.* (1965). These levels are at the $\sigma$ values of 0.991, 0.926, 0.811, 0.664, 0.500, 0.336, 0.189, 0.074, and 0.009.

(ii) $C_d = 0.004$ over land and 0.001 over sea and sea ice.

(iii) $D_w = 0.25$ over land and snow and 1.0 over ocean and sea ice.

(iv) Nonconvective and convective condensation is assumed to occur as the relative humidity exceeds 100%.

(v) Horizontal diffusion of $\zeta$, $D$, $T$, and $M$ with $K_h = 2.5 \times 10^5\ \text{m}^2\ \text{sec}^{-1}$ applied for $l > J$; in the topmost two model levels the horizontal diffusion is applied for all $l$.

(vi) Linearized vertical diffusion of $\zeta$, $D$, $T$, and $M$ with $\mu = 30$ meters for $\sigma \leqslant 0.5$ and zero for $\sigma > 0.5$.

(vii) Specification of the sea surface temperature according to the climatic mean.

(viii) Specification of a continental snow line and the latitudinal extent of sea ice from the poles.

(ix) A full radiative transfer calculation.

(x) Specification of the continental surface and sea ice temperature via a diagnostic heat balance equation.

### B. Radiative Transfer Calculation

The temperature change due to radiative transfer in the model atmosphere is calculated employing the scheme described by Manabe and Strickler (1964) and Manabe and Wetherald (1967). The scheme is identical to that incorporated in the global model at the GFDL described in Holloway and Manabe (1971). The computational schemes takes into account the vertical distribution of the atmospheric absorbers, water vapor, carbon dioxide, ozone, and cloud; the calculation separates the radiative transfer in terms of a long-wave (infrared) component and a short-wave (solar) component.

The distribution of clouds and ozone are specified in terms of zonally averaged distributions, i.e., as functions of latitude and height but not of longitude. The mixing ratio of carbon dioxide is assumed to have a constant value of 0.0456% by weight throughout the atmosphere. The distribution of water vapor is defined by the moisture mixing ratio prognostic of the model. The distribution of ozone has been determined from Dopplick (1974) and is representative of the December–February season. Three categories of clouds are employed in the model–high, middle, and low. This cloud distribution has been derived from London (1957) for the Northern Hemisphere and from Sasamori *et al.* (1972) for the Southern Hemisphere.

The insolation at the top of the model atmosphere is fixed for the duration of the model integration and is for the mean solar declination and mean earth–sun distance for January from the tables given by List (1951). A value of $1.39 \times 10^3$ W m$^{-2}$ has been used for the solar constant. The diurnal variation of solar insolation is eliminated by use of an effective mean zenith angle and daylight duration for each latitude following the definition of Manabe and Möller (1961).

The albedo over ocean and land areas is specified in terms of zonal averages from the charts of Posey and Clapp (1964). The recommendations, reported in Appendix A of Manabe and Wetherald (1975) have been incorporated; other parameter values follow Manabe and Strickler (1964).

### C. Lower Boundary Conditions

Over continents and sea ice a diagnostic heat balance equation is employed to specify the surface temperature.

Climatological specification for mean January conditions of lower boundary conditions are as follows: (i) climatological sea surface temperature from the U.S. Hydrological Atlas; (ii) snow-covered continents poleward of 42°N and over Antarctica; (iii) ice-covered ocean poleward of 73.3°N with no sea ice assumed in the summer south polar regions.

With the diurnal cycle of insolation eliminated it is assumed that there is no net heat conduction into the soil so that the equation for heat balance is

$$S\downarrow + D\downarrow = \sigma_s T_*^4 + c_p \eta_* + L\beta_* + Q_i$$

where $S\downarrow$ and $D\downarrow$ are the net downward insolation and long-wave radiation, respectively, at the surface, $\sigma_s$ is the Stefan–Boltzman constant, and $\eta_*$ and $\beta_*$ as defined in Eq. (35), denote the sensible and latent heat fluxes, respectively, at the surface, and $L$ is the latent heat of water vapor.

The quantity $Q_i$ represents heat conduction into or out of sea ice if present. Following Holloway and Manabe (1971) a constant ice thickness of 2 meters, a thermal conductivity of 0.523 W $m^{-1}$ $°K^{-1}$, and a sub-ice sea water temperature of 271.2 °K are specified. If, over sea ice and snow-covered continents, the solution to the heat balance equation yields a temperature in excess of freezing, the temperature is reset to the freezing point; this implies an assumption that the excess heat would contribute to melting. In solving the heat balance equation over land or sea ice, following Holloway and Manabe (1971), a minimum wind condition is specified, i.e., the $|\mathbf{V}_1|$ of Eq. (35) is not allowed to decrease below 1 m $sec^{-1}$. In the absence of this minimum wind condition the iterative solution of the heat balance equation may fail to converge or may generate unrealistically high surface temperatures.

## D. Mean January Simulation

A January simulation is currently in progress at the time of preparation of this article, and the results presented in the following represent an intermediate assessment of the calculations.

This global simulation for mean January conditions was initialized from a global analysis for 3/11/69. The calculation has been integrated to 86 days on a Cyber 7600 computer via the semi-implicit algorithm with time steps of $\Delta t = 1800$ sec. The calculation requires ~13 min. of CPU time per day of simulation; the radiative transfer calculation is repeated every 12 hrs. The simulation has generated, as is common with many models (see Holloway and Manabe, 1971), an excessive intensity and shear in the stratospheric westerlies of the winter hemisphere; as a result, the model time step had to be reduced from the initial value of 3600 sec to that of 1800 sec. Several changes have been made to the model in the course of the integration. In particular at day 50 the relative humidity threshold for condensation was increased from 80% to 100%; prior to this the global mean precipitation was diagnosed as excessive. In addition, the wetness factor $D_w$ over snow-covered land was reduced from 1.0 to 0.25 to overcome an excess of precipitation in polar regions. Subsequently the model configuration has been unmodified and the time-averaged response of the model for days 52 to 86 are discussed in the following; it should, of course, be noted that the simulation is not considered to be in statistical equilibrium as yet. The 34-day averaging samples the model state at 6 hr intervals.

1. *Temperature*

Figure 8 displays the 34-day mean latitude height distribution of the zonally averaged temperature distribution compared with mean observed distribution for December–February as given by Newell *et al.* (1972).

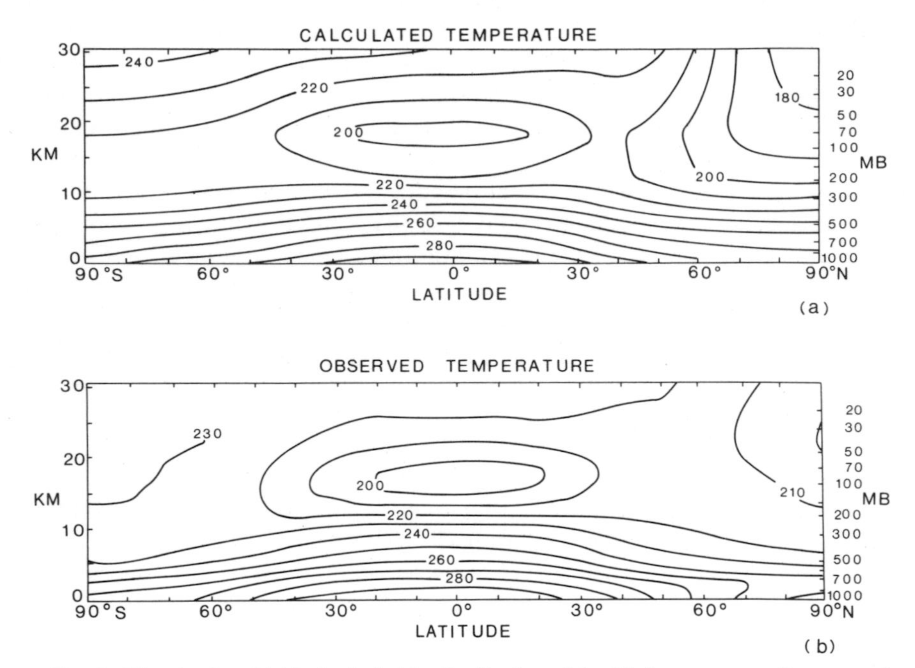


FIG. 8. The simulated (a) latitude–height distribution of the 34-day mean zonally averaged temperature distribution compared with that observed (b), as given by Newell *et al.* (1972); units in degrees Kelvin.

The atmospheric temperature distribution as computed is in reasonable agreement with the observed distribution in the tropics and subtropics. The most noticeable deficiency in the simulation is the excessively cold winter polar stratosphere.

## 2. *Zonal Wind Distribution*

Figure 9 displays the 34-day mean latitude height distribution of the zonally averaged zonal component of the wind compared with the observed distribution for December–February as given by Newell *et al.* (1972).

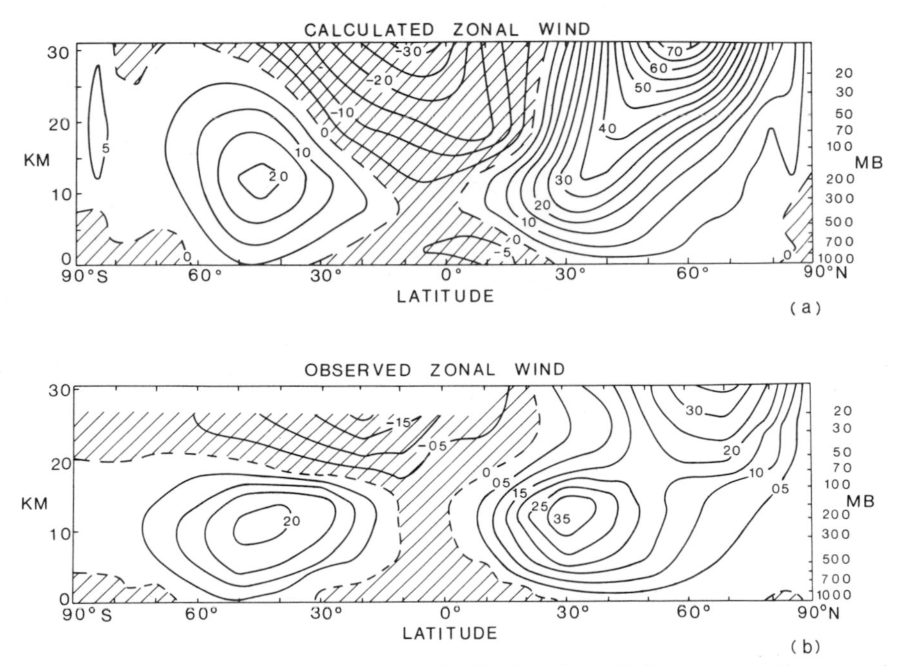


FIG. 9. The simulated (a) latitude height distribution of the 34-day mean zonally averaged zonal component of the wind compared with that observed (b), as given by Newell *et al.* (1972); units in meters per second. The hatched areas denote easterlies.

The calculated wind distribution is in good agreement with the observed in delineating the regimes of tropical easterlies and midlatitude westerlies. The tropical easterlies are captured throughout the depth of the atmosphere, but with the stratospheric maximum in the summer subtropics displaced slightly poleward. The height, latitude, and intensity of the summer mid-latitude jet is well captured. The winter hemisphere is dominated by the overintense and poorly positioned stratospheric westerlies; the tropospheric winter midlatitude westerlies, although not defined as a jet core, are seen to be will simulated in terms of intensity and position. The polar surface easterlies in the model correspond reasonably with those observed.

## 3. *Mean Sea-Level Pressure*

Figure 10 displays the 34-day mean of the MSL pressure compared with the observed as given by Schutz and Gates (1971). The summer hemisphere displays good agreement with the observed distribution with regard to the intensity of low pressures over the continents and the intensity of the subtropical ridge; the centers of the mid-oceanic high pressure ridge are well positioned latitudinally although longitudinally less well positioned. The circumpolar low pressure belt in the Southern Hemisphere is well positioned although lacking a little in intensity; the south polar pressure is additionally somewhat overintense.

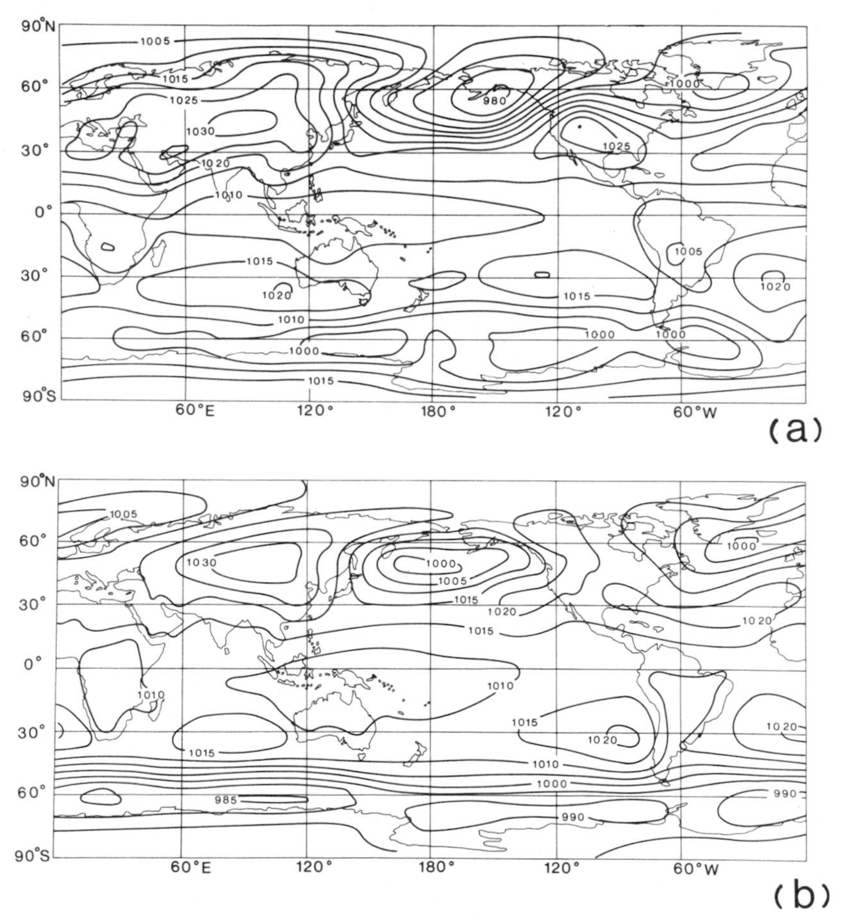


FIG. 10. The 34-day mean of the simulated (a) MSL pressure compared with that observed (b), as given by Schutz and Gates (1971); units in millibars.

The simulated winter hemisphere captures well the intensity and position of the Siberian high, although the Aleutian low is overintense and displaced rather markedly eastward.

The combination of somewhat higher than observed pressures over mid-North America and the displaced overintense Aleutian low have produced rather strong pressure gradients in the northeast Pacific. The Icelandic low is well simulated with respect to intensity, although displaced a little westward of the observed position.

Some guide to the comparative performance of this global simulation can be gained by reference to Appendix B of the U.S. National Academy of Sciences Report on Understanding Climatic Change (U.S. Committee for the GARP, 1975). In that report several large-scale simulations are compared with respect to their ability to simulate mean January conditions.

4. *Moisture*

Figure 11 displays the 34-day mean latitude height distribution of the zonally averaged moisture mixing ratio compared with the mean observed distribution from 10°S to 75°N for January as given by Oort and Rasmussen (1971). The simulated distribution is seen to be in reasonable agreement with that observed.

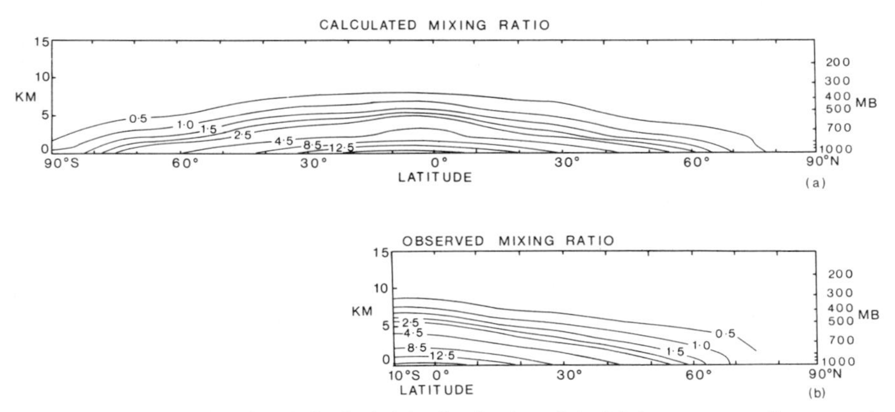


FIG. 11. The simulated (a) latitude–height distribution of the 34-day mean zonally averaged moisture mixing ratio compared with that observed (b), as given by Newell *et al.* (1972); units are grams per kilogram.

The behavior of the model with respect to the global mean values of precipitable water content, evaporation, and precipitation are shown in Fig. 12. The precipitation and evaporation in the spectral model have been diagnosed via summation in grid-point space while the precipitable water content is determined spectrally. Throughout the 34-day period precipitation

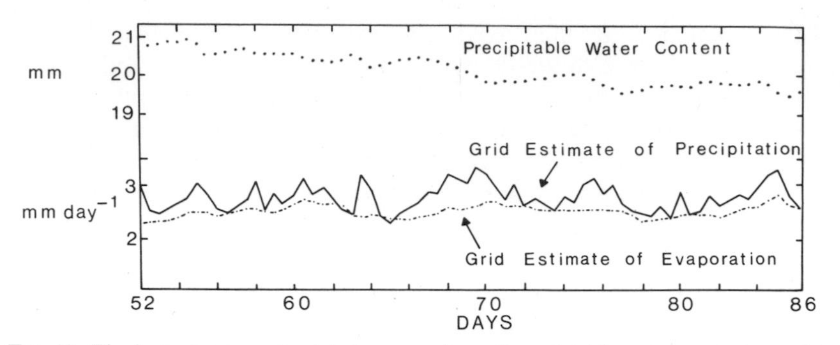


FIG. 12. The instantaneous simulated mean values of preciptable water content, precipitation, and evaporation at 12 hr intervals during the integration from day 52–86; units are, respectively, millimeters, millimeters per day, and millimeters per day.

is generally in excess of evaporation and the precipitable water content of the atmosphere is accordingly declining slowly. The precipitation and evaporation that are sensed by the model in spectral space are not necessarily exactly as given by grid-point evaluation. These diagnostic grid-point summations are given in terms of integrals over the mass field of the change in the moisture mixing ratio induced by nonconvective and convective adjustment and evaporation; the model dynamics are, however, sensitive only to the spectral representation of the changes themselves. The mean precipitation and evaporation rates may be expressed, however, more exactly in spectral space; this more detailed analysis of the model moisture budget and comparison with observations is currently in progress.

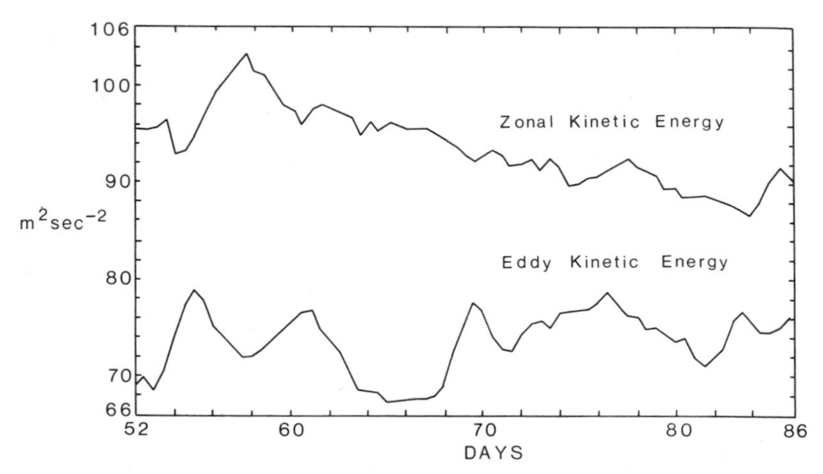


FIG. 13. The instantaneous simulated values of the zonal and eddy kinetic energy at 12 hr intervals during the integration from days 52–86; units are square meters per square second.

### 5. *Kinetic Energy*

The instantaneous values of the zonal and eddy kinetic energy in the global simulation for days 52–86 are shown in Fig. 13. The zonal and eddy kinetic energy are here defined as the zonal and nonzonal contributions to the volume integral in $\sigma$ coordinates of $\mathbf{V} \cdot \mathbf{V}/2$; these may be spectrally evaluated quite simply. The ratio of eddy to zonal kinetic energy is commonly one aspect of interest in assessing general circulation simulations. As deduced from observations, such a ratio is typically of order 1.0 (Lorenz, 1967); here the ratio is seen from Fig. 13 to be $\sim$0.8.

## E. General Comments

As with the operational prediction studies, the general circulation simulations are of a quality that may not be expected at the relatively low resolution of $J = 15$. While there may have been some doubt with respect to the appropriateness of semi-implicit time integration for general circulation studies, it does appear to be quite suitable; substantial time economies are of course, afforded, especially in the spectral model. To some extent the impact of the semi-implicit algorithm is reduced in the global simulation due to the inadequate representation of the winter stratosphere. This problem appears to have plagued many general circulation simulations, and it seems highly desirable for its cause to be clarified and indeed eliminated.

Throughout the 86-day integration the global simulation has conserved mass to a high degree. The mean surface pressure decreased in this period by only 0.018 mb. This conservation is not guaranteed by the spectral formulation as is the case, for example, in models cast in the conservative flux form as commonly used in finite difference models.

The consideration of extended prediction as alluded to in Section IV, E is closely related to the development of the general circulation model described here. The extended simulation is viewed as a means of establishing the systematic behavior of the spectral model and in particular as a means of improving and exploring the parameterizations of physical processes considered of relevance to prediction over time scales of 4–5 days.

# VI. Conclusion

The present contribution outlines the spectral method in general and an extension of the multilevel spectral model formulation reported earlier by Bourke (1974). In particular, the suitability of the spectral model for operational numerical weather prediction and for general circulation simulation has been illustrated.

In principle, the spectral method affords advantages as described in Section I, C over the conventional finite difference methods. The present contribution has not been concerned with comparing finite difference methods and the spectral method, but rather is intended to illustrate the capability of the spectral approach.

While achieving a substantial enhancement of computational efficiency, the transform method of spectral modeling has enabled the incorporation of highly nonlinear physical processes such as, for example, moist convective adjustment and a radiative transfer calculation. The influence of these processes as evaluated in grid-point space may not necessarily be exactly preserved subsequent to a spectral representation that only resolves at most quadratic nonlinearity without aliasing. This effect is, however, an explicit manifestation of a related problem in finite difference models where even the quadratic interaction is misrepresented by aliasing. A related requirement in the spectral model is that of distinguishing between the influence of a particular process as perceived in grid-point space and as it in fact contributes to model dynamics in spectral space. The influence of topography is also in the category of being a rather explicit problem in the spectral model formulation. There is an explicit requirement in the spectral model to predetermine the variance of topography that is compatible with the model resolution; this requirement is less explicit in a finite difference model although just as necessary. The accurate representation of highly nonlinear physical parameterizations and the description of topography in large-scale atmospheric models, whether spectral or finite difference, clearly warrants further attention.

An implication in the pioneering application of the spectral method by Silberman in 1954 was that it may provide a useful alternative to the finite difference methods so successfully applied by Charney *et al.* (1950). The substantial assertion of the present contribution is that this possibility is now realized for the current generation of large-scale numerical models of atmospheric flow. The rapid development and rejuvenation of interest in spectral models subsequent to the emergence of the enhanced capability of the method in 1970 suggests that the full potential for spectral modeling remains to be explored.

## Appendix

### A. Matrix $G$

Extraction of linear components from the term $\dot{\sigma}\gamma + (RT/c_p)\bar{D}$ in the thermodynamic equation yields the term

$$F = \gamma_0(\bar{D} - \bar{D}^\sigma) + \sigma(\partial T_0/\partial\sigma)\bar{D},$$

where the zero subscripts denote global means. As indicated in Section III, B the finite difference representations of $\bar{D}^\sigma$ at level $j$, and $\bar{D}$ are

$$\bar{D}_j^\sigma = D_1\,\Delta\sigma_1 + \sum_{k=1}^{j-1} (D_k + D_{k+1})\frac{\Delta\sigma_{k+1}}{2}$$

$$\bar{D} = D_1\,\Delta\sigma_1 + \sum_{k=1}^{N-1} (D_k + D_{k+1})\frac{\Delta\sigma_{k+1}}{2} + D_N\,\Delta\sigma_{N+1}$$

and hence

$$(\bar{D} - \bar{D}^\sigma)_j = \sum_{k=j}^{N-1} (D_k + D_{k+1})\frac{\Delta\sigma_{k+1}}{2} + D_N\,\Delta\sigma_{N+1}$$

where $N$ is the number of levels and $\Delta\sigma_k$'s are shown in Fig. 3 for $N = 7$.

The required term at level $j$ can now be written as

$$F_j = (\gamma_0)_j\left\{\sum_{k=j}^{N-1} (D_k + D_{k+1})\frac{\Delta\sigma_{k+1}}{2} + D_N\,\Delta\sigma_{N+1}\right\}$$
$$+ \sigma_j\left(\frac{\partial T_0}{\partial\sigma}\right)_j\left\{D_1\,\Delta\sigma_1 + \sum_{k=1}^{N-1} (D_k + D_{k+1})\frac{\Delta\sigma_{k+1}}{2} + D_N\,\Delta\sigma_{N+1}\right\}$$

Regrouping, this reduces to

$$F_j = \sigma_j\left(\frac{\partial T_0}{\partial\sigma}\right)_j\left\{\Delta\sigma_1 D_1 + \sum_{k=1}^{j-1}\frac{\Delta\sigma_{k+1}}{2}(D_k + D_{k+1})\right\}$$
$$+ \left\{(\gamma_0)_j + \sigma_j\left(\frac{\partial T_0}{\partial\sigma}\right)_j\right\}\left\{\sum_{k=j}^{N-1}\frac{\Delta\sigma_{k+1}}{2}(D_k + D_{k+1}) + D_N\,\Delta\sigma_{N+1}\right\}. \quad \text{(A1)}$$

The expression $F$ can also be written in matrix form as

$$\mathbf{F} = \mathbf{GD}$$

The elements of **G** can be easily derived from expression (A1).

### B. Matrices V and V′

The inclusion of vertical diffusion requires the addition of the terms

$$F_v = \frac{g}{p_*}\frac{\partial}{\partial\sigma}\left[\rho^2\frac{g}{p_*}K_v\frac{\partial S}{\partial\sigma}\right]$$

to the tendency equation for $S$, where $S$ represents $\zeta$, $D$, or $M$ and $K_v = \rho(g/p_*)\mu^2|\partial\mathbf{V}/\partial\sigma|$. These terms are linearized by using hemispheric mean values of **V** and $T$ in evaluating $K_v$ and $\rho$. Substitution for $K_v$ and $\rho$ yields

$$F_v = (g/R)^3(\partial/\partial\sigma)[(\sigma/T_0)^3\mu^2|\partial V_0/\partial\sigma|\partial S/\partial\sigma]$$

where $V_0$ is the hemispheric mean of $[(U^2 + V^2)/\cos^2\phi]^{1/2}$ at each level.

$F_v$ can be written in its finite difference form at level $j$ as

$$(F_v)_j = [1/(\sigma_j^h - \sigma_{j-1}^h)][B_{j-1}S_{j-1} - (B_{j-1} + B_j)S_j + B_jS_{j+1}] \quad \text{(A2)}$$

$\sigma_j^h$ denotes the half-level $\sigma$ value immediately above the full level $j$ (see Fig. 3). Also,

$$B_j = A_j^h/(\sigma_{j+1} - \sigma_j)$$

where $A_j^h$ is the value of $(g/R)^3(\sigma/T_0)^3\mu^2|\partial\mathbf{V}_0/\partial\sigma|$ at the half level immediately above full level $j$. $F_v$ can now be written in the matrix form

$$\mathbf{F} = \mathbf{VS}$$

with the elements of the tridiagonal matrix **V** determined by expression (A2) and the prognostic equation for **S** becomes

$$\partial\mathbf{S}/\partial t = \mathbf{Z} + \mathbf{VS}$$

where **Z** represents the remaining terms.

In the thermodynamic equation the quantity being diffused is the potential temperature and the conversion from potential temperature to temperature $T$ results in the equation

$$\partial\mathbf{T}/\partial t = \mathbf{Z} + \mathbf{V'T}$$

where $V'$ is defined by a modified form of expression (A2)

$$(F_v)'_j = \frac{1}{\sigma_j^h - \sigma_{j-1}^h}\left[\left(\frac{\sigma_j}{\sigma_{j-1}}\right)^k B_{j-1}T_{j-1} - (B_{j-1} + B_j)T_j + \left(\frac{\sigma^j}{\sigma_{j+1}}\right)^k B_jT_{j+1}\right] \quad \text{(A3)}$$

with $k = R/c_p$.

ACKNOWLEDGMENT

The authors thank Dr. S. Manabe for making available his computer code for the radiative transfer calculation.

## References

Baer, F., and Platzman, G. W. (1961). *J. Meteorol.* **18**, 393–401.

Bourke, W. (1972). *Mon. Weather Rev.* **100**, 683–689.

Bourke, W. (1974). *Mon. Weather Rev.* **102**, 688–701.

Bourke, W., Puri, K., and Thurling, R. (1974). *Proc. Int. Symp. Spectral Methods Numerical Weather Prediction, 1974*, GARP Working Group on Numerical Experimentation, Rep. No. 7, pp. 22–42.

Charney, J. G., Fjörtoft, R., and von Neumann, J. (1950). *Tellus* **2**, 237–254.

Daley, R., Simmonds, I., and Henderson, J. (1974). *Proc. Int. Symp. Spectral Methods Numerical Weather Prediction, 1974*, Rep. No. 7, pp. 43–45. GARP Working Group on Numerical Experimentation.

Dickinson, R. E., and Williamson, D. L. (1972). *J. Atmos. Sci.* **29**, 623–640.

Dopplick, T. G. (1974). "The General Circulation of the Tropical Atmosphere," Vol. 2, pp. 1–25. MIT Press, Cambridge, Massachusetts.

Doron, E., Hollingsworth, A., Hoskins, B. J., and Simmons, A. (1974). *Q. J. R. Meteorol. Soc.* **100**, 371–383.

Eliasen, E. and Machenhauer, B. (1974). *Proc. Int. Symp. Spectral Methods Numerical Weather Prediction, 1974*, Rep. No. 7, pp. 83–93. GARP Working Group on Numerical Experimentation.

Eliasen, E., Machenhauer, B., and Rasmussen, E. (1970). "On a Numerical Method for Integration of the Hydrodynamical Equations with a Spectral Representation of the Horizontal Fields," Rep. No. 2. Institut for Teoretisk Meteorologi, Köbenhavns Universitet, Denmark.

Elsaesser, H. W. (1966). *J. Appl. Meteorol.* **5**, 246–262.

Gauntlett, D. J., Seaman, R. S., Kinninmonth, W. R., and Langford, J. C. (1972). *Aust. Meteorol. Mag.* **20**, 61–82.

Gordon, T., and Stern, W. (1974). *Proc. Int. Symp. Spectral Methods Numerical Weather Prediction, 1974*, Rep. No. 7, pp. 46–82. GARP Working Group on Numerical Experimentation.

Global Atmospheric Research Program. (1974). *Proc. Int. Symp. Spectral Methods Numerical Weather Prediction, 1974*. Rep. No. 7. GARP Working Group on Numerical Experimentation (WGNE).

Haurwitz, B. (1940). *J. Mar. Res.* **3**, 254–267.

Holloway, J. L., and Manabe, S. (1971). *Mon. Weather Rev.* **99**, 335–370.

Hoskins, B. J., and Simmons, A. J. (1975). *Q. J. R. Meteorol. Soc.* **101**, 637–655.

Kubota, S., Hirose, M., Kikuchi, Y., and Kurihara, Y. (1961). *Pap. Meteorol. Geophys.* **12**, 199–215.

Lee, W. H. K., and Kaula, W. M. (1967). *J. Geophy. Res.* **72**, 753–758.

List, R. J. (1951). "Smithsonian Meteorological Tables," Smithson. Publ. No. 4014. Smithson. Inst., Washington, D.C.

London, J. (1957). "A Study of the Atmospheric Heat Balance," Report. Dept. Meteorol. and Oceanogr. New York Univ., New York.

Lorenz, E. N. (1960). *Tellus* **12**, 243–254.

Lorenz, E. N. (1967). "Nature and Theory of the General Circulation of the Atmosphere." World Meteorol. Organ., Geneva.

Machenhauer, B., and Daley, R. (1972). "A Baroclinic Primitive Equation Model with a Spectral Representation in Three Dimensions," Rep. No. 4. Institut for Teoretisk Meteorologi, Köbenhavns Universitet, Denmark.

Machenhauer, B., and Rasmussen, E. (1972). "On the Integration of the Spectral Hydrodynamical Equations by a Transform Method," Rep. No. 3. Institut for Teoretisk Meteorologi, Köbenhavns Universitet, Denmark.

Maine, R. (1972). *J. Appl. Meteorol.* **11**, 7–15.
Manabe, S., and Holloway, J. L. (1975). *J. Geophy. Res.* **80**, 1617–1649.
Manabe, S., and Möller, F. (1961). *Mon. Weather Rev.* **89**, 503–532.
Manabe, S., and Strickler, R. F. (1964). *J. Atmos. Sci.* **21**, 361–385.
Manabe, S., and Wetherald, R. T. (1967). *J. Atmos. Sci.* **24**, 241–259.
Manabe, S., and Wetherald, R. T. (1975). *J. Atmos. Sci.* **32**, 3–15.
Manabe, S., Smagorinsky, J., and Strickler, R. F. (1965). *Mon. Weather Rev.* **93**, 769–798.
Marchuk, G. I. (1974). "Numerical Methods in Weather Prediction." Academic Press, New York.
Merilees, P. E. (1968). *J. Atmos. Sci.* **25**, 736–743.
Merilees, P. E. (1972). *Atmosphere* **10**, 1–9.
Merilees, P. E. (1973). *Atmosphere* **11**, 13–20.
Miyakoda, K. (1973). *Proc. R. Ir. Acad., Sect. A* **73**, 99–130.
Newell, R. E., Kidson, J. W., Vincent, D. G., and Boer, G. J. (1972). "The General Circulation of the Tropical Atmosphere," Vol. 1, pp. 17–130. MIT Press, Cambridge, Massachusetts.
Oort, A. H., and Rasmusson, E. M. (1971). "Atmospheric Circulation Statistics," NOAA Prof. Pap. No. 5. U.S. Dept. Commerce, Rockville, Maryland.
Orszag, S. A. (1970). *J. Atmos. Sci.* **27**, 890–895.
Orszag, S. A. (1971). *Stud. Appl. Math.* **50**, 293–327.
Orszag, S. A. (1974). *Mon. Weather Rev.* **102**, 56–75.
Phillips, N. A. (1957). *J. Meteorol.* **14**, 184–185.
Phillips, N. A. (1959). *Mon. Weather Rev.* **87**, 333–345.
Platzman, G. W. (1960). *J. Meteorol.* **17**, 635–644.
Posey, J. W., and Clapp, P. F. (1964). Geofis. Int. **4**, 33–48.
Puri, K., and Bourke, W. (1974). *Mon. Weather Rev.* **102**, 333–347.
Rasmussen, E. (1974). "An Investigation of the Truncation Errors in a Barotropic Primitive Equations Model on Spectral Form," Rep. No. 5. Institut for Teoretisk Meteorologi, Köbenhavns Universitet, Denmark.
Robert, A. J. (1966). *J. Meteorol. Soc. Jpn.* **44**, 237–245.
Robert, A. J. (1969). *Proc. WMO/IUGG Symp., Jpn. Meteorol. Agency, 1968.* pp. VII-19–VII-24.
Robert, A. J. (1970). *Proc. Stanstead Semin., 8th, 1969.* Publ. Meteorol. No. 97. McGill University, Montreal.
Robert, A. J., Henderson, J., and Turnbull, C. (1972). *Mon. Weather Rev.* **100**, 329 and 335.
Sasamori, T., London, J., and Hoyt, D. V. (1972). *Meteorol. Monogr.* **13**, 9–23.
Schutz, C. and Gates, W. L. (1971). "Global Climatic Data for Surface, 800 mb: January," Rep. No. R-915-ARPA. Rand Corp., Santa Monica, California.
Shapiro, R. (1970). *Rev. Geophys. Space Phys.* **8**, 359–387.
Shuman, F. G., (1972). "The Research and Development Program at the National Meteorological Center." *NOAA NMC Office Note*, No. 72. U.S. Dept. Commerce. Suitland, Maryland.
Silberman, I. S. (1954). *J. Meteorol.* **11**, 27–34.
Simmons, A. J., and Hoskins, B. J. (1975). *Q. J. R. Meteorol. Soc.* **101**, 551–565.
Teweles, S., and Wobus, H. (1954). *Bull. Am. Meteorol. Soc.* **35**, 455–463.
U.S. Committee for the Global Atmospheric Research Program (1975). "Understanding Climatic Change." U.S. Natl. Acad. Sci. Washington, D.C.
Weigle, W. F. (1972). "Energy Conservation and the Spectrally Truncated One-Level Primitive Equations." Ph.D. Dissertation. University of Michigan, Ann Arbor.

# Author Index

Numbers in italics refer to the pages on which the complete references are listed.

## A


Abarbanel, S., 34, *66*
Anderson, P. A., 34, *60*
Arakawa, A., 32, 36, 41, 43, 52, *60*, *61*, *65*, 70, *108*, 111, *169*, 193, 194, 195, 196, 207, 262, 264, *264*, *265*
Asselin, R., 71, *108*, 164, *169*
Atkinson, G. D., 99, *108*


## B


Baer, F., 44, 50, *61*, *64*, 269, *323*
Ball, F. K., 76, *108*
Barry, R. G., 166, *169*, *172*
Batten, E. S., 43, *62*
Baumhefner, D. P., 57, 58, *61*, 166, *169*, *170*, *171*
Bengtsson, L., 56, *61*
Benoit, R., 120, 165, *169*
Bettge, T. W., 166, *169*
Betts, A. K., 76, 79, 80, 81, *108*
Bleck, R., 16, *61*
Boer, G. J., 99, *109*, 314, 315, 317, *324*
Bourke, W., 45, 48, *61*, 270, 271, 279, 281, 283, 285, 288, 302, 310, 319, *323*, *324*
Bretherton, F. P., 25, 36, *61*
Browning, G. L., 35, 42, *66*, 112, 113, 137, 164, 166, *172*
Bryan, K., 23, 36, *61*, *63*, 69, 71, *108*, 130, *169*
Bryson, R. A., *62*
Budyko, M. I., 130, *169*
Burstein, S. Z., 34, *61*


## C


Cahn, A., 185, 188, *265*
Carson, D. J., 76, 79, 80, 81, *108*
Ceselski, B., 162, *170*
Charney, J. G., 2, 3, 8, 11, *61*, 165, *169*, 268, 320, *323*
Charnock, H., 76, 77, *108*
Chervin, R. M., 59, *61*, 166, *169*, *170*
Chi, C. N., 165, *170*
Clapp, P. F., 87, *109*, 121, *171*, 312, *324*
Clarke, R. H., 42, *64*, 83, 84, *108*
Corby, G. A., 43, 52, 53, *61*, 67, 87, 94, 95, *108*, *109*
Courant, R., 33, *61*
Crowley, W. P., 34, *61*
Cullen, M. J. P., 47, *61*


## D


Daggupaty, S. M., 128, 165, *171*
Daley, R., 45, 51, *63*, 271, *323*
Deardorff, J. W., 81, *108*, 120, 123, 165, *170*
De Santo, G., 125, 165, *171*
De Vries, G., 47, *64*
Dey, C. H., 42, *61*
Dickinson, R. E., 166, *170*, *172*, 281, *323*
Dopplick, T. G., 93, *108*, 312, *323*
Doron, E., 45, *61*, 281, *323*
Douglas, J., Jr., 34, 47, *61*, *65*
Downey, P., 58, *61*, 166, *169*
Druyan, L. M., 52, 58, *61*, *65*
Dupont, T., 47, *65*


## E


Egger, J., 54, *61*
Eilon, B., 34, *61*
Eliasen, E., 45, *61*, 269, 270, 277, 278, 279, 285, *323*
Eliassen, A., 14, 16, *61*
Ellison, T. H., 76, 77, *108*
Ellsaesser, H. W., 44, *62*, 269, 270, 273, 274, *323*

Elvius, T., 48, *62*
Ertel, H., 5, *62*

### F

Fels, S. B., 165, *170*
Fiedler, F., 84, *109*
Fjørtoft, R., 3, 36, *61*, *62*, 191, *265*, 268, 320, *323*
Flattery, T. W., 45, *62*
Fleisher, A., *65*
Fornberg, B., 36, 50, *62*
Francis, P. E., 51, *62*, 75, *109*
Friedrichs, K. O., 33, *61*
Fromm, J. E., 34, *62*

### G

Gary, J. M., 40, 42, 47, 53, *62*
Gates, W. L., 42, 43, *62*, 166, *169*, 316, *324*
Gauntlett, D. J., 304, *323*
Gerrity, J. P., Jr., 35, 50, *62*, *65*
Gilchrist, A., 43, 52, 53, *61*, 67, 68, 73, 87, 94, 95, 107, *108*, *109*
Gordon, T., 271, *323*
Gottlieb, D., 34, *61*, *62*
Gourlay, A. R., 34, *62*
Grammeltvedt, A., 39, 40, *62*
Grimmer, M., 39, 42, 43, *62*, 67, 71, *109*
Gustafsson, B., 48, *62*

### H

Halem, M., 52, *65*
Halpern, P., 47, *65*
Haltiner, G. J., 28, 54, *62*
Hanna, S. R., 81, *109*
Hansen, J. E., 52, *65*
Haurwitz, B., 274, *323*
Hembree, G. D., 53, 58, *64*, *65*
Hembree, L., *62*
Henderson, J., 48, *65*, 271, 295, *323*, *324*
Henrici, P., 32, *62*
Hinkelmann, K. H., 7, *62*
Hinsman, D. E., 47, *62*
Hirose, M., 44, *63*, 269, *323*
Hogan, J. S., 52, *65*
Hollingsworth, A., 45, *61*, 281, *323*
Holloway, J. L., Jr., 25, 40, 42, 43, 53, *62*, *63*, *65*, 73, 86, *109*, 111, 124, 125, 129, 130, *170*, *171*, 268, 312, 313, *323*, *324*
Horn, L. H., *62*
Hoskins, B. J., 45, 48, *61*, *62*, 271, 281, 285, *323*, *324*
Houghton, D. D., 34, *62*, 112, 166, *170*
Hovermale, J. B., 4, *65*
Hoyt, D. V., 113, *171*, 312, *324*
Hunt, G. E., 68, 93, *109*

### I

Isaacson, E., 40, *65*

### J

Julian, P. R., 57, *61*, *62*, 166, *169*

### K

Kahle, A. B., 43, *62*
Kanamitsu, M., 127, 161, 162, *170*
Kao, S.-K., *62*, 165, *170*
Kaplan, L. D., 165, *170*
Karweit, M., 36, *61*
Kasahara, A., 12, 16, 25, 26, 34, 40, 42, 45, 52, 53, 56, *62*, *63*, *64*, *65*, 111, 112, 113, 114, 115, 122, 123, 124, 125, 131, 137, 148, 152, 165, 166, 167, *170*, *171*, *172*
Katayama, A., 53, *63*
Kaula, W. M., 298, *323*
Kida, H., 42, *63*
Kidson, J. W., 99, *109*, 314, 315, 317, *324*
Kikuchi, Y., 44, 51, 53, *63*, 269, *323*
Kinninmonth, W. R., 304, *323*
Kreiss, H.-O., 35, 36, 40, 46, 50, 51, *63*, 126, *170*
Krishnamurti, T. N., 113, 127, 161, 162, *170*
Kubota, S., 44, *63*, 269, *323*
Kuo, H. L., 113, 127, 128, 161, 162, *170*
Kurihara, Y., 31, 32, 40, 41, 42, 44, 53, *63*, 69, *109*, 124, *170*, 269, *323*
Kutzbach, J. E., 166, *170*
Kwizak, M., 48, *63*

## L


Lacis, A. A., 52, *65*
Lamb, H., 5, *63*
Langford, J. C., 304, *323*
Lax, P. D., 34, *63*
Lee, W. H. K., 298, *323*
Leith, C. E., 24, 25, 34, 57, *63*, 75, *109*, 111, *170*
Lewy, H., 33, *61*
Lilly, D. K., 32, 36, 40, *63*, 70, 75, *109*, 165, *170*
List, R. J., 312, *323*
Llewellyn, R. A., 166, *170*
London, J., 113, 131, 165, *170*, *171*, 312, *323*, *324*
Lorenz, E. N., 23, 45, 57, *63*, 216, 217, 222, *265*, 269, 275, 319, *323*


## M


McClintock, M., 166, *170*
Machenhauer, B., 45, 51, *61*, *63*, 269, 270, 271, 277, 278, 279, 285, *323*
MacPherson, A. K., 47, *64*
McPherson, R. D., 35, 48, 56, *62*, *63*
Maine, R., 305, *324*
Manabe, S., 23, 25, 40, 42, 43, 53, *61*, *62*, *63*, *65*, 73, 86, 93, 99, *109*, 111, 124, 125, 129, 130, *170*, *171*, 268, 292, 311, 312, 313, 322, *323*, *324*
Marchuk, G. I., 34, 48, *63*, 301, *324*
Mason, B. J., 105, 106, *109*
Masson, B. W., 34, *65*
Mathur, M., 162, *170*
Matsuno, T., 31, 32, *63*
Mattingly, S. R., 93, *109*
Merilees, P. E., 43, 46, *63*, 275, 281, 284, *324*
Milne, W. E., 32, *64*
Mintz, Y., 41, 43, 52, *61*, *65*, 111, 129, *170*, 207, *265*
Mirin, A., 34, *61*
Miyakoda, K., 32, 42, 58, *64*, 287, *324*
Möller, F., 312, *324*
Molenkamp, C. R., 35, *64*
Monin, A. S., 11, *64*, 83, 84, *109*
Morris, J. L., 34, *62*
Morton, K. W., 30, 33, 35, *64*, 75, *109*
Moxim, W. J., 113, 127, *170*
Moyer, R. W., 42, *64*
Murray, F., 160, *170*


## N


Ndefo, E., 34, *65*
Nelson, A. B., 43, *62*
Newell, R. E., 99, *109*, 314, 315, 317, *324*
Newson, R. L., 43, 52, 53, *61*, 67, 68, 87, 94, 95, *108*, *109*
Norrie, D. H., 47, *64*


## O


Obukhov, A. M., 11, *64*, 84, *109*
Oliger, J. E., 25, 35, 36, 40, 42, 46, 50, 51, *63*, *64*, 113, 122, 126, 148, 152, *170*, *171*
Oort, A. H., 317, *324*
Orszag, S. A., 36, 45, 46, 50, *64*, 269, 270, 275, 278, 282, 284, *324*
Otto-Bliesner, B., 165, *171*


## P


Pacanowski, R. C., 23, *61*
Panofsky, H. A., 84, *109*
Park, J. H., 165, *170*
Peaceman, D. W., 34, *64*
Phillips, N. A., 3, 7, 15, 40, 49, 50, 52, 53, *64*, 70, *109*, 111, *171*, 193, 207, 235, *265*, 281, 285, 297, *324*
Platzman, G. W., 2, 31, 44, 50, *61*, *64*, 269, 270, 275, *323*, *324*
Polger, P. D., 35, *62*
Posey, J. W., 87, *109*, 121, *171*, 312, *324*
Price, G. V., 47, *64*
Priestley, C. H. B., 84, *109*
Puri, K., 281, 288, 302, 310, *323*, *324*


## Q


Quirk, W. J., 25, 52, 58, *61*, *65*


## R


Rachford, H. H., Jr., 34, *64*
Randall, D. A., 264, *265*
Rasmusson, E. M., 45, *61*, *63*, 269, 270, 277, 278, 279, 281, 317, *323*, *324*

Rekustad, J.-E., 16, *61*
Richardson, L. F., 2, 12, 32, *64*, 112, 117, 118, 124, 129, 138, 142, 143, 145, 146, 149, *171*
Richtmyer, R. D., 30, 33, 34, 35, 49, *64*, 75, *109*
Ridley, C., *62*
Riegel, C. A., 42, *62*
Robert, A. J., 45, 48, 50, 58, *63*, *64*, *65*, 164, *171*, 269, 270, 271, 279, 295, *324*
Roberts, K. V., 35, *65*
Rodgers, C. D., 93, 94, 96, *110*
Rowntree, P. R., 68, 87, 94, *110*
Russell, G., 52, *65*

## S

Sadler, J. C., 99, *108*
Sadourny, R., 40, 41, *65*
Saltzman, B., *65*
Sangster, W. E., 53, *65*
Sankar-Rao, M., 42, 43, *65*
Sasamori, T., 26, *63*, 112, 113, 121, 126, 152, 154, 165, *170*, *171*, 312, *324*
Schenk, H. A., 165, *171*
Schlesinger, M. E., 264, *265*
Schneider, S. H., 59, *61*, 166, *169*, *170*
Schubert, W. H., 264, *265*
Schutz, C., 316, *324*
Seaman, R. S., 304, *323*
Shapiro, M. A., 16, *65*
Shapiro, R., 75, *110*, 163, *171*, 298, *324*
Shaw, D. B., 39, 42, 43, *62*, 67, 71, *109*
Sheppard, P. A., 81, *110*
Shulman, I., 58, *64*
Shuman, F. G., 4, 39, 42, 50, 58, 60, *65*, 69, *110*, 306, *324*
Silberman, I. S., 44, *65*, 268, 269, 272, 320, *324*
Simmonds, I., 271, *323*
Simmons, A. J., 45, 48, *61*, *62*, 271, 281, 285, *323*, *324*
Simons, T. J., 51, *65*
Smagorinsky, J., 3, 25, 53, *65*, 111, 124, 125, 129, *170*, *171*, 292, 311, *324*
Smith, P. J., 166, *169*
Somerville, R. C. J., 25, 52, 58, *61*, *65*
Spelman, M. J., 23, 40, 43, *62*, *63*
Stambler, H., 42, *64*
Starr, V. P., 12, *65*
Stern, W., 271, *323*
Stoker, J. J., 40, *65*
Stone, H. M., 125, *170*
Stone, P. H., 25, 52, *65*
Strang, G. W., 34, *65*
Strickler, R. F., 42, 53, 58, *64*, *65*, 93, *109*, 111, 129, *170*, 292, 311, 312, *324*
Suchman, D., 166, *170*
Sundqvist, H., 53, *65*
Sundström, A., 48, *62*
Sutcliffe, R. C., 14, *65*

## T

Takigawa, Y., 53, *63*
Taylor, T. D., 34, *65*
Tenenbaum, J., 52, *65*
Tennekes, H., 76, *110*
Terpstra, T. B., 99, *109*
Teweles, S., 305, *324*
Thiel, L. G., 120, *171*
Thompson, P. D., 3, 28, 57, *65*
Thurling, R., 302, *323*
Tiedtke, M., 39, 43, *65*
Tsay, C.-Y., 165, *171*
Turkel, E., 34, *62*, *65*
Turnbull, C., 48, *65*, 295, *324*

## U

Umscheid, L., Jr., 42, 43, *65*

## V

Vanderman, L. W., 43, *65*
Vincent, D. G., 99, *109*, 314, 315, 317, *324*
von Neumann, J., 3, 33, 48, 49, *61*, 268, 320, *323*

## W

Wang, H.-H., 47, *65*

Washington, W. M., 25, 26, 40, 42, 53, 59, *61*, *62*, *63*, *64*, *65*, 112, 113, 120, 122, 123, 124, 125, 128, 130, 131, 137, 148, 152, 165, 166, 167, *170*, *171*, *172*
Weigle, W. F., 280, *324*
Weiss, N. O., 35, *65*
Wellck, R. E., 25, 42, *64*, 113, 122, 125, 148, 152, 165, *171*
Wendell, L. L., *62*
Wendroff, B., 34, *63*
Wetherald, R. T., 312, *324*
Wiin-Nielsen, A., *65*
Williams, J., 26, *65*, 165, 166, *171*, *172*
Williams, R. T., 54, *62*
Williamson, D. L., 35, 40, 41, 42, 43, *65*, *66*, 112, 113, 137, 164, 165, 166, *170*, *171*, *172*, 281, *323*
Winninghoff, F. J., 180, *265*
Wobus, H., 306, *324*

## Y

Young, J. A., 32, *66*

## Z

Zilitinkevich, S. S., 83, *109*
Zwas, G., 34, *61*, *66*

# Subject Index


## A

Accuracy (numerical), 31, 39, 53, 194, 281
Albedo, 18, 87, 121, 129, 312
Aliasing, 23, 46, 75, 125, 278
Amplification factor, 29

## B

Barotropic equations, 27, 36, 271, 279
Boundary conditions
  lateral, 54
  lower, 15, 119, 137, 165, 208, 286, 312
  upper, 14, 26, 59, 119, 135, 208, 286
Boundary layer, 21, 76, 91, 120, 176, 288

## C

Carbon dioxide, 19, 154, 312
Climate, 58, 68, 96, 111, 165, 311
Clouds, 19, 25, 85, 128, 158, 312
Computers, 60, 68, 305, 312
Condensation, *see* Precipitation
Conservation laws, 3, 14, 27, 72, 115, 202, 211, 220, 236, 244
Convective exchange, 77, 87, 128, 161
Coordinate systems, 12
  horizontal, 37, 51, 69, 132, 176, 193, 290
  vertical, 12, 52, 70, 115, 131, 175, 207, 285
Coriolis force, 3, 38, 71, 116, 141, 242
Courant–Friedrich–Lewy condition, 11

## D

Diagnostic equations, *see* Prognostic equations
Difference schemes, 137, 143, 213, 236, 246, 261
  energy-conserving, 37, 70, 194, 204, 241, 275
  finite
    backward Euler (Matsuno), 30, 192, 261
    box method, 70
    Crank–Nicholson, 30, 47, 51, 192
    explicit, 10, 30
    fractional step (splitting operator), 34, 48, 301
    implicit, 10, 30, 47
    Lax–Wendroff, 34, 192
    leap-frog, 30, 51, 71, 192, 261
    semi-implicit, 48, 295
Diffusion, 17, 75, 84, 123, 147, 287
Dispersion (numerical), 184
Diurnal effect, 80

## E

Eddy coefficients, 20, 25, 75, 84, 124, 287
Emissivity, 93
Enstrophy, 191
Enthalpy, 215, 225
Entropy, 16, 212, 225
Equation of state (ideal gas), 9, 117, 146, 209
Equations of motion
  primitive, 3, 27, 70, 115, 209, 237, 289
  shallow water, 27, 37, 180, 202
  vorticity, 28, 44, 190, 271, 286
Error, *see* Accuracy (numerical); Truncation error

## F

Filtering (smoothing), 43, 50, 75, 82, 162, 252, 298
Finite element method, 47
Four-dimensional data assimilation, 55
Fourier transform, 185, 278
Free atmosphere, 21, 91
Friction, 3, 10, 17, 59, 82, 116, 122, 150, 213

## G

Galerkin approximation, 44
Geopotential, 6, 74, 81, 292
Geostrophic adjustment, 3, 180, 187
Gravitational potential, 5

## H

Homogeneous incompressible flow, 27, 180
Humidity, 91, 116, 127, 145
Hydrostatic equation, 11, 27, 71, 117, 143, 209, 226, 287

## I

Ice, 86, 114, 292, 312
Inertia-gravity wave, 183
Initialization (data), 54, 166, 304
Initial value problem, 2, 32, 185
Insolation, 18, 313
Instability, 40, 46, 71, 193, 256, 280
Integral constraints, 37, 52, 73, 176, 191, 213, 275
Integration (time), 10, 30, 47, 71, 192, 261, 295
Inversion, 80, 92

## J

Jacobian interaction coefficient, 274
Jacobian operator, 191

## L

Land, 86, 114, 292
Latent heat, 75, 84, 118, 127, 160, 259
Legendre transform, 291

## M

Mixing ratio, 20, 212, 251, 256, 287, 296
Moisture, 130, 141, 151, 317
Mountains, *see* Orography

## N

Nonlinear effects, 3, 48, 58, 302

## O

Oceans, 23, 58, 114, 120
Optical thickness, 19
Orography, 17, 52, 73, 113, 134, 297
Ozone, 19, 68, 154, 212, 251, 258, 312

## P

Parameterization processes, 18, 60, 82, 113, 120, 165, 287
Persistency (weather), 4, 57
Poles, 41, 124, 132, 246
Potential temperature, 9, 22, 78, 84, 123, 148, 218, 225, 289
Precipitation, 19, 76, 85, 90, 118, 127, 161, 259, 292
Predictability (atmospheric), 56, 177
  $S_1$ score, 4, 310
Pressure system, 14, 53, 207
Primitive equations, 3, 27, 70, 115, 179, 209, 237
Prognostic equations, 10, 115, 174, 272, 286, 291
Pseudospectral method, 45, 284

## Q

Quadratic conserving scheme, 35, 50, 70, 194
Quasi-geostrophic motion, 3, 14, 180, 190

## R

Radiation (solar), 17, 92, 121, 152, 159, 312
Rainfall (simulation), 104
Resolution (numerical), 59, 67, 112, 167, 176, 293, 303, 311
Reynolds stress, 21, 122
Rossby similarity theory, 83

## S

Sensible heat, 17, 22, 84, 118, 147
Shallowness approximation (*see also* Equations of motion), 7
Snow cover, 130
Solar zenith angle, 159
Sound speed, 11, 47
Spectral method, 43, 269, 276, 280
Speed of computation, 60, 68, 313
Sponge layer, 175
Stability conditions, 11, 30, 41, 250
von Neumann, 33, 49
Stefan's constant, 156
Stream function, 272, 289
Subgrid-scale processes, 24, 74, 79, 122, 147, 287
$\sigma$ System, 15, 53, 70, 175, 207, 285
s System, 12, 52, 115, 134
z System, 12, 53, 115, 131

## T

Thermodynamic equations, 9, 16, 71, 212, 225
Topography, *see* Orography
Truncation error, 24, 41, 52, 235, 281
Turbulence, 21

## V

Vorticity, *see* Equations of motion

## W

Water vapor, 20, 76, 117, 123, 147, 153
Wave energy, 229
Weather forecasting, 3, 20, 25, 52, 111, 120, 166, 302

# Contents of Previous Volumes


**Volume 1: Statistical Physics**

The Numerical Theory of Neutron Transport
*Bengt G. Carlson*

The Calculation of Nonlinear Radiation Transport by a Monte Carlo Method
*Joseph A. Fleck, Jr.*

Critical-Size Calculations for Neutron Systems by the Monte Carlo Method
*Donald H. Davis*

A Monte Carlo Calculation of the Response of Gamma-Ray Scintillation Counters
*Clayton D. Zerby*

Monte Carlo Calculation of the Penetration and Diffusion of Fast Charged Particles
*Martin J. Berger*

Monte Carlo Methods Applied to Configurations of Flexible Polymer Molecules
*Frederick T. Wall, Stanley Windwer, and Paul J. Gans*

Monte Carlo Computations on the Ising Lattice
*L. D. Fosdick*

A Monte Carlo Solution of Percolation in the Cubic Crystal
*J. M. Hammersley*

AUTHOR INDEX—SUBJECT INDEX

**Volume 2: Quantum Mechanics**

The Gaussian Function in Calculations of Statistical Mechanics and Quantum Mechanics
*Isaiah Shavitt*

Atomic Self-Consistent Field Calculations by the Expansion Method
*C. C. J. Roothaan and P. S. Bagus*

The Evaluation of Molecular Integrals by the Zeta-Function Expansion
*M. P. Barnett*

Integrals for Diatomic Molecular Calculations
*Fernando J. Corbató and Alfred C. Switendick*

Nonseparable Theory of Electron-Hydrogen Scattering
*A. Temkin and D. E. Hoover*

Estimating Convergence Rates of Variational Calculations
*Charles Schwartz*

AUTHOR INDEX—SUBJECT INDEX

**Volume 3: Fundamental Methods in Hydrodynamics**

Two-Dimensional Lagrangian Hydrodynamic Difference Equations
*William D. Schulz*

Mixed Eulerian-Lagrangian Method
*R. M. Frank and R. B. Lazarus*

The Strip Code and the Jetting of Gas between Plates
*John G. Trulio*

CEL: A Time-Dependent, Two-Space-Dimensional, Coupled Eulerian-Lagrange Code
*W. F. Noh*

The Tensor Code
*G. Maenchen and S. Sack*

Calculation of Elastic-Plastic Flow
*Mark. L. Wilkins*

Solution by Characteristics of the Equations of One-Dimensional Unsteady Flow
*N. E. Hoskin*

The Solution of Two-Dimensional Hydrodynamic Equations by the Method of Characteristics
*D. J. Richardson*

The Particle-in-Cell Computing Method for Fluid Dynamics
*Francis H. Harlow*

The Time-Dependent Flow of an Incompressible Viscous Fluid
*Jacob Fromm*

AUTHOR INDEX—SUBJECT INDEX

**Volume 4: Applications in Hydrodynamics**

Numerical Simulation of the Earth's Atmosphere
*Cecil E. Leith*

Nonlinear Effects in the Theory of a Wind-Driven Ocean Circulation
*Kirk Bryan*

Analytic Continuation Using Numerical Methods
*Glenn E. Lewis*

Numerical Solution of the Complete Krook-Boltzmann Equation for Strong Shock Waves
*Moustafa T. Chahine*

The Solution of Two Molecular Flow Problems by the Monte Carlo Method
*J. K. Haviland*

Computer Experiments for Molecular Dynamics Problems
*R. A. Gentry, F. H. Harlow, and R. E. Martin*

Computation of the Stability of the Laminar Compressible Boundary Layer
*Leslie M. Mack*

Some Computational Aspects of Propeller Design
*William B. Morgan and John W. Wrench, Jr.*

Methods of the Automatic Computation of Stellar Evolution
*Louis G. Henyey and Richard D. Levée*

Computations Pertaining to the Problem of Propagation of a Seismic Pulse in a Layered Solid
*F. Abramovici and Z. Alterman*

AUTHOR INDEX—SUBJECT INDEX

**Volume 5: Nuclear Particle Kinematics**

Automatic Retrieval Spark Chambers
*J. Bounin, R. H. Miller, and M. J. Neumann*

Computer-Based Data Analysis Systems
*Robert Clark and W. F. Miller*

Programming for the PEPR System
*P. L. Bastien, T. L. Watts, R. K. Yamamoto, M. Alston, A. H. Rosenfeld, F. T. Solmitz, and H. D. Taft*

A System for the Analysis of Bubble Chamber Film Based upon the Scanning and Measuring Projector (SMP)
*Robert I. Hulsizer, John H. Munson, and James N. Snyder*

A Software Approach to the Automatic Scanning of Digitized Bubble Chamber Photographs
*Robert B. Marr and George Rabinowitz*

AUTHOR INDEX—SUBJECT INDEX

**Volume 6: Nuclear Physics**

Nuclear Optical Model Calculations
*Micheal A. Melkanoff, Tatsuro Sawada, and Jacques Raynal*

Numerical Methods for the Many-Body Theory of Finite Nuclei
*Kleber S. Masterson, Jr.*

Application of the Matrix Hartree-Fock Method to Problems in Nuclear Structure
*R. K. Nesbet*

Variational Calculations in Few-Body Problems with Monte Carlo Method
*R. C. Herndon and Y. C. Tang*

Automated Nuclear Shell-Model Calculations
*S. Cohen, R. D. Lawson, M. H. Macfarlane, and M. Soga*

Nucleon-Nucleon Phase Shift Analyses by Chi-Squared Minimization
*Richard A. Arndt and Malcolm H. MacGregor*

AUTHOR INDEX—SUBJECT INDEX

**Volume 7: Astrophysics**

The Calculation of Model Stellar Atmospheres
*Dimitri Mihalas*

Computational Methods for Non-LTE Line-Transfer Problems
*D. G. Hummer and G. Rybicki*

Methods for Calculating Stellar Evolution
*R. Kippenhahm, A. Weigert, and Emmi Hofmeister*

Computational Methods in Stellar Pulsation
*R. F. Christy*

Stellar Dynamics and Gravitational Collapse
*Michael M. May and Richard H. White*

AUTHOR INDEX—SUBJECT INDEX

**Volume 8: Energy Bands of Solids**

Energy Bands and the Theory of Solids
*J. C. Slater*

Interpolation Schemes and Model Hamiltonians in Band Theory
*J. C. Phillips and R. Sandrock*

The Pseudopotential Method and the Single-Particle Electronic Excitation Spectra of Crystals
*David Brust*

A Procedure for Calculating Electronic Energy Bands Using Symmetrized Augmented Plane Waves
*L. F. Mattheiss, J. H. Wood, and A. C. Switendick*

Interpolation Scheme for the Band Structure of Transition Metals with Ferromagnetic and Spin-Orbit Interactions
*Henry Ehrenreich and Laurent Hodges*

Electronic Structure of Tetrahedrally Bonded Semiconductors: Empirically Adjusted OPW Energy Band Calculations
*Frank Herman, Richard L. Kortum, Charles D. Kuglin, John P. Van Dyke, and Sherwood Skillman*

The Green's Function Method of Korringa, Kohn, and Rostoker for the Calculation of the Electronic Band Structure of Solids
*Benjamin Segall and Frank S. Ham*

AUTHOR INDEX—SUBJECT INDEX

**Volume 9: Plasma Physics**

The Electrostatic Sheet Model for a Plasma and Its Modification to Finite-Sized Particles
*John M. Dawson*

Solution of Vlasov's Equation by Transform Methods
*Thomas P. Armstrong, Rollin C. Harding, Georg Knorr, and David Montgomery*

The Water-Bag Model
*Herbert L. Berk and Keith V. Roberts*

The Potential Calculation and Some Applications
*R. W. Hockney*

Multidimensional Plasma Simulation by the Particle-in-Cell Method
*R. L. Morse*

Finite-Size Particle Physics Applied to Plasma Simulation
*Charles K. Birdsall, A. Bruce Langdon, and H. Okuda*

Finite-Difference Methods for Collisionless Plasma Models
*Jack A. Byers and John Killeen*

Application of Hamilton's Principle to the Numerical Analysis of Vlasov Plasmas
*H. Ralph Lewis*

Magnetohydrodynamic Calculations
*Keith V. Roberts and D. E. Potter*

The Solution of the Fokker-Planck Equation for a Mirror-Confined Plasma
*John Killeen and Kenneth D. Marx*

AUTHOR INDEX—SUBJECT INDEX

**Volume 10: Atomic and Molecular Scattering**

Numerical Solutions of the Integro-Differential Equations of Electron–Atom Collision Theory
*P. G. Burke and M. J. Seaton*

Quantum Scattering Using Piecewise Analytic Solutions
*Roy G. Gordon*

Quantum Calculations in Chemically Reactive Systems
*John C. Light*

Expansion Methods for Electron–Atom Scattering
*Frank E. Harris and H. H. Michels*

Calculation of Cross Sections for Rotational Excitation of Diatomic Molecules by Heavy Particle Impact: Solution of the Close-Coupled Equations
*William A. Lester, Jr.*

Amplitude Densities in Molecular Scattering
*Don Secrest*

Classical Trajectory Methods
*Don L. Bunker*

AUTHOR INDEX—SUBJECT INDEX

**Volume 11: Seismology: Surface Waves and Earth Oscillations**

Finite Difference Methods for Seismic Wave Propagation in Heterogeneous Materials
*David M. Boore*

Numerical Analysis of Dispersed Seismic Waves
*A. M. Dziewonski and A. L. Hales*

Fast Surface Wave and Free Mode Computations
*F. A. Schwab and L. Knopoff*

A Finite Element Method for Seismology
*John Lysmer and Lawrence A. Drake*

Seismic Surface Waves
*H. Takeuchi and M. Saito*

AUTHOR INDEX—SUBJECT INDEX

**Volume 12: Seismology: Body Waves and Sources**

Numerical Methods of Ray Generation in Multilayered Media
*F. Hron*

Computer Generated Seismograms
*Z. Alterman and D. Loewenthal*

Diffracted Seismic Signals and Their Numerical Solution
*C. H. Chapman and R. A. Phinney*

Inversion and Inference for Teleseismic Ray Data
*Leonard E. Johnson and Freeman Gilbert*

Multipolar Analysis of the Mechanisms of Deep-Focus Earthquakes
*M. J. Randall*

Computation of Models of Elastic Dislocations in the Earth
*Ari Ben-Menahem and Sarva Jit Singh*

AUTHOR INDEX—SUBJECT INDEX

**Volume 13: Geophysics**

Signal Processing and Frequency-Wavenumber Spectrum Analysis for a Large Aperture Seismic Array
*Jack Capon*

Models of the Sources of the Earth's Magnetic Field
*Charles O. Stearns and Leroy R. Alldredge*

Computations with Spherical Harmonics and Fourier Series in Geomagnetism
*D. E. Winch and R. W. James*

Inverse Methods in the Interpretation of Magnetic and Gravity Anomalies
*M. H. P. Bott*

Analysis of Geoelectromagnetic Data
*S. H. Ward, W. J. Peeples, and J. Ryu*

Nonlinear Spherical Harmonic Analysis of Paleomagnetic Data
*J. M. Wells*

Harmonic Analysis of Earth Tides
*Paul Melchior*

Computer Usage in the Computation of Gravity Anomalies
*Manik Talwani*

Analysis of Irregularities in the Earth's Rotation
*D. E. Smylie, G. K. C. Clarke, and T. J. Ulrych*

Convection in the Earth's Mantle
*D. L. Turcotte, K. E. Torrance, and A. T. Hsui*

AUTHOR INDEX—SUBJECT INDEX

**Volume 14: Radio Astronomy**

Radioheliography
*N. R. Labrum, D. J. McLean, and J. P. Wild*

Pulsar Signal Processing
*Timothy H. Hankins and Barney J. Rickett*

Aperture Synthesis
*W. N. Brouw*

Computations in Radio-Frequency Spectroscopy
*John A. Ball*

AUTHOR INDEX—SUBJECT INDEX

**Volume 15: Vibrational Properties of Solids**

The Calculation of Phonon Frequencies
*G. Dolling*

The Use of Computers in Scattering Experiments with Slow Neutrons
*R. Pynn*

Group Theory of Lattice Dynamics by Computer
*John L. Warren and Thomas G. Worlton*

Lattice Dynamics and Related Properties of Point Defects
*R. F. Wood*

Lattice Dynamics of Surfaces of Solids
*F. W. de Wette and G. P. Alldredge*

Vibrational Properties of Amorphous Solids
*R. J. Bell*

Lattice Dynamics of Quantum Crystals
*T. R. Koehler*

Methods of Brillouin Zone Integration
*G. Gilat*

Computer Studies of Transport Properties in Simple Models of Solids
*William M. Visscher*

AUTHOR INDEX—SUBJECT INDEX

**Volume 16: Controlled Fusion**

Numerical Magnetohydrodynamics for High-Beta Plasmas
*Jeremiah U. Brackbill*

Waterbag Methods in Magnetohydrodynamics
*David Potter*

Solution of Continuity Equations by the Method of Flux-Corrected Transport
*J. P. Boris and D. L. Book*

Multifluid Tokamak Transport Models
*John T. Hogan*

Icarus—A One-Dimensional Plasma Diffusion Code
*M. L. Watkins, M. H. Hughes, K. V. Roberts, P. M. Keeping, and J. Killeen*

Equilibria of Magnetically Confined Plasmas
*Brendan McNamara*

Computation of the Magnetohydrodynamic Spectrum in Axisymmetric Toroidal Confinement Systems
*Ray C. Grimm, John M. Greene, and John L. Johnson*

Collective Transport in Plasmas
*John M. Dawson, Hideo Okuda, and Bernard Rosen*

Electromagnetic and Relativistic Plasma Simulation Models
*A. Bruce Langdon and Barbara F. Lasinski*

Particle-Code Models in the Nonradiative Limit
*Clair W. Nielson and H. Ralph Lewis*

The Solution of the Kinetic Equations for a Multispecies Plasma
*John Killeen, Arthur A. Mirin, and Marvin E. Rensink*

AUTHOR INDEX—SUBJECT INDEX

A 7
B 8
C 9
D 0
E 1
F 2
G 3
H 4
I 5
J 6